MANUFACTURING SCIENCE-II

By.

S.A. RIZVI
Ph.D. (Pursuing) Welding
Mechanical Engineering Department
IIT (BHU), Varanasi
Faculty Member, JMI, New Delhi

Dr. S.P. TEWARI
Ph.D. (Welding)
Prof. Department of Mechanical Engineering
IIT (BHU), Varanasi

WAJAHAT ALI
B.E. (Mechanical), J.M.I.
Mechanical Engineering Department
SSRIET, CCS University, MEERUT (UP)

S.K. KATARIA & SONS®

Educational Publishers
4885/109, Prakash Mahal, Dr. Subhash Bhargav Lane,
Opposite Delhi Medical Association, Daryaganj, New Delhi-110002 (INDIA)
Phone: +91-11- 43551243; *Mobile:* +91-9871775858
e-mail: katariabook@gmail.com; katariabooks@yahoo.com
Website: www.skkatariaandsons.com
(e-books, subscriptions and bulk buying also available on our website)

Published by:

S.K. Kataria & Sons®

4885/109, Prakash Mahal, Dr. Subhash Bhargav Lane,
Opposite Delhi Medical Association, Daryaganj, New Delhi-110002 (INDIA)
Phone: +91-11- 43551243; *Mobile:* +91-9871775858
e-mail: katariabook@gmail.com; katariabooks@yahoo.com
Website: www.skkatariaandsons.com

First Edition: 2011; Third Edition: 2012; Reprint: 2023

Price: ₹ 465/-

Laser Typesetting by: S.K. Computers, Delhi-32 *Printed at:* Aapte Udhyog Noida

Preface

P

This **'Manufacturing Science-II'** has been written exclusive to meet the full requirement of **B.Tech. Mechanical Engineering** and other related branch of Vth semester students. This subject is taught as compulsory at Graduate level in various Professional Colleges for Engineering students. Basic aim of this book is that a student should understand various Manufacturing processes such as **metal cutting, surface finishing**, **unconventional machining**, **welding** and **drilling** etc. The book written in very simple language so that it is easily understandable. At the end of each chapter some additional **objective questions** are given. These questions are very helpful in various competitive examinations like **GATE**, **IES**, **UPSC**, **SAIL**, **HPCL** etc.

It is hoped that the book in its present form shall be preferred by the reader over other books available in the market. Any suggestions for improvement would be highly appreciated.

S.A. Rizvi
Wajahat Ali

Acknowledgements

I am grateful to my mother (Smt. **Amna Begum Rizvi**) and all my family members and colleagues for the encouragement given by them to write this book and for their valuable suggestion.

I feel greatly indebted to my wife Saba Rizvi and my kids Alian Zehra, Khadija-tul Kubra and Syed Mohd. for their constant support and valuable guidance from time to time.

My special thanks are due to Mr. Sanjeev Kumar Kataria of M/s. S.K. Kataria & Sons, New Delhi publisher for his encouragement and the confidence he reposed in this work.

S.A. Rizvi

Dedicated to my wife
Saba Zafar Rizvi

– S.A. Rizvi

&

Dedicated to my wife
Dr. Mamata Tewari

– Dr. S.P. Tewari

Contents

UNIT–I METAL CUTTING

1. THEORY OF METAL CUTTING 3–70

1.1 Introduction 3
- 1.1.1 Various Metal Removal Process 3

1.2 Cutting Tool 4
- 1.2.1 Single Point Cutting Tool 5
- 1.2.2 Multi Point Cutting Tools 10
- 1.2.3 ASA System (American System) 11

1.3 Chip Formation Mechanism 12

1.4 Types of Chips 14
- 1.4.1 Discontinuous Chip 14
- 1.4.2 Continuous Chips 15
- 1.4.3 Continuous Chip with Built-up Edge 15
- 1.4.4 Differentiate Between the Continuous and Discontinuous Chip 16

1.5 Types of Metal Cutting Process 17
- 1.5.1 Orthogonal Cutting 17
- 1.5.2 Oblique Cutting 17
- 1.5.3 Differentiate between Orthogonal and Oblique Cutting 18

1.6 Determination of Shear Angle 19

1.7 Earnest Merchant's Model for Orthogonal Cutting 20
- 1.7.1 Assumptions 21
- 1.7.2 Chip Velocity 23

1.8 Lee and Shaffer's Theory 24
- 1.8.1 Assumptions 24

1.9 Cutting Fluids 25
- 1.9.1 Main Functions of a Cutting Fluid 26
- 1.9.2 Requirements of a Good Cutting Fluid 27
- 1.9.3 Composition of a Cutting Fluid 27
- 1.9.4 Functions of a Cutting Fluid 27
- 1.9.5 Cutting Fluid Effects in Machining 30
- 1.9.6 Cutting Fluid Selection Criteria 30
- 1.9.7 Cutting Fluid Maintenance and Disposal 31

1.10 Cutting Tool Materials 31
- 1.10.1 Types of Cutting Tool Materials 31
- 1.10.2 Comon Cutting Tool Materials 33

1.11 Cemented Carbides Steel 37
- 1.11.1 Ceramics – Alumina 38
- 1.11.2 Diamond 38
- 1.11.3 Cubic Boron Nitride (CBN) 39

1.12 Effect of Alloying Elements 39
- 1.12.1 Vanadium 39
- 1.12.2 Cobalt 39
- 1.12.3 Carbon 39
- 1.12.4 Silicon 40
- 1.12.5 Manganese 40
- 1.12.6 Phosphorus 40
- 1.12.7 Chromium 40
- 1.12.8 Tungsten 40

1.13 Tool Wear 40
- 1.13.1 Mechanism of Tool Wear 41
- 1.13.2 Abrasion 41
- 1.13.3 Adhesion 41
- 1.13.4 Diffusion 41
- 1.13.5 Chemical Reactions (Corrosion) 41
- 1.13.6 Plastic Deformation (Fracture) 42
- 1.13.7 Flank Wear 42

1.14 Machinability 42
- 1.14.1 Factors Affecting the Machinability 43
- 1.14.2 Tool Life 43

1.15 Machining Economics 44
1.16 Basic Terms 46
Problems 60
Objective Type Questions 63
Answers 69
Questions 69

UNIT–II MACHINE TOOLS

2. LATHE MACHINE 73–111

2.1 Lathe Machine 73
2.1.1 Selection of Lathe Machine for Metal Cutting 73
2.1.2 Types of Lathe Machines 75
2.1.3 Specification of a Lathe Machine 79
2.1.4 Common parts of Lathe Machine 80
2.2 Guide Ways 81
2.2.1 Apron 81
2.2.2 Compound Slide Rest Consisting of the Cross Slide and Top Slide 82
2.2.3 Carriage 82
2.3 Head Stock 83
2.3.1 Headstock Spindle 84
2.4 Backgear 85
2.5 Leadscrew 86
2.6 Tailstock 87
2.7 Counter Shaft 88
2.8 Work Holding Devices 89
2.8.1 Coolet Chucks 89
2.8.2 Push Out Type 89
2.8.3 Draw in Type 90
2.8.4 Dead Length Type 90
2.8.5 Three Jaw Chuck 90
2.8.6 Four Jaw Chuck 90
2.9 Lathe Operations 91
2.9.1 Facing 91
2.9.2 Turning 91
2.9.3 Drilling 96
2.9.4 Reaming 96
2.9.5 Boring 96
2.9.6 Knurling 97
2.9.7 Thread Cutting 97
2.10 Capstan Lathe 99
2.10.1 Different Parts of Capstan Lathe 99
2.10.2 The Turret Saddle and Auxiliary Slide 101
2.11 Turret Lathe 101
2.11.1 Different Parts of Turret Lathe 102
2.12 Difference Between Capstan Lathe and Turret Lathe 104
2.12.1 Comparison Between a Capstan Lathe, Turret and an Engine Lathe 104
2.13 Capstan and Turret Lathe Mechanism 105
2.13.1 Turret Indexing Mechanism 105
2.13.2 Bar Feeding Mechanism of Capstan Lathe 106
2.14 Safety Guideline for Working on Lathe Machine 107
Objective Type Questions 108
Answers 110
Questions 111

3. SHAPER, SLOTTER, PLANNER MACHINE 112–129

3.1 Shaper Machine 112
3.1.1 Principal Parts of a Shaper 112
3.1.2 Additional Components 117
3.2 Shaper Drive Mechanisms 117
3.2.1 Crank and Slotted Link Mechanism 117
3.2.2 Whitworth Quick Return Mechanism 119
3.3 Components Shapes Produced by Shaper 120
3.3.1 Slotter Machine 120
3.3.2 Main Parts of a Slotter 121
3.3.3 Specification of a Slotter Machine 122
3.3.4 Types of Slotter 122
3.3.5 Drive Mechanism of a Slotter 122
3.3.6 Work Holding Devices 123
3.3.7 Slotter Cutting Tools 123
3.4 Planer 124
3.4.1 Classification of Planer 125
3.4.2 Construction 126
3.5 Difference Between Shaper and Planer 127
Objective Type Questions 128
Answers 129
Questions 129

4. MILLING 130–175

4.1 Milling 130
4.1.1 Knee-Type 131
4.1.2 Universal Horizontal Milling Machine 132

4.1.3 Ram-Type Milling Machine 133
4.1.4 Universal Ram-Type Milling Machine 133
4.1.5 Swivel Cutter Head Ram-Type Milling Machine 133
4.2 Tools for Milling Machine 134
4.2.1 Milling Cutters 134
4.2.2 Milling Cutter Nomenclature 134
4.2.3 Types of Teeth 135
4.3 Classification of Milling Cutters 136
4.3.1 Helical Milling Cutters 136
4.3.2 Metal Slitting Saw Milling Cutter 137
4.3.3 Side Milling Cutters 137
4.3.4 End Milling Cutters 138
4.3.5 T-Slot Milling Cutter 138
4.3.6 Woodruff Key Slot Milling Cutters 139
4.3.7 Angle Milling Cutters 139
4.3.8 Concave and Convex Milling Cutters 140
4.3.9 Corner Rounding Milling Cutter 140
4.3.10 Special Shaped-Formed Milling Cutter 140
4.4 Selection of Milling Cutters 140
4.5 Size of Milling Cutter 141
4.5.1 Arbors 141
4.5.2 Standard Milling Machine Arbor 142
4.5.3 Screw Arbor 143
4.6 Collets, Spindle Adapters, and Quick-change Tooling 144
4.6.1 Collets 145
4.6.2 Spindle Adapters 145
4.6.3 Chuck Adapter 145
4.6.4 Quick-Change Tooling 146
4.7 Vises 147
4.8 Indexing Fixture 147
4.8.1 High-Speed Milling Attachment 147
4.8.2 Vertical Spindle Attachment 147
4.8.3 Universal Milling Attachment 148
4.8.4 Rotary Table or Circular Milling Attachment 148
4.8.5 Offset Boring Head 148
4.8.6 Offset Boring Head and Tools 148
4.9 Mounting and Indexing Work 149
4.9.1 Methods of Mounting Workpieces 149
4.9.2 Clamping a Workpiece to the Angle Plate 150
4.9.3 Clamping Workpieces in Fixtures 150
4.9.4 Holding Workpieces Between Centers 150
4.9.5 Holding Workpieces in a Chuck 150
4.9.6 Indexing 151
4.10 Feeds for Milling 155
4.11 Plain Milling 157
4.12 Angular Milling 158
4.13 Straddle Milling 159
4.13.1 Milling a Hexagon 159
4.14 Face Milling 160
4.14.1 General 160
4.14.2 Mounting the Workpiece 161
4.14.3 Depth of Cut 161
4.15 Gang Milling 162
4.16 Form Milling 162
4.17 Fly Cutting 163
4.18 Key Way Milling 164
4.18.1 Woodruff Key 164
4.18.2 Square-End Machine Key 165
4.19 T-slot Milling 167
4.20 Sawing and Parting 168
4.21 Helical Milling 169
4.22 Conventional Milling–"Up" Milling 170
4.23 Climb Milling— "Down" Milling 170
4.24 Chip Thickness in Milling Operation 171

Objective Type Questions 173
Answers 174
Questions 174

5. DRILLING AND BORING 176–192

5.1 Drill Machine 176
5.1.1 Types of Drill Machine 177
5.2 Twist Drills 181
5.2.1 Definition 181
5.2.2 General Classifications 181
5.2.3 Nomenclature of Twist Drills and Other Terms Relating to Drilling 183

5.3 Combined Drills and Countersinks 187
5.3.1 Definition 187
5.3.2 General Classifications 188
5.3.3 Nomenclature of Combined Drills and Countersinks 188
5.3.4 Process Parameters 189
Objective Type Questions 190
Answers 192
Questions 192

UNIT–III GRINDING AND SUPER FINISHING

6. GRINDING 195–219
6.1 Grinding 195
6.2 Grinding Wheels 195
6.3 Types of Grinding Wheels 197
6.3.1 Straight Grinding Wheels 197
6.3.2 Cylinder or Wheel Ring 197
6.3.3 Tapered Grinding Wheels 197
6.3.4 Straight Cup 197
6.3.5 Dish Cup 198
6.3.6 Saucer Grinding Wheels 198
6.3.7 Diamond Grinding Wheels 198
6.4 Grinding Wheel Components 198
6.4.1 Grade of Grinding Wheel 200
6.4.2 Bonds 200
6.5 Abrasive Grain Types 201
6.5.1 Abrasive Grain Properties 201
6.5.2 Abrasive Manufacturing 202
6.5.3 Abrasive Grain Types 202
6.6 Grade/Hardness 203
6.7 Structure 203
6.8 Mechanics of Grinding 203
6.8.1 Surface Grinding 203
6.8.2 Maximum Chip Thickness (t_m) in Surface Grinding 205
6.8.3 Mechanism of Grinding Wheel Wear 206
6.9 Selection of a Grinding Wheel 207
6.9.1 Specification of Selection Factors 208
6.9.2 Grinding Operations 209
6.10 Standard Wheel Shapes 209
6.11 Wheel Care and Safety 209
6.12 Types of Grinding Machine 210
6.13 Surface Grinding 210
6.14 Centerless Grinding 213
Objective Type Questions 215
Answers 218
Questions 218

7. SUPER FINISHING PROCESS 220–226
7.1 Introduction 220
7.2 Honing Process 220
7.2.1 Honing Machine 221
7.2.2 Advantages of Honing 222
7.2.3 Limitations of Honing 222
7.3 Lapping 223
7.3.1 Lapping Speed 223
7.3.2 Method of Lapping 224
7.3.3 Advantages of Lapping Process 225
7.3.4 Limitations of Lapping 225
7.3.5 Applications of Lapping 225
7.4 Polishing 225
Questions 226

8. LIMITS, FITS TOLERANCE AND SURFACE ROUGHNESS 227–241
8.1 Introduction 227
8.2 FITS 228
8.2.1 Terminology for Limits and Fits 229
8.2.2 Classification of Fits 230
8.2.3 Fit Between Mating Parts 230
8.3 Basic Hole System 231
8.4 Surface Roughness or Surface Texture 237
8.4.1 Specification 238
Objective Type Questions 239
Answers 240
Questions 240

UNIT–IV METAL JOINING (WELDING)

9. METAL JOINING 245–315
9.1 Welding 245
9.1.1 History of Arc Welding 245
9.1.2 Classification of Welding Process 247

9.1.3 Metals that Can be Welded 247
9.1.4 Advantages of Welding 249
9.1.5 Limitations of Welding 249
9.1.6 Comparison of Welding with Other Joining Process 249
9.1.7 Applications of Welding 250

9.2 Gas Welding 251
9.2.1 Principle of Oxy-acetylene Welding 251
9.2.2 Availability of O_2 and C_2H_2 Gases 253
9.2.3 Oxygen 254
9.2.4 Acetylene 254
9.2.5 Acetylene Generator 254
9.2.6 Safety/Precautions of Acetylene Cylinder 255
9.2.7 Backfire/Popping 255
9.2.8 Flashback 256
9.2.9 Arrangement of Oxy-acetylene Gas Welding 256
9.2.10 Welding Torch 257
9.2.11 Gas Welding Technique 258
9.2.12 Edge Preparations for Butt Joint with Left-Hand Welding 258
9.2.13 Right-Hand Welding Technique 259
9.2.14 Advantages of Gas Welding 260
9.2.15 Limitations of Gas Welding 260
9.2.16 Applications of Gas Welding 260
9.2.17 Oxy-acetylene/Gas-cutting 261
9.2.18 Operating of Cutting Torch 261

9.3 Arc Welding Process 262
9.3.1 Functions of Flux Used for Coating of Electrode 264
9.3.2 Edge Preparation 264
9.3.3 Striking the Arc 264
9.3.4 Arc Blow 264
9.3.5 Arc Length 266
9.3.6 Welding Equipments Required for Arc Welding 266
9.3.7 Comparison Between AC and DC Arc Welding 268
9.3.8 Advantages of Arc Welding 268
9.3.9 Limitations of Arc Welding 269
9.3.10 Applications of Arc Welding 269

9.4 Metal Inert Gas Welding (MIG) 269
9.4.1 Introduction 269
9.4.2 Principle of Operation 270
9.4.3 Equipments Required for GMAW 270
9.4.4 Different Type of Materials can be Welded by MIG 272
9.4.5 Advantages of MIG Welding 272
9.4.6 Metal Transfer in MIG 273

9.5 Tungsten Inert Gas Welding (TIG) 274
9.5.1 Principle of Operation 274
9.5.2 Advantages of GTAW 276
9.5.3 Limitations 276
9.5.4 Applications of TIG Welding 276
9.5.5 TIG Welding Bead 276
9.5.6 Comparision Between TIG (GTAW) and MIG (GMAW) Welding 277

9.6 Resistance Welding 277
9.6.1 Variables of Resistance Welding 278
9.6.2 Types of Resistance Welding 280
9.6.3 Seam Welding 282
9.6.4 Projection Welding 284
9.6.5 Percussion Welding 286
9.6.6 Flash Butt Welding 287

9.7 Atomic Hydrogen Welding 290
9.7.1 Introduction 290
9.7.2 Procedure 291
9.7.3 Equipments 291
9.7.4 Advantages of Atomic Hydrogen Welding 291
9.7.5 Disadvantages of Atomic Hydrogen Welding 292
9.7.6 Applications 292
9.7.7 Submerged Arc Welding 292

9.8 Electroslag Welding 295
9.8.1 Working Principle 295
9.8.2 Advantages 297
9.8.3 Limitations 297
9.8.4 Electrogas Welding 297

9.9 Solid State Welding 299
9.9.1 Friction Welding 299
9.9.2 Welding Parameters (Variable) 301
9.9.3 Advantages of Friction Welding 301

9.9.4 Limitations of Friction Welding 301
9.9.5 Materials that Can be Welded 301
9.10 Flux Cored Arc Welding (FCAW) 302
9.10.1 Definition and Concept 302
9.10.2 Principle of Operation 302
9.10.3 Welding Equipments 302
9.10.4 Weldable Metals 304
9.10.5 FCAW Advantages and Applications 304
9.10.6 FCAW Disadvantages 304
9.11 Soldering and Brazing 305
9.11.1 Soldering 305
9.11.2 Brazing 308
Questions 313

UNIT–V UNCONVENTIONAL MACHINING PROCESSES AND WELDING

10. METALLURGICAL ASPECT IN WELDING 319–353

10.1 Metallurgy of Fusion Weld 319
10.1.1 Heat Affected Zone (HAZ) 320
10.1.2 Effect of Heat on Welding and Heat Affect Zone (HAZ) 321
10.2 Welding Stresses 324
10.2.1 Residual Stresses 325
10.2.2 Method for Reducing/Relieving the Stresses 325
10.2.3 Chief Factors Responsible for Residual Stress 325
10.2.4 Measurement and Determination of Residual Stresses in Weldments 326
10.2.5 Solidification Shrinkage 327
10.2.6 Contraction of Solid Metal 327
10.2.7 Longitudinal Shrinkage 327
10.2.8 Transverse Shrinkage 328
10.3 Main Causes for Distortion 329
10.3.1 Main Types of Distortion 329
10.3.2 How Much Shall we Allow for Weld Shrinkage? 331
10.3.3 Transverse Shrinkage 331
10.3.4 Longitudinal Shrinkage 331
10.3.5 Factors Affecting the Distortion 331
10.4 Welding Defect 331
10.4.1 General Sources of Welding Defects 332
10.4.2 Excessive Penetration Bead 332
10.4.3 Lack of Penetration 332
10.4.4 Undercut 333
10.4.5 Cracks 333
10.4.6 Hot Cracking 334
10.4.7 Factors Promoting Hot Cracking 334
10.4.8 Cold Cracking 335
10.4.9 Cold Cracking—Causes and Remedies 335
10.4.10 Factor Responsible for Cold Cracking 335
10.4.11 Distortion 336
10.4.12 Creater 336
10.4.13 Porosity 336
10.4.14 Slag Inclusion 337
10.4.15 Incomplete Fusion and Penetration 337
10.4.16 Inclusions 338
10.4.17 Spatters 339
10.4.18 Poor Weld Bead Appearance 339
10.5 Weld Decay in Steel HAZ of Stainless Steel Weld 339
10.5.1 Preventing of Weld Decay 340
10.5.2 Work-Hardened Metals 340
10.6 Factors that Affect the Weldability 341
10.6.1 Weldability Testing 341
10.7 Welding Joints 342
Objective Type Questions 344
Answers 351
Questions 351

11. UNCONVENTIONAL MACHINING PROCESS (UCM) 354–391

11.1 Introduction 354
11.2 Classifications of UCM Process 354
11.3 Non-conventional Machining Process are Followings 356
11.4 Common UCM Process 356
11.4.1 Comparison of Unconventional Machining Process 356

11.5 Advantages of UCM Process 357
11.6 Limitations of UCM Process 357
11.7 Electro-Chemical Machining (ECM) 357
11.7.1 Introduction 357
11.7.2 Principle of ECM 358
11.7.3 Characteristic of ECM 359
11.7.4 Metal Removal Rate in ECM 359
11.7.5 Advantages of ECM 360
11.7.6 Limitations of ECM Process 360
11.7.7 Electrolyte Used in ECM 361
11.7.8 Applications of ECM 361
11.7.9 Power Needed for ECM 362
11.8 Electrical Discharge Machining (EDM) 363
11.8.1 Working Principle of EDM 364
11.8.2 Advantages of EDM 365
11.8.3 Limitations of EDM Process 365
11.8.4 Dielectric Fluid Used in EDM 365
11.8.5 Condenser Circuit and Metal Removal Rate 366
11.8.6 EDM Wire Cut Machine 367
11.9 Ultrasonic Machining Process (USM) 368
11.9.1 Introduction 368
11.9.2 Working Principle 368
11.9.3 Elements of USM 369
11.9.4 Advantages of USM Process 370
11.9.5 Limitations of Process 370
11.9.6 USM Characteristics 370
11.9.7 Material Removal Rate 371
11.9.8 Applications 372
11.10 Abrasive Jet Machining 372
11.10.1 Introduction 372
11.10.2 Elements of AJM 373
11.10.3 Advantages of AJM 374
11.10.4 Limitations of AJM 375
11.10.5 Applications of AJM 375
11.11 Laser Beam Machining (LBM) 375
11.11.1 Introduction 375
11.11.2 Working Principle 375
11.11.3 Characteristic of LBM 377
11.11.4 Advantages of LBM 377
11.11.5 Limitations of LBM 378
11.11.6 Applications of LBM 378
11.11.7 Common Application of LASER 378
11.12 Electron Beam Machining 378
11.12.1 Introduction 378
11.12.2 Working Principle 379
11.12.3 Characteristic of EBM 379
11.12.4 Electron Beam Machining–Process 380
11.12.5 Advantages of EBM 380
11.12.6 Limitations of EBM 380
11.12.7 Applications of EBM 380
11.12.8 Generation of Contract of Electron Beam 381
11.12.9 Electron Beam Machining–Equipment 381
11.13 Ion Beam Machining 382
11.13.1 Ion Source 383
11.13.2 Plasma Source 383
11.13.3 Extraction Grid 383
11.13.4 Ground Electrodes 384
11.13.5 Power Supply 384
11.13.6 Focus of Ion Beam 384
11.13.7 Application of Ion Beam 384
11.13.8 Advantage of Ion Beam Machining 385
11.13.9 Limitation of ion Beam Machining 385
Objective Type Questions 385
Answers 389
Questions 389

12. UNCONVENTIONAL WELDING PROCESS 392–411

12.1 Laser Beam Welding 392
12.1.1 Working Principle 393
12.1.2 Types of Laser 394
12.1.3 Advantages of Laser Beam Welding 395
12.1.4 Disadvantages 395
12.1.5 Applications 395
12.2 Ultrasonic Welding 396
12.2.1 Welding Equipments 397
12.2.2 Working Principle of Ultrasonic Welding 398
12.2.3 Materials Suitable for Ultrasonic Welding (Application) 398
12.2.4 Applications of Ultrasonic Welding 398

12.2.5 Advantages of Ultrasonic Welding 399
12.2.6 Limitations of Ultrasonic Welding 399
12.3 Electron Beam Welding 399
12.3.1 Definition 400
12.3.2 Electron Gun 400
12.3.3 Vacuum Chamber 401
12.3.4 Procedure 401
12.3.5 Advantages of EBW 401
12.3.6 Limitations of EBW 402
12.3.7 Applications of EBW 402
12.4 Plasma Arc welding 402
12.4.1 Non-transferred Arc 403
12.4.2 Uses of Equipment 403
12.4.3 Process 404
12.4.4 Advantages of Plasma Arc Welding 405
12.4.5 Limitations of PAW 405
12.4.6 Applications 405
12.4.7 Difference between the Plasma Arc Welding (PAW) and TIG Welding 406
12.5 Explosive Welding 406
12.5.1 Development of Explosive Welding 407
12.5.2 Procedure 407
12.5.3 Salient Feature About Explosive Welding 408
12.5.4 Weld Interface and Weld Parameter 410
12.5.5 Terms Used in Explosive Welding 410
12.5.6 Advantages of Explosive Welding 410
12.5.7 Limitations 410
12.5.8 Applications: Main Applications 411

EXPERIMENTS 413–441

BIBLIOGRAPHY 442–443

PREVIOUS YEAR EXAMINATION PAPER 444–445

INDEX 446–450

UNIT-I

Metal Cutting

Chapter 1

Theory of Metal Cutting

1.1 INTRODUCTION

Metal cutting or machining is a cutting/machining process in which the unwanted material is removed from the surface of workpiece and produced the required shape and size of object. In metal cutting operations a ***single point*** *or* ***multi point*** **cutting** tool is used as a cutting tool and cutting tool is always hardend as compared to workpiece. One of the major drawback of metal cutting operations that material losses take place in term of ***chips*** which are very difficult to dispose off.

Machining of metal basically depends upon the following factors:

(i) Type of work piece or object material (hard or soft)

(ii) Cutting tool material

(iii) Cutting fluid is used or not

(iv) Feed

(v) Speed

(vi) Depth of cut

(vii) Cutting tool geometry.

1.1.1 Various Metal Removal Process

Different types of metal removal processes are given below:

(i) Turning process (on lathe machine)

(ii) Drilling process (on drill machine)

(iii) Milling process (on milling machine)

(iv) Grinding process (on grinding machine)

(v) Boaring process (on boring machine)

(vi) Gear cutting

(vii) Shaping and planing

(viii) Unconventional machine process

(ix) Broaching process etc.

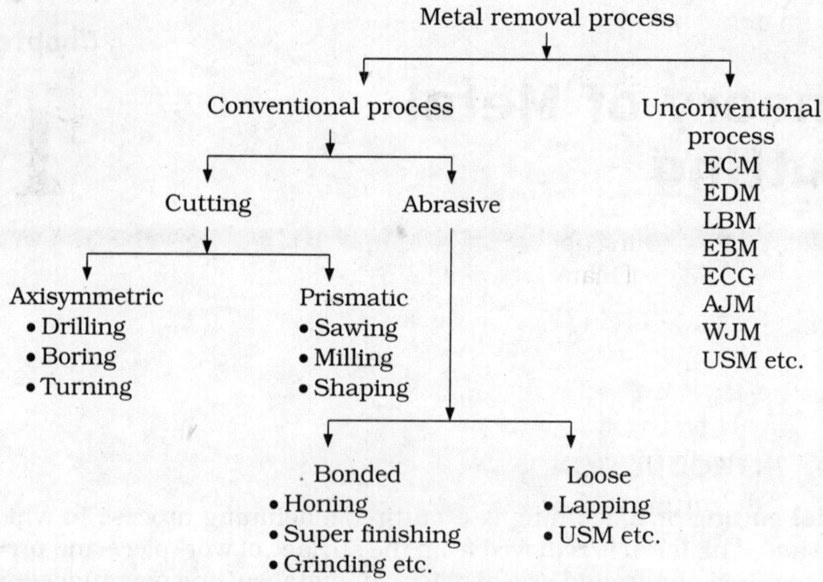

Fig. 1.1 Block diagram of a metal removal process

Chip Formation: As it is known that during the metal cutting operation chips are produced and this chip are various type. Basically generation of chip depends upon the various factor such as *type of material machined, feed, speed, depth of cut* and many more, dispose of the same chips are also very important task.

1.2 CUTTING TOOL

Machining was carried out by plain carbon steel or hardened alloy steel till 1900. Machining speed was very low at that time as cutting tool was made out of less hardened steel. In between **Taylor** discovered high speed as a cutting tool material. Alloy cast tool was discovered in 1915. They are having higher hot hardness and wear resistance. Sintered tungsten carbide tools was discovered in 1925 by using power metallurgy. Cemented oxide tool material was discovered in 1960.

Cutting tool is a item or device which is used to remove the unwanted materials from the surface of workpiece.

Principal categories of cutting tools include single point lathe tools, multi point milling tools, drills, reamers, and taps etc. All of these tools may be standard catalog items or tooling designed and custom-built for a specific manufacturing need.

The number one error when selecting tooling is calculating monetary savings based on lowest cost per tool, rather than on maximized productivity and extended tool life.

To effectively select tools for machining, a machinist or engineer must have specific information about:

- starting and finished part shape
- workpiece hardness
- material's tensile strength

- material's abrasiveness
- type of chip generated
- work holding setup
- power and speed capacity of the machine tool.

Changes in any of these conditions may require a thorough review of any cutting tool selection.

Different machining applications require different cutting tool materials. The ideal cutting tool material should have all of the following characteristics:

- harder than the work that to be machined
- high temperature stability
- resists wear and thermal shock
- impact resistant
- chemically inert to the work material and cutting fluid.

No single cutting tool material incorporates all these qualities. Instead, trade-offs occur among the various tool materials. For example, ceramic cutting tool material has high heat resistance, but has a low resistance to shock and impact. Every new and evolving tool development has an application where it will provide superior performance over others. Many newer cutting tool materials tend to reduce, but not eliminate the applications of older cutting tool materials.

Cutting tools are basically two types:

(i) Single point cutting tool

(ii) Multi point cutting tool.

1.2.1 Single Point Cutting Tool

A single point cutting tool consist only one cutting edge. In metal cutting operation single point tool used in **Lathe**, **Shaper, Planer** and **Slotter** machines.

Various parts of single point cutting tool

A single point cutting tool having the following parts:

1. **Shank:** It is the main body of cutting tool and this portion is usually gripped in the tool post at the starting end of the tool. There is no provision of cutting angles.
2. **Nose radius:** This is also important element of single point cutting tool. It is the point where the side cutting edge and end cutting edge intersect. Nose radius affect the surface roughness of the parts (workpiece). Usually it is represented by *r* and its values lies from 0.2 to 1.4 mm.
3. **Flank:** It is the surface below cutting edge.
4. **Heel:** It is the intersection of flank and base of the cutting tool.
5. **Cutting edge:** It is the edge of cutting tool face which removes the unwanted material from the workpiece.

Cutting edge are two types:

(i) Side cutting edge

(ii) End cutting edge

(i) ***Side cutting edge:*** It is formed by the intersection of flank and side flank. It is used in cutting.

(ii) ***End cutting edge:*** It is formed by the intersection of face and flank.

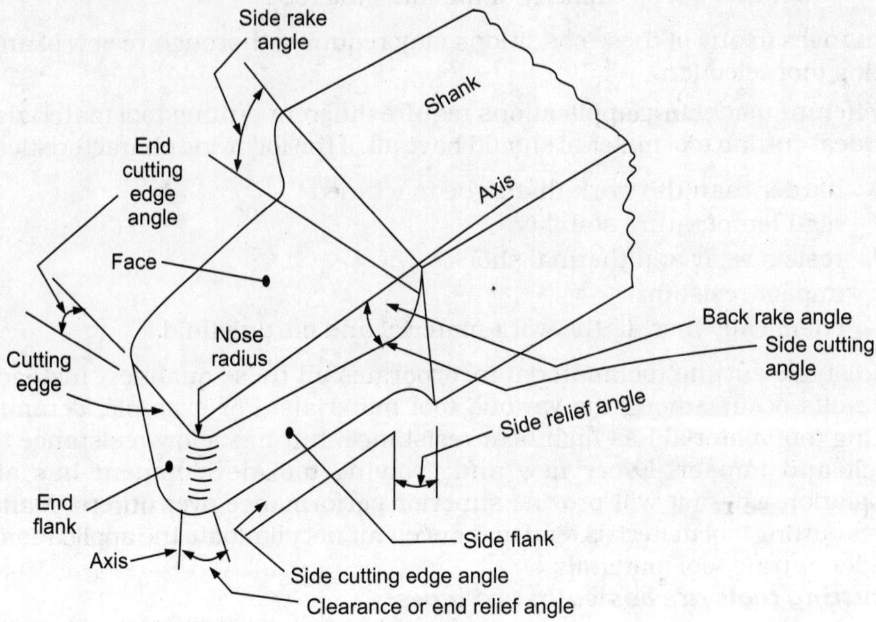

Fig. 1.2 Element of a single point cutting tool.

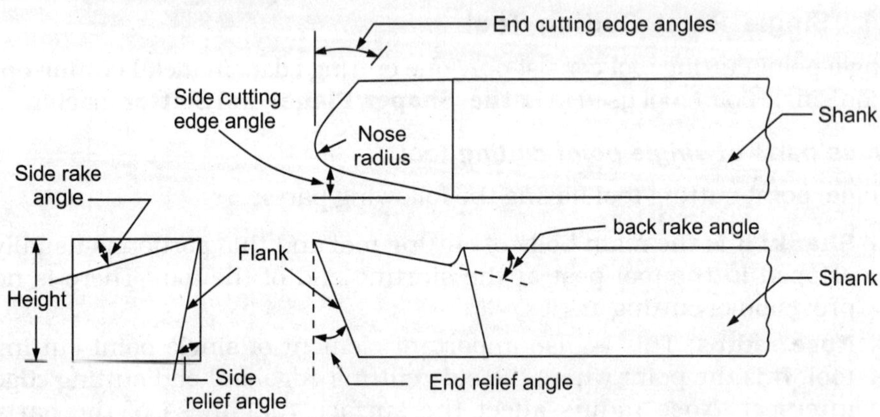

Fig. 1.3 Various angles of a single point cutting tool.

1.2.1.1 Designation of a Single Point Cutting Tool

Designation of a cutting tools are **seven** and given below:

(i) Side rake angle : It is denoted by α_s, and this angle is measured the slope of the top surface of the tool to the side, in a direction perpendicular to longi-tudinal axis.

(ii) End relief angle : It is denoted by θ_e and it is the angle between the portion of the front flank below the cutting edge and a line drawn through the cutting edge perpendicular to the base.

(iii) Side relief angle : It is denoted by θ_s and it is the angle between the flank and a vertical plane through the side cutting edge.

(iv) End cutting edge angle : It is denoted by C_e and it is the angle between the end cutting edge and a line perpendicular to the longitudinal axis of the straight shank tool.

(v) Back rake angle : It is denoted by d_s and it is the angle between the tool face and line parallel to the base of tool and it is measured in plane (perpendicular) through the side cutting edge.

(vi) Side cutting edge angle: It is denoted by C_s and it is the angle between the side cutting and longitudinal axis of the tool.

(vii) Nose radius : Junction of the side cutting edge and end cutting edge is called nose and its radius is called ***nose radius***. It is denoted by *r*.

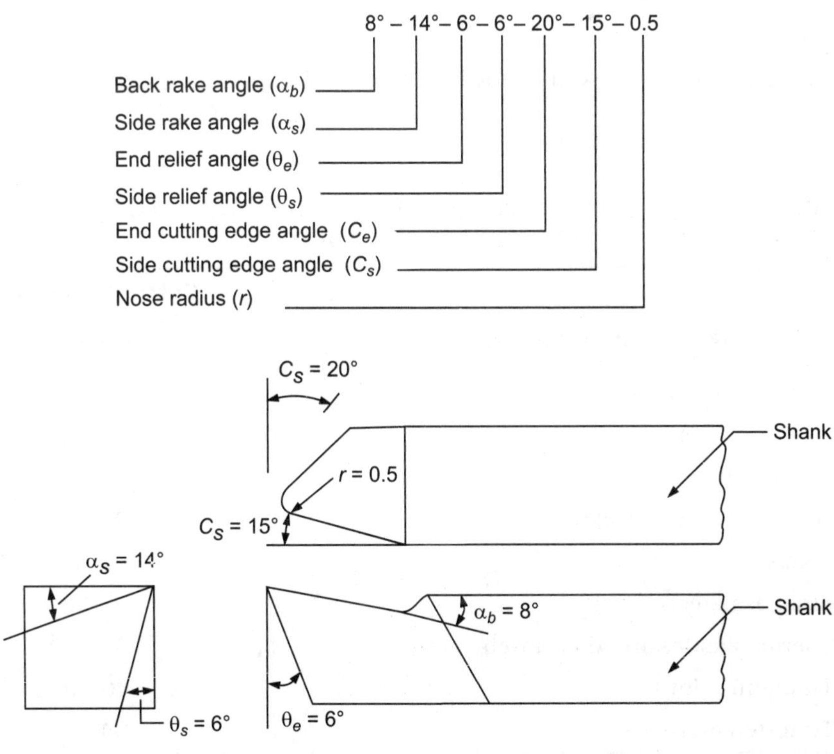

Fig. 1.4 Three views of single point cutting tool

1.2.1.2 Back rake angle

It is the angle between the tool face and a line parallel to the base of tool and it is measured in a plane (perpendicular) through the side cutting edge. Rake angle may be **zero**, **positive** or **negative** for different metals.

Positive rake angle for soft material while on the other hand **negative** and zero rake angle may be for very hard and brittle material (like ceramics etc.).

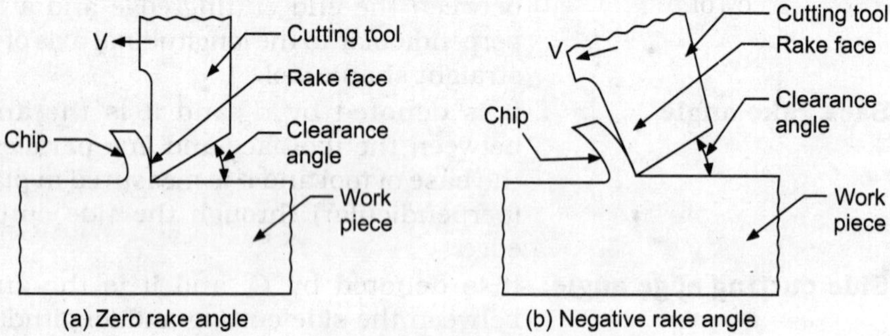

Fig. 1.5 Rake angle [19]

Positive Rake Angle:

- Better surface finish can be produced by positive rake angle.
- By positive rake angle power consumption can be reduced while tool life can be increases.
- Less cutting forces are required due to sharpness of cutting edge.

Negative Rake Angle: More creater produce.

Table 1.1 Approximate ranges of recommended cutting speed in turning operation

Workpiece material	**Cutting speeds** *m/min*
Aluminium alloys	200–1000
Cast iron, gray	60–900
Copper alloys	50–700
High-temperature alloys	20–400
Steels	50–500
Stainless steels	50–300
Thermoplastics and thermosets	90–240
Titanium alloys	10–100
Tungsten alloys	60–150

Indexable carbide Inserts are available now in the wide variety of shapes such as **square**, **Triangular**, **rectangular**, **rhombus**, **hexagonal**, **pentagonal** etc. ANSI introduce a special numbering system for inserts and tool holder.

For example: TNMG323E

T = Triangular shape of insert

N = relief angle (N = 0°)

M = accuracy of point location for M – ±0.002″ to ±0.004″

G = type of insert

3 = size of insert (9.53 mm)

2 = thickness of 1/16″(2 = 1/8″ = 3.18 mm)

3 = cutting point radius in 64 of an inch (3 = 3/64″ = 1.19 m)

E = other (E = un round insert)

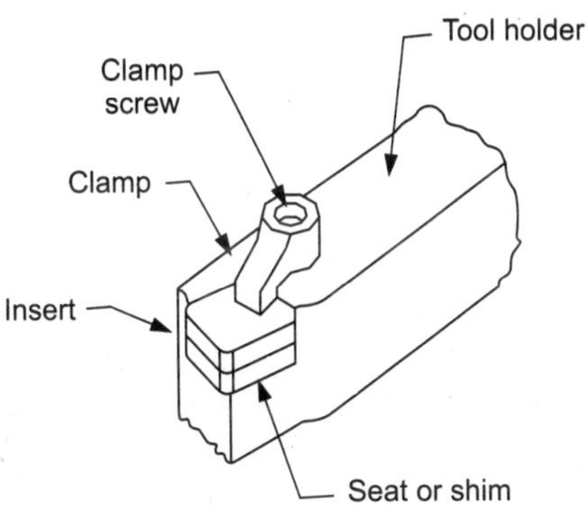

Fig. 1.6 (*a*) Square insert in tool holder

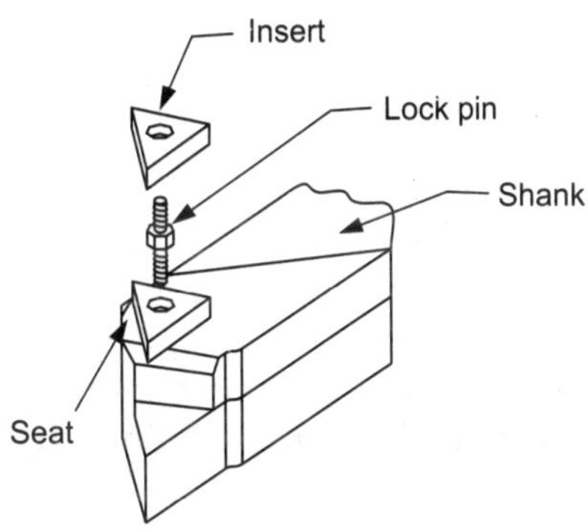

Fig. 1.6 (*b*) Trangular insert in tool holder

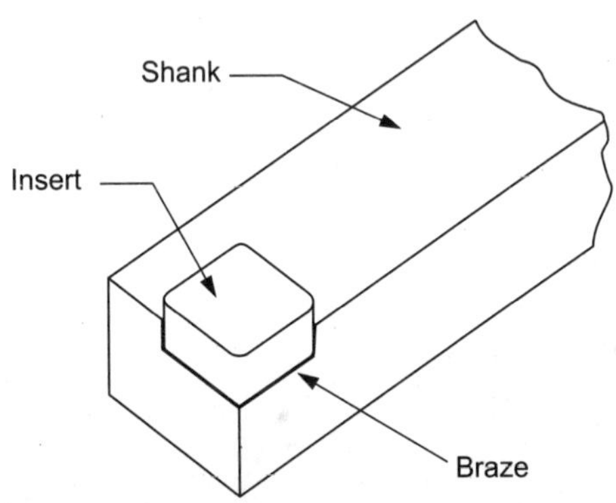

Fig. 1.6 (*c*) Insert brazed with tool holder

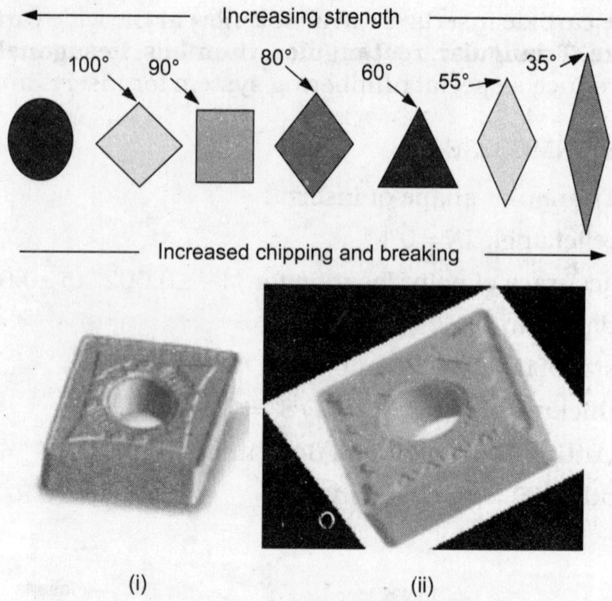

Fig. 1.6 (d) Different Inserts

1.2.2 Multi Point Cutting Tools

Multi point cutting tool is a device which is used to remove the unwanted material from the workpiece to convert it into a desired shape and size. Multi point cutting tools are differ with single point cutting tools in manner of cutting edge. Single point cutting tools have only *one cutting* edge where as multi point cutting tools consist *more* than *one cutting edge*. In metal cutting operation multi point cutting is used generally in ***Drill, milling, hobbing, broaching, tapping, reaming,*** etc. Process geometry of multi point cutting tool is quite different with single point cutting tool.

Multi pint cutting tool may be out of *high speed steel (HSS), crome steel* etc. depending upon the material to be machined.

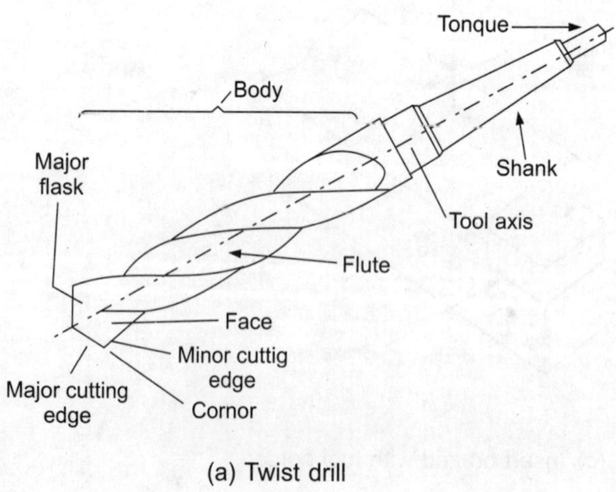

(a) Twist drill

(b) Tap tool

(c) Hobbing tool

(d) Milling cutters

Fig. 1.7 Multi-point cutting tools

1.2.3 ASA System (American System)

According to ASA system, geometry of the rake face of a cutting tool is expressed in term of back rake angle and side rake angle (Fig. 1.8). It is a quite easy to measure the setting angles for grinding fixture in terms of back rake angle and side rake angle, but the main draw back of this system, is that special fied angle are not related to actual position of cutting edge.

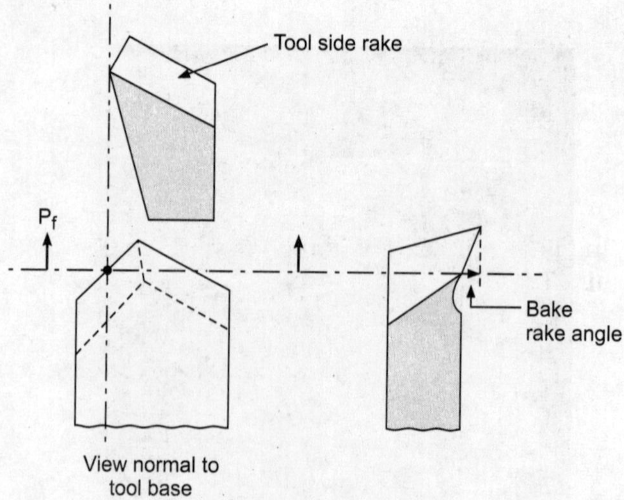

Fig. 1.8 ASA system

1.3 CHIP FORMATION MECHANISM

Chip formation mechanism is shown in Fig. 1.9. A wedge shape cutting tool is used in this process as the tool comes contact with the workpiece, it exerts a pressure on workpiece, resulting in the compression of the metal near the tool tip as tool travel in advance to workpiece due to friction, plastic deformation of workpiece take place near the cutting edge of tool. This sharp edge region of cutting tool is refer as ***shear zone***. This shear plane is actually a narrow zone (0.025 mm).

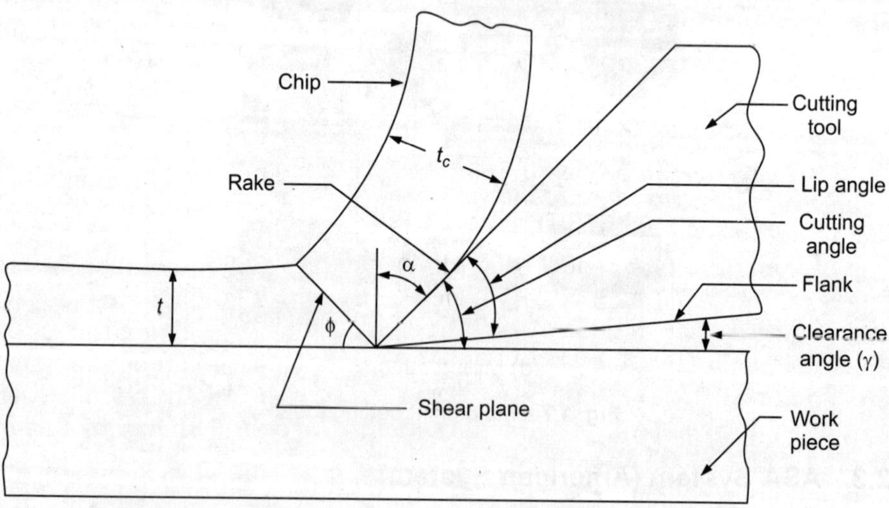

t = Uncut chip thickness (mm)

t_c = Cut chip thickness (mm)

α = Rake angle

ϕ = Shear angle

γ = Relief angle

Fig. 1.9 Mechanism of chip formation

Shear angle value usually depends upon the following factors:

(i) Type of workpiece material
(ii) Cutting tool material
(iii) Cutting condition
(iv) Cutting tool geometry.

Metal cutting is a complex phenomenon first of all it should very clear that what is chip. To remove the unwanted material from the workpiece is known as **chip**. This is one of the most ***drawback*** of metal cutting process that material losses take place in the form of chip Fig. 1.10 shows the mechanism of chip formation. During the metal cutting operation chips produced due to the **plastic deformation** of the material, the zone from which metal removed is refer as shear zone. Separation of material from workpiece start as a yielding or fracture.

In metal cutting operation (turning) a wedge shape tool is used, having the two intersecting surfaces to form the cutting edge. Surface along which the chip flows as known *as **rake face*** and surface which is relived to clear the newely surface is refer *as **flank***.

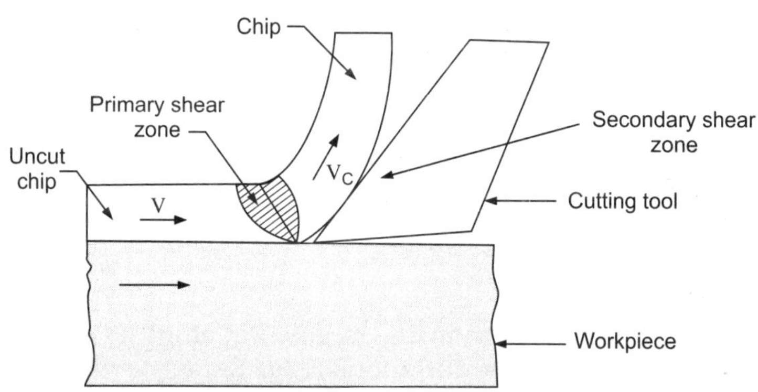

Fig. 1.10 Plastic deformation of material during a turning operation

As applying the force the sharp edge of cutting tool gradually comes in contact with workpiece by applying the continuous pressure on the tool, the value of stress acting on workpiece increases. Initially it reach the elastic limit and then finally plastic limit. After cross the plastic limit the deformation of workpiece starts and finally chips produced. So during the metal cutting operation chips produced due to ***compressive force*** and ***shearing action***.

To calculate shear strain with the help of Piispanen's model shown in Fig. (1.11) shear strain γ is given by

$$\gamma = \frac{\Delta S}{\Delta Y} = \frac{AB}{CD} = \frac{AD}{CD} + \frac{DB}{CD} = \tan\phi + \cot(\phi - \alpha)$$

$$\gamma = \frac{\cos\alpha}{\sin\phi\cos(\phi-\alpha)} = \frac{V_S}{V\sin\phi}$$

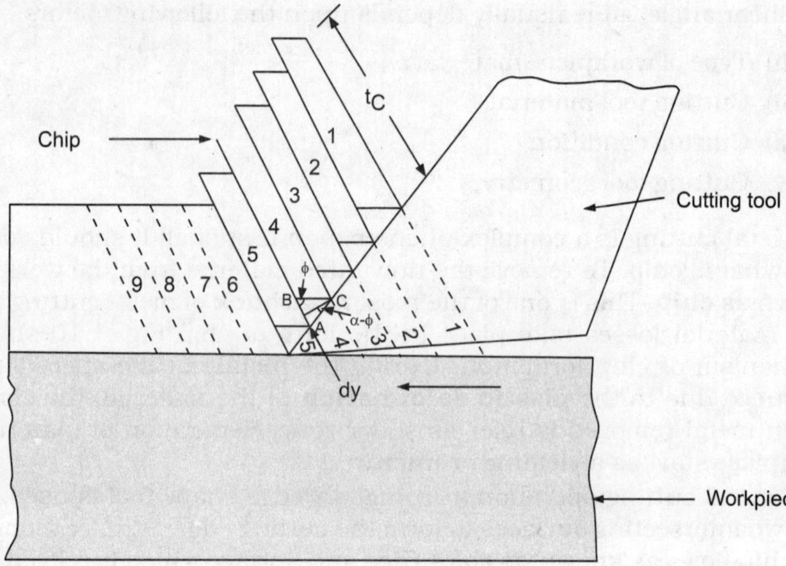

Fig. 1.11 Piispanen's model of chip formation [13]

1.4 TYPES OF CHIPS

During the machining operation usually three types of chips are produced:

(i) Discontinuous chip
(ii) Continuous chip
(iii) Continuous chip with built up edge.

1.4.1 Discontinuous Chip

This type of chips are also known as segmental chips as they generate in the form of small pieces. Discontinuous chips usually generate during the machining of brittle material such as *Cast iron, ceramics, bronze,* etc. This type of chips are not longer in size (length) and dispose off this type chips are very easy. No need to used chip dispose off device, chip break.

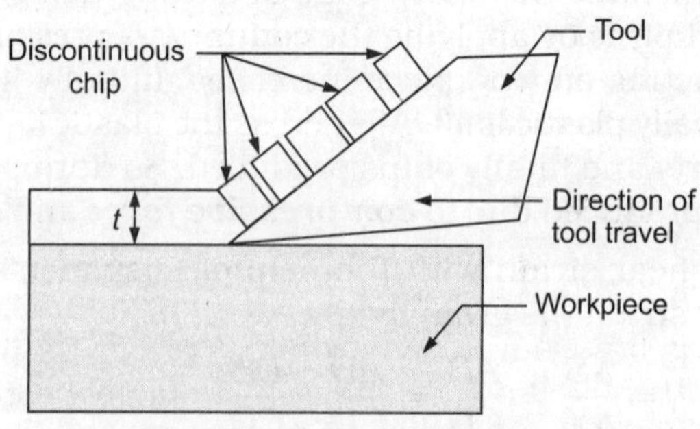

Fig. 1.12 Discontinuous chip

1.4.1.1 *Favourable Condition for Discontinuous Chip*

- Brittle and hard material
- Lower cutting speed
- Small rake angle
- Higher depth of cut
- High friction between tool and work piece interface.

1.4.2 Continuous Chips

Such type of chips are in the form of *long ribbon* of uniform thickness as shown in Fig. 1.13. These type of chips are produced during the machining of ductile material like *Mild steel, Aluminium, Copper, Low Carbon steel* etc., at high speed. This type of chips are in a single piece and are in the form of spiral, thickness of this type chips are through out uniform and dispose off this type chips are difficult as compared to discontinuous chip.

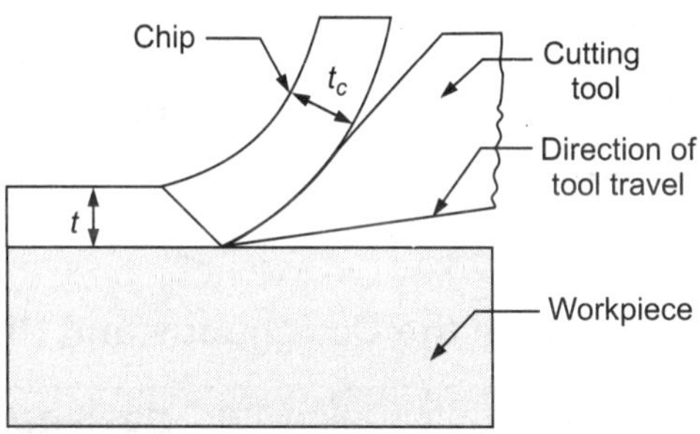

Fig. 1.13 Continuous chip

1.4.2.1 *Following Factors are Responsible for Continuous Chips*

- Ductile material (such as M.S, Al, Cu etc.)
- Low coefficient of friction between the tool and work interface
- High cutting speed
- Large rake angle
- Coolent used during process
- Sharp cutting edge
- Force feed.

1.4.3 Continuous Chip with Built-up Edge

Such type of chips are also long but not smooth as (continuous chips) shown in Fig. 1.14. The basic difference between the continuous chip and continuous chips with built up edge that continuous chips with built-up edge have a sharp **burr** on the edge of chip and the dispose off same is very difficult as if no regular dispose off the chip then it welded with the tool due to high friction this may responsible for tool wear.

BUE chips are produced due to the high local temperature and extreme pressure in the cutting zone.

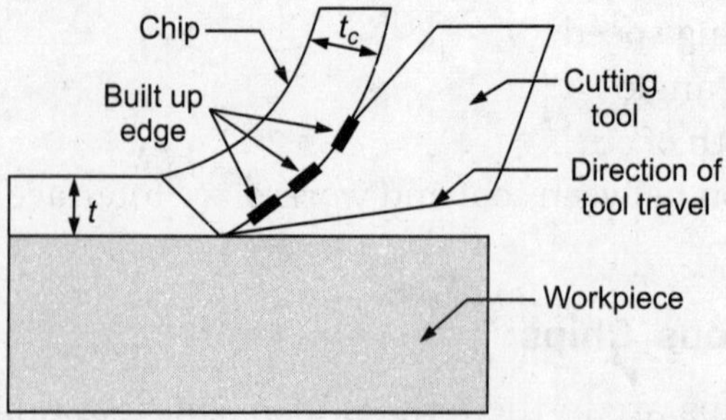

Fig. 1.14 Continuous chip with built-up (BUE) edge.

1.4.3.1 *Following Factors are Responsible for BUE Chip*

- Ductile material (such as M.S, Al etc.)
- Low rake angle
- Coarse feed
- High friction between the tool and work interface
- Low cutting speed
- Insufficient use of cutting fluid.

1.4.4 Differentiate Between the Continuous and Discontinuous Chip

Continuous chips	*Discontinuous chips*
(i) 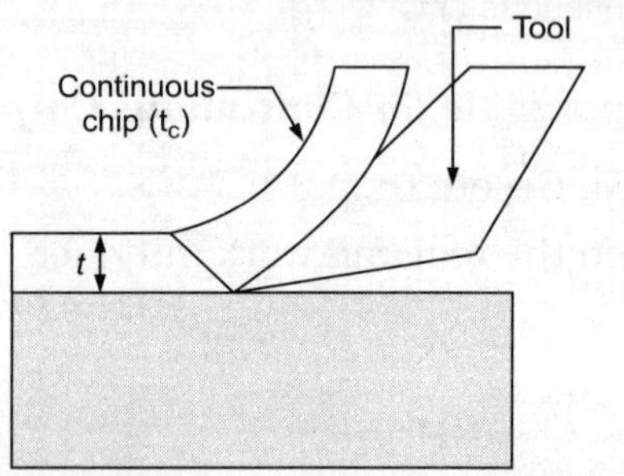	(i) 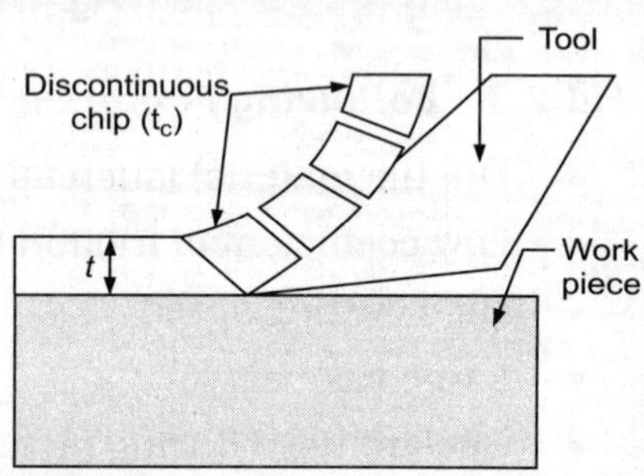
(ii) This type of chips are produced during the machining of ductile materials like M.S, Al etc.	(ii) This type of chips are produced during the machining of brittle materials like-cast iron etc.
(iii) Large rake angle may because of continuous chips.	(iii) Small rake angle may because of discontinuous chips.
(iv) Factors responsible for continuous chips. (a) Low friction (b) Fine feed (c) High cutting speed (d) Ductile material.	(iv) Factors responsible for discontinous chips. (a) High friction (b) Coarse feed (c) Low cutting speed (d) Brittle materials.

1.5 TYPES OF METAL CUTTING PROCESS

Metal cutting process **broadly** classified into two groups:

(i) Orthogonal cutting (two dimensional cutting process)

(ii) Oblique cutting (three dimensional cutting process).

1.5.1 Orthogonal Cutting

In orthogonal cutting operation cutting edge is straight, parallel to the original plane surface of workpiece and perpendicular to the direction of cutting or in simple words, a cutting process in which the cutting edge of tool is arrange to be perpendicular to the direction of relative work, tool motion is known as orthogonal cutting.

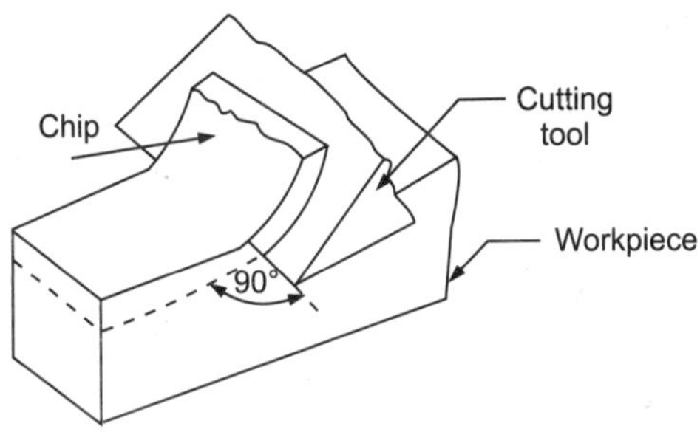

Fig. 1.15 Orthogonal cutting

1.5.1.1 Requirements for Orthogonal Cutting Operation

- Tool should be perfectly sharp
- Continuous chip should be produced without any built up edge
- Depth of cut is constant and smaller than the width of cut
- Chip should not flow either side and direction of chip flow velocity is normal to the cutting edge
- Width of cutting edge is greater than the width of workpiece.

1.5.2 Oblique Cutting

In oblique cutting operation, the cutting edge is placed at an angle to normal to the direction of tool travel. In oblique cutting chip flow freely. All modern machining operations consist **oblique cutting**.

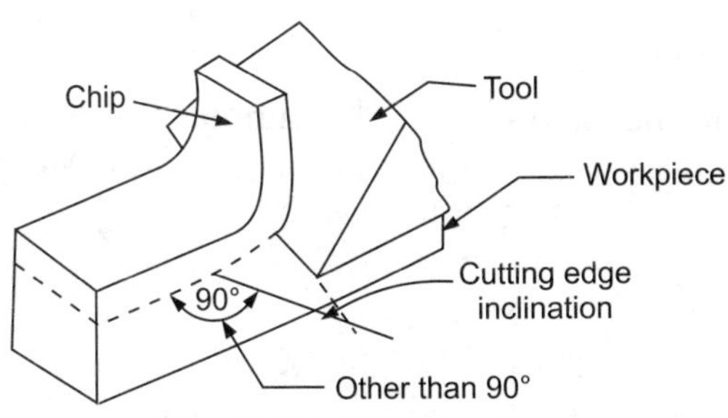

Fig. 1.16 Oblique cutting

1.5.2.1 *Requirements of Oblique Cutting*

- Cutting edge should be inclind at other than 90° angle with the normal to the direction of tool travel.
- Width of the cutting edge may or may not be greater than width of work piece.
- Three components of forces should be act on the cutting edge and they should be mutually perpendicular to each other.

1.5.3 Differentiate between Orthogonal and Oblique Cutting

Orthogonal cutting	***Oblique cutting***
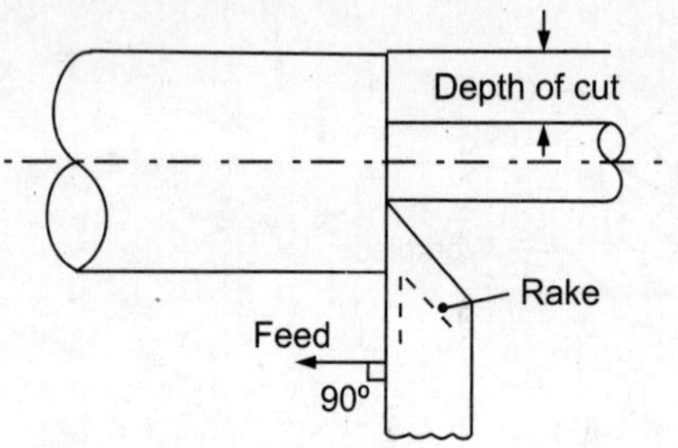	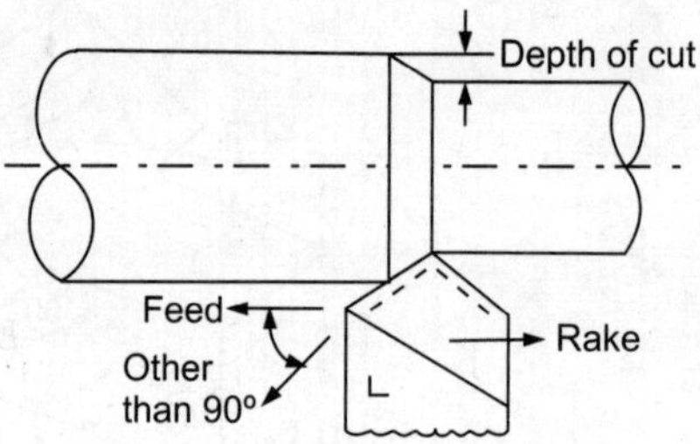
(i) Orthogonal cutting is known as 2-dimensional cutting operation.	(i) Oblique cutting is known as 3-dimensional cutting operation.
(ii) In orthogonal cutting operation tool life is shorter.	(ii) In obleque cutting tool life is more.
(iii) In orthogonal cutting operation 2-component of forces are acting.	(iii) In oblique cutting 3-component of forces are acting.
(iv) In orthogonal cutting operation cutting forces of the tool is making an angle of 90° to the line of action or path of the tool.	(iv) In oblique cutting forces of the tool is making an angle other than 90°.
(v) Maximum chip thickness produce at the middle of chip.	(v) Maximum chip thickness may not produced at middle.
(vi) Chip removed are flats and spiral in shape.	(vi) Chips removed are of curling type.
(vii) In orthogonal cutting operation more heat generated.	(vii) In oblique cutting less heat is generated.
Example: Knief turning, slotting parting.	**Example:** Turning on a Lathe machine and boaring.

1.6 DETERMINATION OF SHEAR ANGLE [UPTU-2010-11]

Before calculating the shear plane angle it is very important to know— What is shear plane angle? ***Shear plane angle*** is the angle between the cutting velocity vector and the plane across, which the shear deformation of workpiece take place. It is usually denoted by ϕ, during the chip formation process there is continuous flow of material, therefore no change in volume of work material. Hence

$$t \cdot b \cdot v = t_c \cdot V_c \cdot b_c$$

where t = uncut chip thickness,

b = width of cut,

v = cutting velocity

Similarly, t_c = cut chip thickness

b_c = width of chip,

V_c = chip velocity.

as $b_c > b$

So above equation can be write like, $t \times v = t_c \times v_c$

or $$\frac{t}{t_c} = \frac{V_c}{V}$$

or $$r = \frac{t}{t_c} = \frac{V_c}{V} = \frac{l_c}{l}$$

where r = Chip thickness ratio

L_c = length of chip formed

L = uncut length on workpiece.

Now from Fig. 1.17, $$\sin\phi = \frac{t}{AC}$$

Depth of cut $t = AC \sin \phi$

Chip thickness $t_c = AC \cos (\phi - \alpha)$

but $$r = \frac{t}{t_c}$$

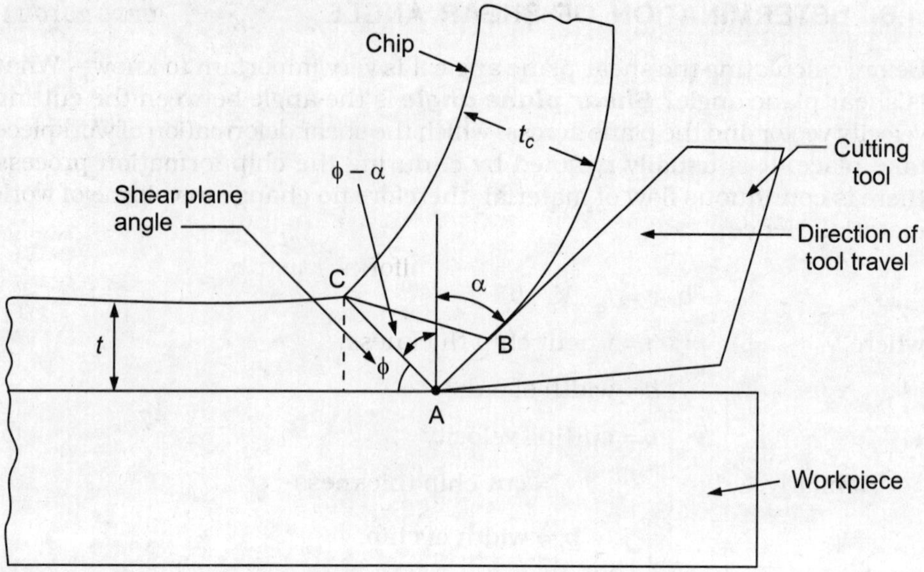

Fig. 1.17 Shear plane and shear plane angle

or $$r = \frac{AC \sin \phi}{AC \cos(\phi - \alpha)}$$

$$r = \frac{\sin \phi}{\cos(\phi - \alpha)}$$

or $$\frac{1}{r} = \frac{\cos\phi \cos\alpha + \sin\phi \sin\alpha}{\sin\phi}$$

$$\frac{1}{r} = \cos\alpha \cdot \cot\phi + \sin\alpha$$

$$\frac{1}{r} = \frac{\cos\alpha}{\tan\phi} + \sin\alpha$$

$$\frac{1}{r} - \sin\alpha = \frac{\cos\alpha}{\tan\phi}$$

$$\frac{1 - r\sin\alpha}{r} = \frac{\cos\alpha}{\tan\phi}$$

or $$\boxed{\tan\phi = \frac{r\cos\alpha}{1 - r\sin\alpha}}$$

1.7 EARNEST MERCHANT'S MODEL FOR ORTHOGONAL CUTTING

[UPTU-2011-12]

Ernest Merchant analysis [1945] is for the orthogonal cutting. This model is based on minimum energy dissipation during the cutting process.

1.7.1 Assumptions

1. Cutting tool have sharp edge.
2. Workpiece material is rigid and perfectly elastic.
3. Uncut chip thickness is uniform throughout.
4. Continuous chip is produced without any built up edge.
5. Stresses on the shear plane are uniformly distributed.
6. There is no uniform distribution of normal and shear stresses on the shear plane.

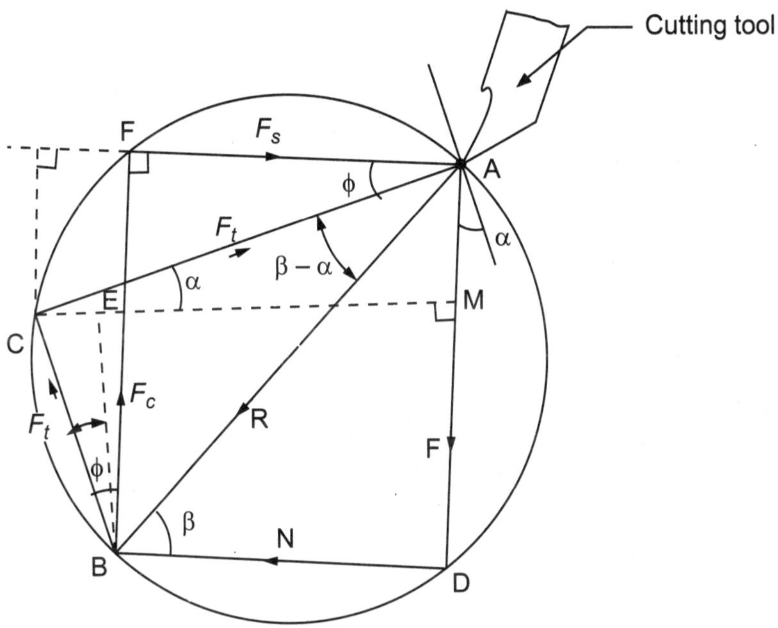

Fig. 1.18 Merchant's force circle

where F_t = tangential force or cutting force

F_f = feed force

F_s = shear force on the shear plane

F_c = compressive force on the shear plane

F = frictional force

N = normal force

α = rake angle

ϕ = shear plane angle

β = angle of friction

but $$F = DH + HA$$

$$= F_f \cos\alpha + F_t \sin\alpha$$

Now, $$N = BH = BD$$

$$= CM - CE$$

$$= F_t \cos\alpha - F_f \sin\alpha$$

$$\text{Ratio of } \frac{F}{N} = \frac{F_f \cos\alpha + F_t \sin\alpha}{F_t \cos\alpha - F_f \sin\alpha}$$

$$= \frac{F_f + F_t \tan\alpha}{F_t - F_f \tan\alpha}$$

But $$\mu = \tan\beta = \frac{F}{N}$$

where μ = coefficient of friction

$$\mu = \tan\beta$$

$$= \frac{F_f + F_t + \tan\alpha}{F_t - F_f \tan\alpha}$$

This friction is not equal to sliding case friction, now shear plane area A is calculated by

$$A_s = \frac{b \times t}{\sin\phi}$$

where b = width of cut

t = uncut chip thickness

Shear force $$F_s = \tau_s \times A_s$$

$$= \frac{\tau_s \times b \times t}{\sin\phi}; \qquad \tau_s = \text{Shear stress}$$

but normal stress is $$\tau_s = \frac{N_s}{A_s}$$

or $$N_s = A_s \times \tau_s$$

$$= \frac{\tau_s \times b \times t}{\sin\phi}$$

Now, $$F_s = F_t \cos\phi - F_f \sin\phi$$

and $$F_c = F_f \cos\phi + F_t \sin\phi$$

and $$F_t = R\cos(\beta - \alpha) \quad \text{and} \quad F_f = R\sin(\beta - \alpha)$$

$$F_s = R\cos(\beta-\alpha)\cos\phi - R\sin\phi\sin(\beta-\alpha)$$

$$= R\cos(\beta-\alpha+\phi)$$

Now

$$\frac{F_t}{F_s} = \frac{R\cos(\beta-\alpha)}{R\cos(\beta-\alpha+\phi)}$$

$$= \frac{\cos(\beta-\alpha)}{\cos(\beta-\alpha+\phi)}$$

$$F_t = \frac{F_s\cos(\beta-\alpha)}{\cos(\beta-\alpha+\phi)}$$

$$= \frac{\tau_s \times b \times t\cos(\beta-\alpha)}{\cos(\beta-\alpha+\phi)\times\sin\phi}$$

On differentiation with regard to ϕ

$$\frac{\partial F_t}{\partial\phi} = \frac{-\tau_s\cdot b\cdot t\cos(\beta-\alpha)\,[\cos\phi\cdot\cos(\beta-\alpha+\phi)+\sin\phi\cdot\sin(\beta-\alpha+\phi)]}{\{\sin\phi\cdot\cos(\beta-\alpha+\phi)\}^2}$$

$$\frac{\partial F_t}{\partial\phi} = \frac{-\tau_s\cdot b\cdot t\cos(\beta-\alpha)\cdot\cos(2\phi+\beta-\alpha)}{\{\sin\phi\cdot\cos(\beta-\alpha+\phi)\}^2}$$

For minimum energy condition

$$\frac{\partial F_t}{\partial\phi} = 0$$

$$\frac{-\tau_s\cdot b\cdot t\cos(\beta-\alpha)\cdot\cos(2\phi+\beta-\alpha)}{\{\sin\phi\cdot\cos(\beta-\alpha+\phi)\}^2} = 0$$

or

$$\cos(2\phi+\beta-\alpha) = 0 = \cos\frac{\pi}{2}$$

$$2\phi+\beta-\alpha = \frac{\pi}{2}$$

or

$$\boxed{\phi = \frac{\pi}{4} - \frac{(\beta-\alpha)}{2}}$$

1.7.2 Chip Velocity

Let us assume that work piece is moving against the cutting tool with a velocity V. Chip velocity V_C along the tool rake face is the vector sum of V. Shear velocity V_S is the velocity of chip relative to the work piece and directed along the shear plane.

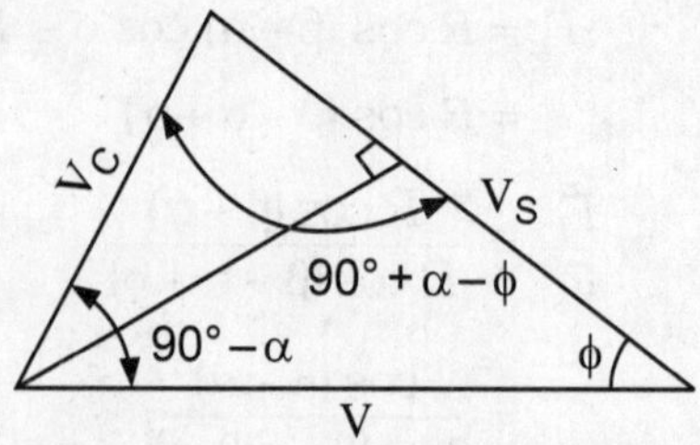

Fig. 1.19

Applying properties of trangle in above Fig. 1.19

$$\frac{V}{\sin(90+\alpha-\phi)} = \frac{V_S}{\sin(90-\alpha)}$$

$$= \frac{V_C}{\sin\phi}$$

$$V_C = \frac{V\sin\phi}{\cos(\phi-\alpha)}$$

and
$$V_S = \frac{V\cos\alpha}{\cos(\alpha-\alpha)}$$

Power required: Power required during the turning process is calculated by

$$P = \frac{V \times f_C}{60} \text{ watts}$$

where f_C = Tangential force (N)

V = Velocity (m/min)

1.8 LEE AND SHAFFER'S THEORY *[UPTU 2011-12]*

Lee and Shaffer's theory for orthogonal cutting operation is based on the slip line field theory.

1.8.1 Assumptions

- Chip produced during the process are stress free.
- Workpiece material is ideal plastic rigid.
- Existence of plane similar to earnest-merchant shear plane.

In Fig. 1.20 direct stress is zero on line AC. Hence slip line are at 45° to AC. Stress field is thus a set of orthogonal straight line and stress distribution is uniform.

Slip line field for orthogonal cutting is shown in Fig. 1.20

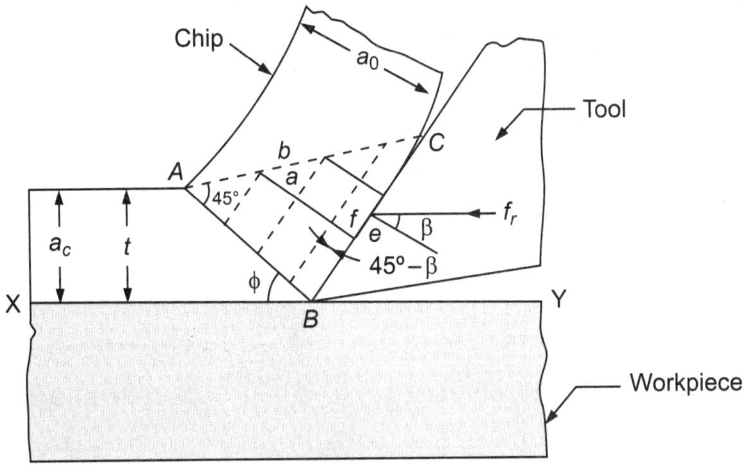

Fig. 1.20 Shear plane model of Lee and Shaffer

R = resultant tool force

β = mean friction angle on tool force

ϕ = shear plane angle

α = rake angle

t = uncut chip thickness

t_C = chip thickness

ΔBCA contain the zone of deformation plane AC is stress free and slip line angle. AC is 45° *AB* is the shear plane and there is velocity discontinuity along AC.

from figure

$$\angle XBA = \phi$$

$$\angle CBY = 180° - 90° - \alpha = 90° - \alpha$$

$$\angle ABC = 45 + \beta$$

$$\phi + \angle ABC = 90° + \alpha$$

$$\phi = 45° - (\beta - \alpha)$$

or

$$\phi + \beta - \alpha = 45°$$

Case (i). For zero rake angle if $\beta = 45°$ and $\alpha = 0°$ then $\phi = 0°$ high friction and low rake angle is responsible for built up edge formation.

Case (ii). Negative rake angle if $\beta = 350°$ and $\alpha = -10°$ than $\phi = 0°$ this solution is again impracticable.

1.9 CUTTING FLUIDS

During the machining operation (turning, drilling, milling etc.) due to the friction between the tool and workpiece interface high heat energy generation take place, which may lead metallurgical change of workpiece/components, hence the cooling of workpiece as well as cutting tool is very essential otherwise tool wear can also occur. So a cutting fluid is used during the machining operation (tool-workpiece chip zone) to improve the machining performance.

So cutting fluid is known as a substance which may be **Liquid, gas** is applied in between the cutting zone of workpiece and cutting tool to cool the workpiece and cutting tool as well as to wash away the chips.

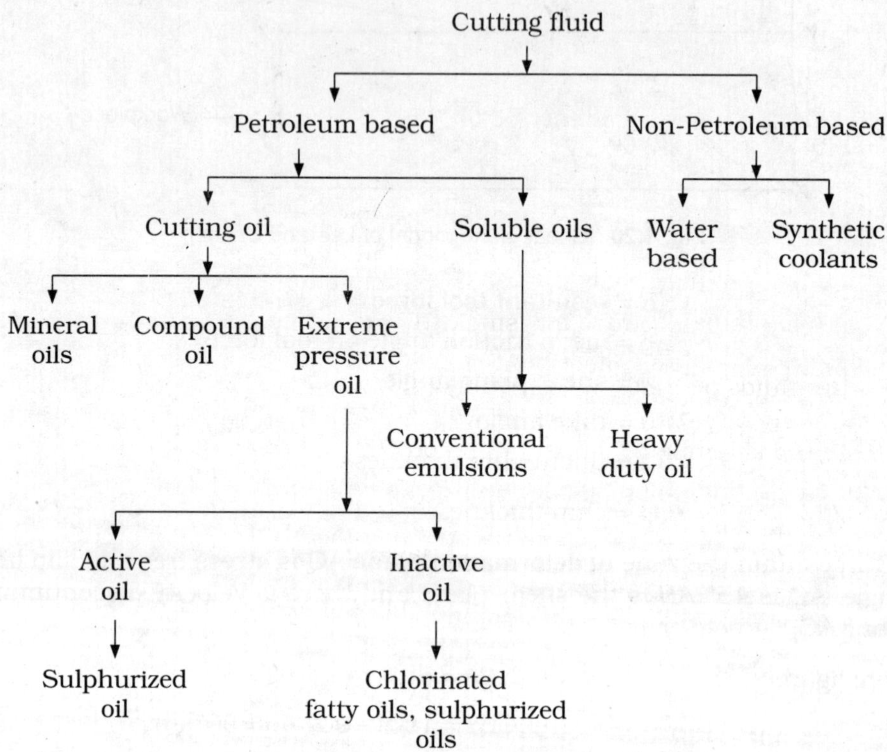

Fig. 1.21 Flow chart of cutting fluids

1.9.1 Main Functions of a Cutting Fluid

- To cool the workpiece
- To cool the cutting tool
- To make a thin film layer between workpiece and cutting tool to reduce the friction
- To prevent the workpiece surface for any rusting
- To improve the surface finish (surface roughness)
- To wash away the chips from cutting zone
- To avoid the built up edge formation
- To reducing work piece thermal deformation.

1.9.2 Requirements of a Good Cutting Fluid

- It should not react with the work piece material as well as cutting tool material
- It should have high heat absorption capacity
- It should have good thermal conductivity
- It should not be toxic
- It should have good lubrication property
- It should not harm full to operation skin as well as eyes
- It should have high boiling and high flash point
- It should be transparent
- It should be easy handle, easy storage for a long time with out changing its properties
- It should be economical
- It should have low viscosity
- It should not produce any smoke during process.

Cutting fluids play a **significant role** in machining operations and impact shop productivity, tool life and quality of work. The **primary function** of cutting fluid is temperature control through cooling and lubrication.

A fluid's cooling and lubrication properties are critical in decreasing tool wear and extending tool life. Cooling and lubrication are also important in achieving the desired size, finish and shape of the workpiece.

A secondary function of cutting fluid is to flush away chips and metal fines from the tool/workpiece interface to prevent a finished surface from becoming marred and also to reduce the occurrence of built-up edge (BUE).

1.9.3 Composition of a Cutting Fluid

Cutting fluid consist the following:

- Base oil
- Emulsifier
- Stabilizing agent
- Corrosion inhibitors
- Foam inhibitors
- Lubricating, anti wear and extreme pressure additive
- Netutralizing agent.

1.9.4 Functions of a Cutting Fluid

(i) Lubrication
(ii) Cooling
(iii) Anti-wetting.

(i) Lubrication: This is one of the unique function of cutting fluid. Basic aim of lubrication is to reduce the friction between the tool and workpiece interface and the mechanism of lubrication is quite different with hydrodynamic lubrication which is used for ball bearing case. In the lubrication system cutting fluid makes a thin film between the workpiece and cutting tool to reduces the friction thus prevent intimate contact of workpiece and cutting tool.

Through a film of lubrication the coefficient of friction between the sliding chip and tool and worn tool flank problem can be reduced. This lubrication is a time dependent with increasing the cutting speed its effect decreases.

(ii) Cooling: It is well knowned that during the cutting process a high degree of heat produced, due to friction between the tool interface and workpiece which could be responsible for the change metallurgy of workpiece and cause for tool wear there fore to avoid this problem cooling is must. For cooling purpose a fluid is used and cutting fluid should so much capable that it can absorb all the heat itself. For cooling purpose merchant presented an idealized picture of the void presenting between the chip and cutting tool during cutting operation. For proper/efficient cooling it is must that the cutting fluid must proper penetrate to the tool.

(iii) Anti-wetting: It of spite the cooling and lubrication action of the cutting fluids, there is always metal to metal contact (in machining) between the tool and work interface, heat generated during process is enough to weld the contacting asperities of the metal. To prevent this compounds of **chlorine, sulphur, phosphorus** etc. are added to

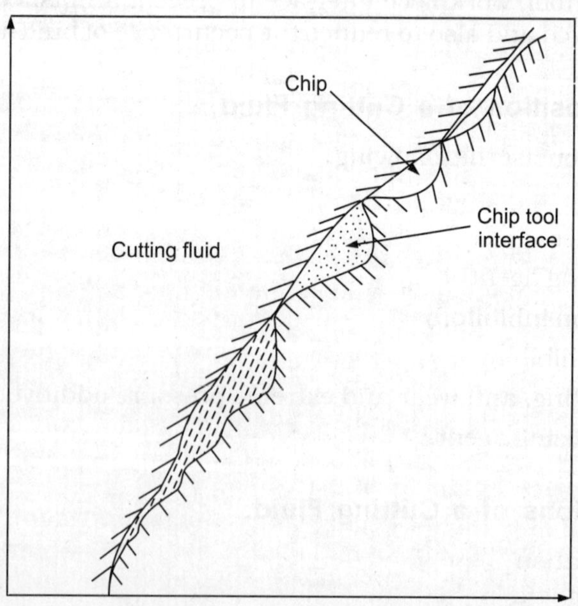

Fig. 1.22 Merchant idealized picture of cutting fluid between tool and chip.

cutting fluids. These compounds give rise to a soapy, metallic film which prevents particles of workpiece from welding.

TABLE 1.2 Comperative performance of various cutting fluids

S. No.	*Type of cutting fluid*	*Lubrication properties*	*Cooling properties*	*Anti wetting properties*
1.	Air	Poor	Poor	Poor
2.	Water	Poor	Excellent	Poor
3.	Fatty Oil	Excellent	Fair	Fair
4.	Emulsion	Fair	Good	Fair
5.	E.P. Cutting Oil	Good	Fairly good	Excellent
6.	Mineral Oil	Good	Fairly good	Fair

Practically all cutting fluids presently in use fall into one of **four categories**:

- Straight oils
- Synthetic fluids
- Soluble oils
- Semi synthetic fluids

Fig. 1.23 Application of cutting fluid

(i) Straight oils are non-emulsiable and are used in machining operations in an undiluted form. They are composed of a base mineral or petroleum oil and often contains polar lubricants such as **fats**, **vegetable oils** and **esters** as well as extreme pressure additives such as **Chlorine**, **Sulphur** and **Phosphorus**. Straight oils provide the best lubrication and the poorest cooling characteristics among cutting fluids.

(ii) Synthetic Fluids contain no petroleum or mineral oil base and instead are formulated from alkaline inorganic and organic compounds along with additives for corrosion inhibition. They are generally used in a diluted form (usual concent ratio = 3 to 10%). Synthetic fluids often provide the best cooling performance among all cutting fluids.

(iii) Soluble Oil Fluids form an emulsion when mixed with water. The concentrate consists of a base mineral oil and emulsifiers to help produce a stable emulsion. They are used in a diluted form (usual concentration = 3 to 10%) and provide good lubrication and heat transfer performance. They are widely used in industry and are the least expensive among all cutting fluids.

(iv) Semi-synthetic fluids are esentially combination of synthetic and soluble oil fluids and have characteristics common to both types. The cost and heat transfer performance of semi-synthetic fluids lie between those of synthetic and soluble oil fluids.

1.9.5 Cutting Fluid Effects in Machining

The **primary functions** of cutting fluids in machining are:

- Lubricating the cutting process primarily at low cutting speeds.
- Cooling the workpiece primarily at high cutting speeds.
- Flushing chips away from the cutting zone.

Secondary Functions Include:

- Corossion protection of the machined surface.
- Enabling part handling by cooling the hot surface.

Process Effects of Using Cutting Fluids in Machining Include:

- Longer Tool Life.
- Reduced Thermal Deformation of Workpiece.
- Better Surface Finish (in some applications).
- Ease of Chip and Swarf handling.

1.9.6 Cutting Fluid Selection Criteria

The principal criteria for selection of a cutting fluid for a given machining operation are:

- Process performance
  - Heat transfer performance
  - Lubrication performance
  - Chip flushing
  - Fluid mist generation
  - Fluid carry-off in chips
  - Corrosion inhibition
  - Fluid stability (for emulsions)
- Cost Performance
- Environmental Performance
- Health Hazard Performance

1.9.7 Cutting Fluid Maintenance and Disposal

Monitoring and maintenance of cutting fluid is required due to contamination and degradation. Eventually, fluids require disposal once their efficiency is lost. Waste management and disposal become a major problem concerning environmental liability. The primary concern is the significant negative effects to worker's health associated with use of the cutting fluids.

Cutting fluid maintenance involves checking the concentration of soluble oil emulsions (using refractometers), ***pH (using a pH meter)***, the quantity of tramp oil (hydraulic oil leaking into the cutting fluid system) and the quantity of particulates in the fluid. Action taken to maintain the fluid includes adding make-up concentrate or water, skimming of tramp oil, adding **biocides** to prevent **bacterial** growth and filtering the particulates by centrifuging.

The cutting fluid within a coolant system degrades with time due to bacterial growth and contamination with tramp oil and fine metal swarf from the machining operation. When it becomes uneconomical to maintain the fluid by regular make-up operations it is dumped. Prior to letting the fluid flow into a sewer system, it should be treated to bring the fluid composition to safe disposal levels.

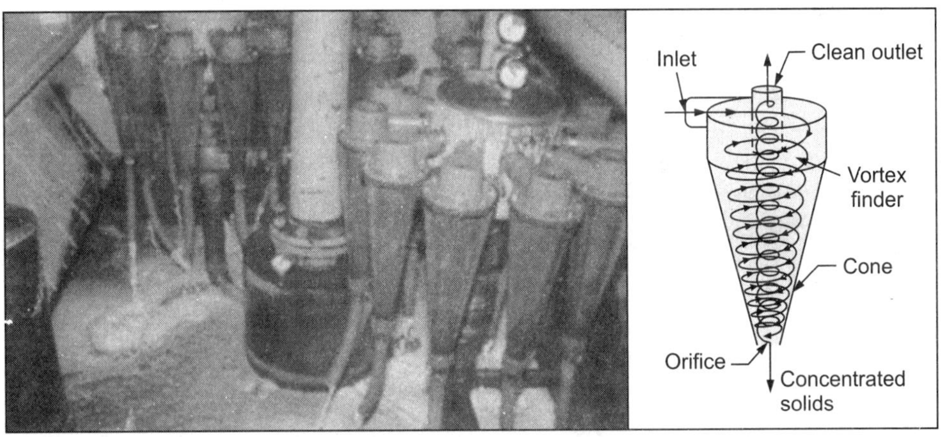

Fig. 1.24 Cutting fluid disease off.

1.10 CUTTING TOOL MATERIALS

1.10.1 Types of Cutting Tool Materials

Various type of cutting toal materials are available for cutting tool purpose and the selection of cutting tool materials depends upon various factor such as type of **material machined**, **surface roughness required**, **cutting speed** etc. before 1900 only plain carbon steel tool was used for cutting purpose but now different and some special characteristic tools are used for cutting purpose.

As rates of metal removal have increased, so the need of heat resistant cutting tools are more. The result has been a progression from high-speed steels to carbide, and on to ceramics and other superhard materials. Developed around 1900, high-speed steel **cuts four times** faster than the carbon steels they replaced. There are over 30 grades of high-speed steel, in three main categories: **tungsten**, **molybdenum**, and **molybdenum-cobalt** based grades. Since the 1960s the development of powdered metal high-speed steel has allowed the production of near-net shaped cutting tools, such as drills, milling cutters and form tools. The use of coatings, particularly **titanium nitride, alloys, high speed steel** tools to cut faster and last longer. **titanium nitride** provides a high surface hardness, resists corrosion, and it minimizes friction.

In industry today, carbide tools have replaced high-speed steels in most applications. These carbide and coated carbide tools cut about **3 to 5 times** faster than high-speed steels. Cemented carbide is a powder metal product consisting of fine carbide particles cemented together with a binder of cobalt. The major categories of hard carbide include **tungsten carbide, titanium carbide**, **tantalum carbide**, and **niobidum carbide**. Each type of carbide affects the cutting tool's characteristics differently. For example, a higher tungsten content increases wear resistance, but reduces tool strength. A higher percentage of cobalt binder increases **strength**, but lowers the **wear resistance**. Carbide is used in solid round tools or in the form of replaceable inserts.

Every manufacturer of carbide tools offers a variety for specific applications.

The proper choice can double tool life or double the cutting speed of the same tool. Shock-resistant types are used for interrupted cutting. Harder, chemically-stable types are required for high speed finishing of steel. More heat-resistant tools are needed for machining the superalloys, like Inconel and Hastelloy.

There are no effective standards for choosing carbide grade specifications so it is necessary to rely on the carbide suppliers to recommend grades for given applications. Manufacturers do use an ANSI code to identify their proprietary carbide product line.

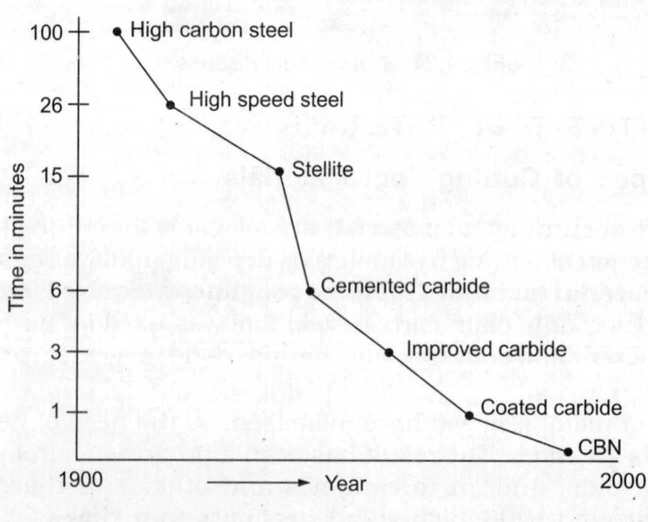

Fig. 1.25 Trend in time of turning a workpiece over last hundred year [8]

Two-thirds of all carbide tools are coated. Coated tools should Parts Produced by EBM.

1.10.1.1 Cutting Tool mostly Possess the Following Properties

(i) Wear resistance: As cutting tool is used for cutting purpose, wear of tool take place and the wear of cutting tool encountered due to the abrasion, adhesion, diffusion. So a cutting tool should have proper wear resistance power, wear resistance is defined as the ability of the tool material to retain its shape and sharpness for a long time during machining of a work piece at a specific cutting speed.

(ii) Hardness: Hardness is the property of a material by virtue of it resist any scratching, cutting, penetration, indentation, abrasion, etc. Hence a cutting tool should have proper hardness to retain its shapes, properties etc. at high temperature while cutting at specific speed, higher the cutting speed, higher the cutting temperature and productivity.

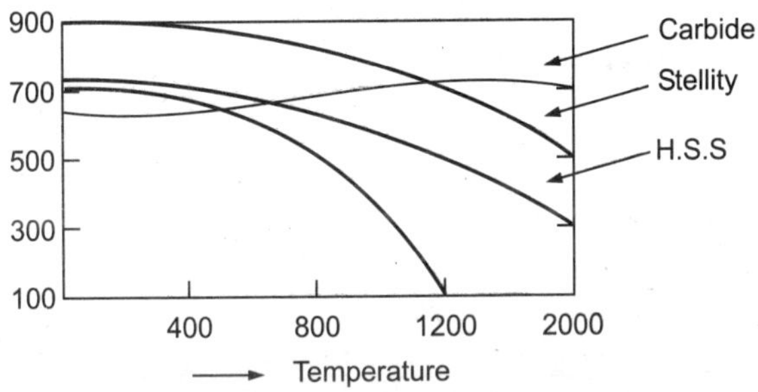

Fig. 1.26 Effect of temperature on the hardness of various tool materials.

(iii) Toughness: Toughness is also referred as impact strength. Toughness of any material is defined as the ability of material to absorbed the energy before the fracture. Tougher material are able with stand impact load.

(iv) Chemical Stability: A cutting tool material should be chemical stable is should not react with any cutting fluid, which a cause of tool wear due to chemical action of fluid.

(v) Other desirable properties: Cutting tool material should not destroyed during high heat generation and its should be easily brazed on tool holder.

(vi) Cutting tool material should have low coefficient of thermal expansion.

1.10.2 Comon Cutting Tool Materials

Following cutting tools material are available:

(i) Plain carbon steel	(ii) High speed steel
(iii) Cemented carbide steel	(iv) Ceramics tool
(v) Diamond tool	(vi) Boron Nitride
(vii) Satellites tool	

1.10.2.1 Plain Carbon Steel

If carbon is present in a steel from 0.6 to 1.5% and some small amount alloys that steel is referred as a plain carbon steel. It is one of the oldest material which was used as a cutting tool and hardness of plain carbon steel varies 60-70 HRC. But if cutting temperature goes beyond the 200°C then their hardness goes to decreasing hence they are very suitable for cutting for materials.

Plain carbon steel can be classified according to their carbon content:

(i) Low carbon or mild steel – 0.05 to 0.30% C

(ii) Medium carbon steel – 0.3 to 0.6% C

(iii) High carbon steel – 0.6 to 1.5 % C

Advantages:

- in expensive
- low hardness 60-70 HRC
- good machinability
- formability to the uncalled state.

Limitations:

- cannot retain their hardness above 200°C
- necessity for rapid quenching in water.

Applications:

- wood working tools
- chisels
- hammers
- kinves and punches etc.

1.10.2.2 High Speed Steel

Abbrivation of high speed steel is **(HSS)** and it is an alloy steel in which some special alloying element such as Tungsten (*W*), cromium (c_r), vanadium (v_d) are added to get the some specific properties.

Specification of HSS

W – 18%,
Cr – 4%.
V_d – 1%

and carbon upto 0.75%. ***HSS*** is the one of the widely material used as a cutting tool due to their high red hardness properties. *HSS* can retain their properties upto 650°C. High speed steel are basically two types one is tungsten based another is molybdenum based.

Advantages:

- Retain their hardness at high temperature.
- high wear resistance power at high temperature.
- good shock absorption properties.
- high degree of toughness as compared to any other cutting tool.
- Low in cost (economical) as compared to diamond and other tools.

TABLE 1.3 Typical composition of HSS

Types of steel	*Chemical composition*					
	C	W	V_d	C_r	C_o	M_b
M1	0.8	1.5	1.0	4.0	1.5	8.0
M6	0.8	4.0	1.55	4.0	12.0	5.0
M30	0.85	2.0	1.25	4.0	5.0	8.0

HIGH-SPEED TOOL STEELS and their requirements are defined by The American Society for Testing and Materials in Specification A600-79 as follows:

High-speed tool steels are so named primarily because of their ability to machine materials at high cutting speeds. They are complex iron-base alloys of **carbon, chromium, vanadium, molybdenum,** or **tungsten,** or combinations thereof, and in some cases substantial amounts of cobalt. The carbon and alloy contents are balanced at levels to give high attainable hardening response, high wear resistance, high resistance to the softening effect of heat, and good toughness for effective use in industrial cutting operations.

Commercial practice has developed two groups of cutting materials:

- The recognized standard high-speed tool steel, which serves almost all applications under mild to severe metal-cutting conditions.
- A smaller group of intermediate steels, which are satisfactory for limited applications under mild to moderate metal-cutting conditions The minimum requirements that must be met to be classed as a standard high-speed tool steel, and those for an intermediate high-speed tool steel, are listed in Table 1.4. To be acceptable for either group, an alloy must meet all of the requirements shown for that group. A chronology of some of the significant developments in high-speed tool steels is given in Table 1.4. The research work in 1903 on a 14% tungsten alloy led to the development of the first high-speed tool steel, which is now designated T_1.

1.10.2.2.1 Molybdenum Based and Tungsten Based Classification

[UPTU 2011-12]

There are presently more than 40 individual classifications of high-speed tool steels, according to the American Iron and Steel Institute (AISI). When these are compounded by the number of domestic manufacturers, the total number of individual steels in the high-speed tool steel category exceeds 150. The AISI established a classification system for the high-speed tool steels many years ago. That system consists of a *T* for those steels that have **tungsten** as one of their primary alloying elements and an *M* for those steels that have **molybdenum** additions as one of their primary alloying elements. In addition, there is a number that follows either the M or the *T*. Thus, there are high-speed tool steels designated M_1, M_2, M_{41}, T_1, T_{15}, and so on. That number does not have any special significance other than to distinguish one from another. For example, *M*, does not mean that it is more highly alloyed than M_2 or has greater hardenability or poorer wear resistance, and so on. It merely separates the types and attempts to simplify selection for the user. Table 1.3 lists the nominal analyses of the common *M* and *T* types.

The *T* (Tungsten based) series contains 12 to 20% tungsten, with chromium, vanadium, and cobalt as the other major alloying elements. The *M* (Molybdenum based) series contains approximately 3.5 to 10% molybdenum, with chromium, vanadium, tungsten, and cobalt as the other alloying elements. All types, whether molybdenum or tungsten, contain about 4% chromium; the carbon and vanadium contents vary. As a general rule, when the vanadium content is increased, the carbon content is usually increased. The tungsten type T_1 does not contain molybdenum or cobalt. Cobalt-base tungsten types range from T_4 through T_{15} and contain various amounts of cobalt. Molybdenum types M_1 through M_{10} (except M_6) contain no cobalt, but most contain some tungsten. The cobalt-base, molybdenum-tungsten, premium types are generally classified in the M_{30} and M_{40} series. Super

TABLE 1.4 Significant year in the development of high-speed tool steels

Year	*Development*
1903	0.70% C, 14% W, 4% Cr prototype of modem high-speed tool steels
1904	0.30% V addition
1906	Introduction of electric furnace melting
1910	Introduction of first 18-4-1 composition (AISI T_1)
1912	3 to 5% Co addition for improved hot hardness
1923	12% Co addition for increased cutting speeds
1939	Introduction of high-carbon high-vanadium super high-speed tool steels (M_4 and T_{15})
1940-1952	Increasing substitution of molybdenum for tungsten
1953	Introduction of sulfurized free-machining high-speed tool steel
1961	Introduction of high-carbon high-cobalt super hard high-speed tool steels (M_{40} series)
1970	Owdered metal high-speed tool steels
1973	Addition of higher silicon/nitrogen content to M_7 to increase hardness
1980	Development of cobalt-free super high-speed tool steels.
1982	Introduction of aluminum-modified high-speed tool steels for cutting tools

high-speed steels normally range from M_{40} upward; they are capable of being heat treated to high hardnesses. The M series steels generally have higher abrasion resistance than the T series steels and less distortion in heat treatment; also, they are less expensive. Tools made of high-speed steel can also be coated with titanium nitride, titanium carbide, and numerous other coatings by physical vapor deposition technique for improved performance and increased tool life.

Molybdenum forms the same double carbide with iron and carbon as tungsten does but has half the atomic weight of tungsten. As a consequence, molybdenum can be substituted for tungsten on the basis of approximately one part of molybdenum, by weight, for two parts of tungsten. The melting point of the molybdenum steels is some what lower than that of the tungsten grades, and thus they require a lower hardening temperature and have a narrower hardening range. The **M-type** high-speed tool steels are **tougher** than the **T-type high-**speed tool steels, but the hot hardness is slightly lower. Compensation for this reduced hot hardness is partially accomplished by the addition of tungsten and, to a lesser extent, vanadium to the plain molybdenum grades. This is one important reason for the popularity of the tungsten-molybdenum grades (like M_2, M_3, M_4). They have good hot hardness, which is so desirable in high-speed tool steels.

1.11 CEMENTED CARBIDES STEEL

Cemented carbides or sintered carbides steel were introduced in the 1930s and have high hardness over a wide range of temperatures, high thermal conductivity, high Young's modulus making them effective tool and die materials for a range of applications. The two groups used for machining are **tungsten carbide** and **titanium carbide**, both types may be coated or uncoated.

Cemented carbide represented a big break through in tool material, they can retain their hardness at a temperature of 1000°, they can operate at much higher speed than H.S.S. cemented carbide are used in the forms of tips and small in rests.

Tungsten carbide particles (1 to 5 micro-m) are bonded together in a cobalt matrix using powder metallurgy. The powder is pressed and sintered to the required insert shape, titanium and niobium carbides may also be included to impart special properties. A wide range of grades are available for different applications. Sintered carbide tips are the dominant type of material used in metal cutting. The proportion of cobalt (the usual matrix material) present has a significant effect on the properties of carbide tools. 3 – 6% matrix of cobalt gives greater hardness while 6 – 15% matrix of cobalt gives a greater toughness while decreasing the hardness, wear resistance and strength. Tungsten carbide tools are commonly used for machining **steels**, **cast irons** and **abrasive non-ferrous materials**.

Titanium carbide has a higher wear resistance than tungsten but is not as tough. With a nickel-molybdenum alloy as the matrix, TiC is suitable for machining at higher speeds than those which can be used for tungsten carbide. Typical cutting speeds are: 30 – 150 m/min or 100 – 250 when coated.

1.11.1 Ceramics – Alumina

Introduced in the early 1950s, and produced by sintering process and the main constituent is Al_2O_3 or Alumina upto 10%. Ceramic tools are brittle by nature and they can retain their hardness at 1400°C they are better resistance to abrasion wear as compared to cemented carbide. This type of tool may be used to machining very hard material i.e., highly abrasive cast **iron**, **heat treated steel** (RC > 65) and monal, stellite (high temperature alloy), for greater strength to the tool tip these is a provision of negative rake angle:

Fine grained high purity **aluminium oxide** (Al_2O_3) and **silicon nitride** (Si_3N_4) are pressed into insert tip shapes and sintered at high temperatures. Additions of titanium carbide and zirconium oxide (ZrO_2) may be made to improve properties. But while **ZrO_2** improves the fracture toughness, it reduces the hardness and thermal conductivity. **Silicon carbide** (SiC) whiskers may be added to give better toughness and improved thermal shock resistance.

The tips have high abrasion resistance and hot hardness and their superior chemical stability compared to HSS and carbides means they are less likely to adhere to the metals during cutting and consequently have a lower tendency to form a built up edge.

Their main weakness is low toughness and negative rake angles are often used to avoid chipping due to their low tensile strengths. Stiff machine tools and work set ups should be used when machining with ceramic tips as otherwise vibration is likely to lead to premature failure of the tip. Due to high cutting speed. chips are produced.

Typical cutting speeds: 150 – 650 m/min.

1.11.2 Diamond

The hardest known substance is diamond. It used as a cutting material in the form of a single crystal structure or polycrystalline compact. Diamond tools are used for light cuts, where these provide high tool life, which can be justify the high cost of diamond. Although single crystal diamond has been used as a tool for machining non-ferrous material such as **Al**, **Cu**, **Brass**, **Bronze** etc. They are **brittle** by nature and need to be mounted at the correct crystal orientation to obtain optimal tool life. Single crystal diamond tools have been mainly replaced by polycrystalline diamond (**PCD**). This consists of very small synthetic crystals fused by a high temperature high pressure process to a thickness of between 0.5 and 1mm and bonded to a carbide substrate. The result is similar to **CBN** tools. The random orientation of the diamond crystals prevents the propagation of cracks, improving toughness. Diamond is not recommenced for cutting ferrous material.

Because of its reactivity, **PCD** is not suitable for machining plain carbon steels or nickel, titanium and cobalt based alloys. they are used for machining glass, reinforced plastic, eutectic alloys etc.

PCD is most suited to light uninterrupted finishing cuts at almost any speed and is mainly used for very high speed machining of aluminium – silicon alloys, composites and other non - metallic materials.

Typical cutting speeds: 200 – 2000 m/min.

1.11.3 Cubic Boron Nitride (CBN)

Introduced in the early 1960s, this is the second hardest material available after diamond. CBN tools may be used either in the form of small solid tips or as a 0.5 to 1 mm thick layer of polycrystalline boron nitride sintered onto a carbide substrate under pressure. In the latter case the carbide provides shock resistance and the CBN layer provides very high wear resistance and cutting edge strength. Cubic boron nitride is the standard choice for machining alloy and tool steels with a hardness of 50 RC or higher.

Typical cutting speeds: 30–310 m/min.

1.12 EFFECT OF ALLOYING ELEMENTS

Alloying elements play an important role in cutting tool material. Basic object of alloying element in a cutting tool is to produced some special properties. Some alloying elements are given below:

1.12.1 Vanadium

Was first added to high-speed tool steels as a scavenger to remove slag impurities and to reduce nitrogen levels in the melting operation, but it was soon found that the element materially increased the cutting efficiency of tools. The addition of vanadium promotes the formation of very hard, stable carbides, which significantly increase wear resistance and, to a lesser extent, hot hardness. An increase in vanadium, when properly balanced by carbon additions, has relatively little effect on the toughness. For this reason, vanadium-beating grades are a very good choice when very fast cutting operations are demanded, as in finishing cuts, or when the surface of the material is hard and scaly. The special characteristics of the high-speed tool steels that are due to high vanadium additions have given rise to several specially developed steels for very severe service requiting high toughness as well as exceptional hot hardness and wear resistance. The T_{15}, M_4, and M_{15} grades are in this category; their vanadium content is 4.88, 4.13, and 5.0%, respectively.

1.12.2 Cobalt

The main effect of cobalt in high speed tool steel is to increase the hot hardness and thus to increase the cutting efficiency when high tool temperatures are attained during the cutting operation. Cobalt raises the heat-treating temperatures because it elevates the melting point. Hardening temperatures for cobalt high-speed tool steels can be 14 to 28°C (25 to 50 °F) higher than would be normal for similar grades without cobalt.

Cobalt additions slightly increase the brittleness of high speed tool steels. Cobalt steels are especially effective on rough or hogging cuts, but they are not usually suited to finishing cuts that do not involve high temperatures.

1.12.3 Carbon

Is by far the most important of the elements and is very closely controlled.While the carbon content of any one high speed tool steel is usually fixed within narrow limits, variations within these limits can cause important changes in the mechanical properties and the cutting ability. As the carbon concentration is increased, the working hardness also rises; the elevated temperature hardness is higher; and the number of hard, stable, complex carbides increases. The latter contribute much to the wear resistance and other properties of the high-speed tool steels.

1.12.4 Silicon

The influence of silicon on high speed tool steel, up to about 1.0%, is slight Increasing the silicon content from 0.15 to 0.45% gives a slight increase in maximum attainable tempered hardness and has some influence on carbide morphology, although there seems to be a concurrent slight decrease in toughness. Some manufacturers produce at least one grade with silicon up to 0.65%, but this level of silicon content requires a lower maximum austenitizing temperature than does a lower silicon level in the same grade, if overheating is to be avoided. In general, however, the silicon content is kept below 0.45% on most grades.

1.12.5 Manganese

Generally, manganese is not high in concentration in high-speed tool steels. This is because of its marked effect in increasing brittleness and the danger of cracking upon quenching.

1.12.6 Phosphorus

Has no effect on any of the desired properties of high-speed tool steels, but because of its well-known effect in causing cold shortness, or room-temperature brittleness, the concentration of phosphorus is kept to a minimum.

1.12.7 Chromium

Is always present in high speed tool steels in amounts ranging from 3 to 5% and is mainly responsible for the hardenability. Generally, the addition is 4% because it appears that this concentration gives the best compromise between hardness and toughness. In addition, chromium reduces oxidation and scaling during heat treatment.

1.12.8 Tungsten

In the high speed tool steels, tungsten is of vital importance. It is found in all T-type steels and in all but two of the M-type steels. The complex carbide of iron, tungsten, and carbon that is found in high speed tool steels is very hard and significantly contributes to wear resistance. Tungsten improves hot hardness, causes secondary hardening, and imparts marked resistance to tempering. When the tungsten concentration is lowered in high-speed tool steels, molybdenum is usually added to.

1.13 TOOL WEAR

Productivity and economy of manufacturing by machining are significantly affected by life of the cutting tools. Cutting tools may fail by **brittle fracture,**

plastic deformation or **gradual wear**. Turning **carbide inserts** having **enough strength**, toughness and **hot hardness** generally fail by gradual wears. With the progress of machining the tools attain crater wear at the rake surface and flank wear at the clearance surfaces, as schematically shown in following Fig. 1.27 due to continuous interaction and rubbing with the chips and the work surfaces respectively. Among the aforesaid wears, the principal flank wear is the most important because it raises the cutting forces and the related problems.

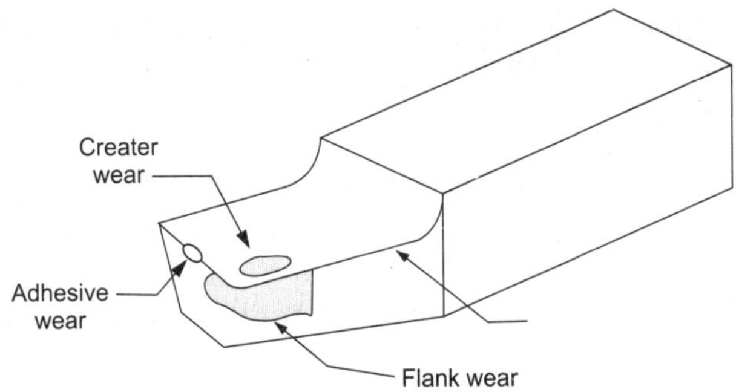

Fig. 1.27 Types of tool wear

Effects of Tool Wear

The wear on a tool causes the following effects.

- The cutting force increases
- The dimensional accuracy of the work decreases
- The surface roughness of the work increases
- The tool-work system may start vibrating
- The work piece may get damaged or tool may break ultimately.

1.13.1 Mechanism of Tool Wear

To know the right mechanism of tool wear and its reasons, the researchers all over the world conducted lots of experiments. Due to the inabilities of the researchers to observe the wear actually taking place on different places of a tool, the bulk of the knowledge is based primarily on theory supported by limited investigations. In general there are different types of wear that affect a cutting tool:

1.13.2 Abrasion

Mechanical wearing, hard particles in workpiece removes small portions of the tool, that cause flank and crater wear. This is the dominant cause of flank wear.

1.13.3 Adhesion

Two metals contact under high pressure and temperature that cause welding between the materials.

1.13.4 Diffusion

Atoms on the boundry of workpiece and tool changes place. This is the principle cause for crater wear.

1.13.5 Chemical Reactions (Corrosion)

The high temperatures and clean surfaces at the chip-tool interface in machining at high speeds can result in chemical reactions, in particular, oxidation, on the rake surface of the tool. The oxidized layer, being softer than the parent tool material, is sheared away, exposing new material to sustain the reaction process.

1.13.6 Plastic Deformation (Fracture)

Cutting forces acting on the cutting edge at high temperature cause the edge to deform plastically.

Cursive wear (cutting action of particles in a fluid)

1.13.7 Flank Wear

Progressive or gradual wear that develop on the flank surface of cutting tool is known as 'flank wear:

At the starting tool is sharp and wear load on the flank has zero width. It is noted that the wear land is not uniform width.

During the steady wear phase, flank wear is caused mainly through abrasion, whereas during the rapid wear phase, it is caused by diffusion.

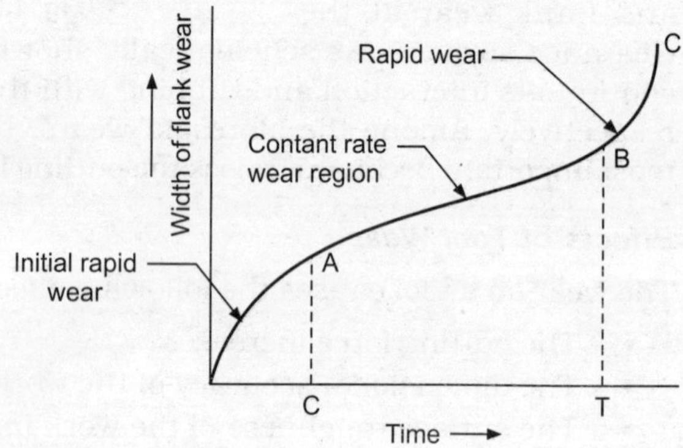

Fig. 1.28 Typical wear cause for a cutting tool.

1.14 MACHINABILITY

To define the machinability, it is not so easy how ever machinability is defined as how much easyness a metal can be machined, i.e., how much easyness any machining operation can be performed on a object or other manner machinability can be defined as the property of material which governs the area case or difficulty with which it can be machined under a specific cutting condition.

Machinability are two types-good machinability and bad machinability. Good machinability means good surface finish, long tool life. Low power consumption and low cutting forces while bad machinability is totally opposite the first.

Machinability always expressed in rating and this is for most ferrous metal, using AISI (American Iron & Steel Institute) as basies of 100% machinability for stainless steel it is upto 60% and for aluminium it is upto 240%.

Machinability usually defined in terms of following factors:

(i) Surface finish and integnity of a part required
(ii) Tool life
(iii) Forces required to remove the material
(iv) Power required to generate the forces
(v) Chip control

1.14.1 Factors Affecting the Machinability

Machinability of any material is affected by several such as:

(i) Cutting condition
- Feed
- Speed
- Depth of cut

(ii) Workpiece material variable
- Capacity of Machine
- Rigidity of Machine

(iii) Workpeice-material variable
- Hardness
- Strength
- Chemical composition
- Rigidity of workpiece
- Strain hardening

(iv) Tool variable
- Tool material
- Tool geometry
- Nature of engagement of tool with work

Excellent Machinbility
- MN alloys
- Al alloys
- Zn alloys

Excellent Machinbility
- Gon metal
- Cu-Al alloys
- Malleable cast iron
- Free cutting steel

Poor Machinability
- Low carbon steel
- Low alloys steel
- Annealed steel

Not Machinable Material
- Stain less steel
- White cast iron
- Stellite (alloys of W, Cr, Co, C)

1.14.2 Tool life

Total time accumulated by a cutting tool before the failure is referred as **'tool life'**. Every tool have its own specified life once tool life over cutting tool became failure. These is no exact definition of tool life however, usually it is defined as tool's life which has been expended when it can no longer in use or other words no effective to produce satisfactory parts under specific cutting condition.

1.14.2.1 Factors Affecting Tool Life

Tool life of a cutting tool is affected by the following factors

(a) Hardness of tool materials like soft or hard
(b) Tool material lake H.S.S, Diamond, CBN etc.
(c) Type of materials being cut like **ferrous** or **non-ferrous** hard or soft material.
(d) Type of finish required like **rough surface, medium surface** and finish surface.
(e) Profile of cutting tool.
(f) Type of machining operation being performed lake turning, milling, boaring, tapping etc.
(g) Cutting speed like low cutting speed, high cutting speed.
(h) Workpiece microstructure.
(i) Temperature generation during the process.
(j) Feed and depth of cut.
(k) Number of components produced.

1. **Cutting condition:** machinability is to much affected by cutting condition which consist the speed, feed and depth of cut. Usually cutting speed is calculated by **Taylor's tool equation**

$$VT^n = C$$

where V = cutting velocity
T = Time (tool life in Hr.)
C = Constant (value depends up to workpiece machine tool material)
n = Index whose value is differ for different material.

Boston later found that feed, depth of cut also effect the surface roughness and he gave the express

$$VT^n = \frac{C_1}{f^x d^y} \quad \text{or} \quad C_1 = VT^n \times f^x \times d^y$$

Normally surface finish, improve with the cutting speed, but if on continuous increasing in cutting speed surface finish (machinability) decreases.

2. **Machining variable:** Performance of any machining operation basically depends upon the rigidity of system consisting of machine tool etc. Machine should be sufficient rigid and should have a sufficient power to withstand the cutting forces, on which the operation to be performed. If a machine tool is not have sufficient power and no sufficient rigid in Design tool life will reduce with affecting the surface finish, and accuracy of objects. Lower the value of cutting speed reduces the machinability hence machinability is directly affected by machine variable.

1.15 MACHINING ECONOMICS

Optimizing cutting speed is formulated by W. Gilbert with respect to Taylor's tool life formula. There are two objectives in this optimization:

- Maximizing production rate
- Minimizing unit cost

Both objectives seek a balanced MRR and tool life.

Maximizing Production Rate

Choose cutting speed to minimize machining time per production unit.

In turning three elements contribute to the total production cycle time for one part:

- Part handling time (loading + unloading + starting the machining) = T_h
- Machining time (actual machining) = T_m

$$\text{Machining time/piece} = T_m = \frac{L}{f \times N} \text{ min}$$

Now, cutting speed

$$V = \frac{\pi DN}{1000} \text{ m/min.}$$

or

$$T_m = \frac{\pi DL}{1000 \times f \times V} \text{ min}$$

where f = feed, N = speed of spindle in r.p.m.
L = Length of bar

- Tool change time (at the end of tool life, the tool must be changed) = T_t

Therefore, total time per unit product for the operation cycle

$$T_c = T_h + T_m + T_t / n_p$$

where n_p = integer number of parts we can produce within the tool life.

Our objective is to minimize T_c, which is the function of the cutting speed.

$$\text{Remember in Turning operation, } T_m = \frac{n \times D \times L}{V \times S_o}$$

Taylor's tool life formula, $VT^n = C$

or

$$T = \left(\frac{C}{V}\right)^{1/n} \qquad \eta_p = \frac{T}{T_m}$$

or

$$n_p = \frac{(C/V)^{1/n} \times V \times S_o}{n \times D \times L} = \frac{C^{1/n} \times S_o}{n \times D \times L \times \left(\frac{1}{n} - 1\right)}$$

So, T_C becomes

$$T_C = \frac{T_h + n \times D \times L}{V \times S_o} + \frac{\{T_t \times n \times D \times V^{\left(\frac{1}{n} - 1\right)}\}}{S_o \times C^{\frac{1}{n}}}$$

To minimize we need to take derivative of T_C w.r.t. to V and equate it to 0

$$\text{Therefore the maximum } V = V_{max} = \frac{C}{\left[\left(\frac{1}{h} - 1\right) \times T_t\right]^n}$$

Choose cutting speed to minimize production cost per unit product. In turning 4 elements contribute to the total production cost for one part (cost rate is $/min)

- Cost of part handling time(cost of the time that operator spends loading and unloading the part) = $C_o \times T_h$
- Cost of machining time= $C_o \cdot T_m$
- Cost of tool change time= $C_o \cdot T_t / n_p$
- Tooling cost= C_t / n_p,

where, C_t = Cost for cutting edge = P_t / n_e
P_t = Price of the tool
n_e = Number of cutting edges
C_o = Cost rate ($/min) for the operator and machine.

If the tool is regrindable, $C_t = P_t / n_g + T_g \times C_g$

where, n_g = umber of tool lifes
T_g = time to grind
C_g = grinding labor cost

Therefore total cost per unit product for the operation cycle,

$$C_C = C_o T_h + C_o T_m + \frac{C_o T_t}{h_p} + \frac{C_t}{h_p}$$

$$= C_o T_h + \frac{C_o \pi DL}{VS_o} + \frac{(C_o T_t + C_t)\left(\pi DLV^{\frac{1}{n}-1}\right)}{S \times C^{1/h}}$$

To minimize the cost we need to take derivative of C_o w.r.t., and equate it to 0.

Therefore the minimum V,

$$V_{\min} = C \times [\{n/(1-n)\} \cdot (C_o/(C_o \cdot T_t + C_t)]^n$$

Means that it is the cost minimizing speed, and the corresponding tool life is

$$T_{\min} = [(1/n) - 1] \cdot (C_o \cdot T_t + C_t) / C_o .$$

1.16 BASIC TERMS

It is necessary that the dimensions, shape and mutual position of surfaces of individual parts of mechanical engineering products are kept within a certain accuracy to achieve their correct and reliable functioning. Routine production processes do not allow maintenance (or measurement) of the given geometrical properties with absolute accuracy. Actual surfaces of the produced parts therefore differ from ideal surfaces prescribed in drawings. Deviations of actual surfaces are divided into four groups to enable assessment, prescription and checking of the permitted inaccuracy during production:

- Dimensional deviations
- Shape deviations
- Position deviations
- Surface roughness deviations

Example 1.1

A mild steel billet of 150 mm diameter was turned with a carbide tool at 30 m/min. Life of tool found 2 hours under this condition at the cutting speed 20 m/min, the tool life was observed to be 5.0 hours. Calculate the value of 'C' and the exponent 'n' in the tool life equation.

Solution: By Taylor's tool equation

$$VT^n = C$$

given $V_1 = 30$ m/min.

$V_2 = 20$ m/min.

as material is same hence

$$T_1 = 2.0 \text{ hr or } 2.0 \times 60 \text{ min} = 120 \text{ min.}$$

$$T_2 = 5.0 \text{ hr or } 5.0 \times 60 \text{ min} = 300 \text{ min.}$$

$$V_1 T_1^{\,n} = V_2 T_2^{\,n}$$

$$30 \times (120)^n = 20 \times (300)^n,$$

$$\frac{30}{20} = \left(\frac{300}{120}\right)^n$$

$$1.5 = (2.5)^n$$

Taking log both side, we get

$$n \log 2.5 = \log 1.5$$

$$n \times 0.916 = 0.405 \quad \text{or} \quad n = \frac{0.405}{0.916}$$

$$n = 0.443$$

Putting the value of n in above equation

$$30 \times (120)^{0.443} = C \qquad \text{or} \quad C = \mathbf{250}$$ **Ans.**

Example 1.2

A mild steel workpeice was turned by a tool at 30 m/min having 1.5 hours tool life. What will be the possible life when engaged on a light finishing cut. Assume n = 0.125 for rough cut and n = 0.12 for finish cut.

Solution: According to Taylor's tool equation

$$VT^n = C$$

given $n_1 = 0.125$

$n_2 = 0.12$

For rough cut $V_1 = 30$ m/min,

$T_1 = 1.5 \text{ hr} = 1.5 \times 60 = 90$ min

$$30 \times (90)^{0.125} = C \qquad \text{or} \qquad C = 52.65$$

$$(T_1)^{0.12} = 1.755$$

$$T_2 = (1.755)^{\frac{1}{0.12}} = 108.35 \text{ min}$$

$$T_2 = \mathbf{108.35\ min} \quad \text{or} \quad T_2 = \mathbf{108.35\ min} \qquad \textbf{Ans.}$$

Example 1.3

Calculate the percentage change in cutting speed required to given 40% reduction in tool life if, n = 0.25.

Solution: According to Taylor's tool equation

$$VT^n = C$$

$$T_2 = (1 - 0.40) \times T_1$$

$$T_2 = 0.60\,T_1$$

$$V_1 T_1{}^n = V_2 T_2{}^n$$

$$\frac{V_2}{V_1} = \left(\frac{T_1}{T_2}\right)^n = \left(\frac{T_1}{0.60 T_1}\right)^n$$

or
$$\frac{V_2}{V_1} = (1.67)^{0.25}$$

$$\frac{V_2}{V_1} = \mathbf{1.135}$$

value of V_2 should be 13.5% greater than V_1 value.

Example 1.4

Either cemented carbide or ceramic (oxide) may be used as tool material while maching on carbon steel. The Taylor's tool life equations are:

(i) for carbide tool $VT^{0.25} = 800$

(ii) for ceramic tool $VT^{1.25} = 8000$.

Calculate the break-even speed above which tool give better tool life.

[UPTU-2002-2003]

Solution:

(i) $VT^{0.25} = 800$. Taking log both sides $\log_e^V + 0.25 \log_e^T = \log_e 800$ ...(I)

(ii) $VT^{1.25} = 8000$

$\log_e^V + 1.25 \log_e^T = \log_e 8000$ ...(II)

Solving equation I & II we get

$$4 \log_e V = 5 \log_e 800 - \log_e 8000$$

or
$$4 \log_e V = 14.52 - 3.91 = 10.61$$

$$\log_e V = 2.6525$$

or
$$V = \mathbf{14.15\ m/min} \qquad \textbf{Ans.}$$

Example 1.5

If Taylor's tool life exponent n = 0.5 and constant C = 400, what will the % age increases in tool life when cutting speed is reduced by half?

[UPTU-2002-2003]

Solution: Taylor's equation $V_1 T_1^{\,n} = C_1$ and $V_2 T_2^{\,n} = C_2$ as $C_1 = C_2 = 400$

hence $$V_1 T_1^{\,n} = V_2 T_2^{\,n}$$

$$V_1 T_1^{\,0.5} = 400 \qquad \text{...(I)}$$

and $$V_2 T_2^{\,0.5} = 400 \qquad \text{...(II)}$$

but $$V_2 = 0.5\, V_1$$

Putting the value of V_2 in equation (II), we have

$$V_1 T_1^{\,0.5} = 400 \qquad \text{...(III)}$$

$$0.5\, V_1 T_2^{\,0.5} = 400 \qquad \text{...(IV)}$$

Taking log both sides, we get

$$\log V_1 + 0.5 \log T_1 = \log 400$$

and $$\log V_1 + 0.5 \log T_2 = \log 800$$

On solving equation

$$0.5\,(\log T_1 \;\; \log T_2) = -0.301$$

or $$\log \frac{T_1}{T_2} = -0.602 \quad \text{or} \quad \frac{T_1}{T_2} = e^{-0.602}$$

or $$T_2 = \mathbf{1.8226\ T_1}$$

Tool life increases upto 82.6%.

Example 1.6

Using the modified formula of tool life $VT^n d^x f^y = C$. Calculate the change in tool life if the depth of cut (d) is reduced 1.5 times, feed is reduced 2.5 times and the cutting speed is made double. Assume the exponent have following values n = 0.15, x = 0.15 and Y = 0.6. ***[UPTU-2005-2006]***

Solution: $$VT^n d^x f^y = C$$

Let $$V = V_1, \quad T = T_1$$

finally $$d_2 = \frac{d_1}{1.5}\ \text{(given)} \qquad d = d_1$$

and $$f_2 = \frac{f_1}{1.5}\ \text{(given)} \quad f = f_1 \quad \& \quad V_2 = 2V_1$$

$$V_1 T_1^n d_1^x f_1^y = V_2 T_2^n d_2^x f_2^y$$

$$\left(\frac{V_1}{V_2}\right)\left(\frac{d_1}{d_2}\right)^x\left(\frac{f_1}{f_2}\right)^y = \left(\frac{T_2}{T_1}\right)^n$$

or $$\left(\frac{1}{2}\right) \times \left(\frac{d_1}{\frac{d_1}{1.5}}\right)^x \times \left(\frac{f_1}{\frac{f_2}{1.5}}\right)^y = \left(\frac{T_2}{T_1}\right)^n$$

put $x = 0.15$, $n = 0.15$ and $y = 0.6$

$$\frac{1}{2} \times (1.5)^{0.15} \times (1.5)^{0.6} = \left(\frac{T_2}{T_1}\right)^{0.15}$$

or $$0.5 \times 1.063 \times 1.275 = \left(\frac{T_2}{T_1}\right)^{0.15}$$

or $$\left(\frac{T_2}{T_1}\right)^{0.15} = 0.6778, \quad \frac{T_2}{T_1} = (0.6778)^{\frac{1}{.15}}$$

$$\frac{T_2}{T_1} = (0.6778)^{6.67} = \mathbf{0.7472}$$

or $$T_2 = \mathbf{0.07472\ T_1}$$ **Ans.**

Example 1.7

The following equation for tool life is given for a M.S block turning

$$VT^{0.15} f^{0.82} d^{0.35} = c$$

A 50 minute tool life was obtained while cutting at v = 40 m/min, f = 0.4 mm/rev and d = 2.5 mm.

Determine the change in tool life if the cutting speed, feed and depth of cut are increases by 25% individually and also taken together.

Solution: $v \times T^{0.15} \times f^{0.82} \times d^{0.35} = c$ ...(I)

or $$c = 40 \times (50)^{0.15} \times (0.4)^{0.82} \times (2.5)^{0.35}$$

$$= 40 \times 1.798 \times 0.472 \times 1.3781 = 40.78$$

(i) As cutting speed increases by 25% therefore $v = 1.25 \times 40 = 50$ m/min putting this value in equation I, we get

$$50 \times T^{0.15} \times 0.4^{0.82} \times 2.5^{0.35} = 40.78$$

or $$T^{.15} = 1.254$$

or $$T = 1.254^{\frac{1}{0.15}} \quad \text{or} \quad T = 4.54 \text{ min}$$ **Ans.**

(ii) As feed is increases by 25% therefore $f = 0.4 \times 1.25 = 0.50$ mm/rev

$$40 \times T^{.15} \times (.5)^{.82} \times 2.5^{0.35} = 40.78$$

or $$T^{.15} \times 31.22 = 40.78$$

$$T^{.15} = (1.306)$$

or $$T = (1.306)^{\frac{1}{0.15}}$$

$$T = 5.94 \simeq 6.0 \text{ min}$$ **Ans.**

(iii) As depth of cut is increases by 25% therefore $d = 2.5 \times 1.25 = 3.125$ mm

$$40 \times T^{.15} \times (0.4)^{.82} \times (3.125)^{0.35} = 40.78$$

$$T^{.15} \times 28.11 = 40.78$$

or $$T^{.15} = 1.45$$

$$T = (1.45)^{\frac{1}{0.15}}$$

$$T = 11.9 \simeq 12 \text{ min} \quad \textbf{Ans.}$$

(iv) As $v = 50\ f = 0.50$, and $d = 3.125$ then

$$50 \times T^{.15} \times (0.50)^{.82} \times (3.125)^{.35} = 40.78$$

$$T^{.15} = \frac{40.78}{42.20} = 0.966$$

$$T = (0.966)^{\frac{1}{0.15}}$$

$$= 0.795 \text{ min} \quad \textbf{Ans.}$$

Example 1.8

A single point cutting tool with 12° rake angle is used to machine a steel workpiece. The depth of cut, i.e., uncut thickness is 0.81 mm and the chip thickness under orthogonal machining condition is 1.8 mm. Find the shear angle. ***[GATE – 2011]***

Solution: $t = 0.81$ mm,

$t_c = 1.8$ mm,

$\alpha = 12°, \quad \phi = ?$

We know, $$\tan \phi = \frac{r \cos \alpha}{1 - r \sin \alpha};$$

$$r = \frac{t}{t_c} = \frac{0.81}{1.8} = 0.45$$

$$\tan \phi = \frac{0.45 \times 0.978}{1 - 0.45 \times 0.207}$$

$$= \frac{0.4401}{0.90685} = 0.4853$$

$$\phi = \tan^{-1}(0.4853) = 25.88$$

or $$\phi = \mathbf{26°}$$ **Ans.**

Example 1.9

During the orthogonal cutting of 80 mm diameter rod to reduces 78.5 mm. If the mean length of the cut chip is 79.5 mm and rake angle is 12°, then Find

(i) cutting ratio

(ii) shear angle.

Solution: $t = l_c = 79.5$ mm, $\alpha = 12°$

Length of uncut chip $t_c = l = \dfrac{\pi(80 + 78.5)}{2} = 248.8 \simeq 249$ mm

$$r = \frac{t}{t_c} = \frac{V_c}{V} = \frac{l_c}{l} = \frac{79.5}{249} = 0.319 \simeq 0.32$$

(ii) Shear angle $\tan\phi = \dfrac{r\cos\alpha}{1 - r\sin\alpha}$

$$\tan\phi = \frac{0.32 \times 0.978}{1 - 0.32 \times 0.207}$$

$$= \frac{0.313}{0.933} = 0.335$$

$$\phi = \tan^{-1}(0.335) = \mathbf{18.53°}$$ **Ans.**

Example 1.10

Following data relates to an orthogonal cutting (turning) operations:

Chip thickness = 0.56 mm,

Feed = 0.2 mm/rev,

rake angle = 10°

Calculate–

(i) Chip thickness ratio,

(ii) Chip reduction coefficient,

(iii) Shear angle,

(iv) Shear strain involved in the deformation process.

[UPTU–2006–07]

Solution: $\alpha = 10°$, $t_c = 0.56$, $t = 0.2$

(i) $r = \dfrac{t}{t_c} = \dfrac{0.2}{0.56} = \mathbf{0.357}$ **Ans.**

(ii) Chip reduction coefficient $= \dfrac{1}{r} = \dfrac{1}{0.357} = \mathbf{2.8}$ **Ans.**

(iii) Shear angle

$$\tan\phi = \frac{r\cos\alpha}{1 - r\sin\alpha} = \frac{0.357 \times 0.985}{1 - 0.357 \times 0.174} = \frac{0.3516}{0.9378} = 0.375$$

$$\phi = \tan^{-1}(0.375) = \mathbf{20.5°}$$ **Ans.**

(iv) Shear strain

$$\varepsilon = \frac{\cos\alpha}{\sin\phi\cos(\phi - \alpha)} = \frac{\cos 10°}{\sin 20.5° \times \cos(20.5 - 10)}$$

$$= \frac{0.9848}{0.3502 \times 0.4832} = \frac{0.9848}{0.3443} = \mathbf{2.86.}$$ **Ans.**

Example 1.11

During the orthogonal machining with a cutting tool having 14° rake angle and chip thickness is measured 0.48 mm and the uncut thickness measured 0.20 mm. Determine,

(i) Chip thickness ratio
(ii) Shear angle
(iii) Shear strain

Solution: $t = 0.20$, and $t_c = 0.48$,

(i) $r = \frac{t}{t_c} = \frac{0.20}{0.48} = \mathbf{0.4}$ **Ans.**

(ii) $\tan\phi = \frac{r\cos\alpha}{1 - r\sin\alpha}$

$$\tan\phi = \frac{0.42 \times 0.97}{1 - 0.42 \times 0.242} = \frac{0.4074}{0.8983} = 0.453$$

$\phi = \tan^{-1}(0.4535) = \mathbf{24.39°}$ **Ans.**

(iii) $\varepsilon = \cot\phi + \tan(\phi - \alpha)$

$= \cot 24.39 + \tan(24.39 - 14) = 0.0114 + 0.1833 = \mathbf{0.1947}$ **Ans.**

Example 1.12

During the orthogonal cutting, the following data was observed:

Cutting force = 2000 N,
Chip thickness ratio = 0.50,
Feed force = 1000 N,
Rake angle = 12°

Determine– (i) Shear force and compression force.
(ii) Coefficient of friction between tool and workpiece interface.

Solution: $F_f = 1000$ N, $F_t = 2000$ N, $r = 0.50$, $\alpha = 12°$

$$\tan\phi = \frac{r\cos\alpha}{1 - r\sin\alpha} = \frac{0.50 \times 0.978}{1 - 0.50 \times 0.207} = \frac{0.488}{0.8960} = 0.5446$$

$\phi = \tan^{-1}(0.5446) = \mathbf{28.57°}$ **Ans.**

We know $F_c = F_f\cos\phi + F_t\sin\phi$ and $F_s = F_t\cos\phi - F_f\sin\phi$

$F_c = 1000 \times \cos 28.57 + 2000 \times \sin 28.57°$

$= 878.20 + 956.46 = \mathbf{1834.6\ N}$ **Ans.**

Now, $F_s = F_t\cos\phi - F_f\sin\phi$

$$F_s = 2000 \cos 28.57° - 1000 \times \sin 28.57°$$

$$= 1756.46 - 478.23 = \mathbf{1278.2\ N} \qquad \textbf{Ans.}$$

(ii) μ = Coefficient of friction

$$\mu = \tan\beta = \frac{F_f + F_t \tan\alpha}{F_t - F_f \tan\alpha} = \frac{1000 + 2000 \times \tan 12°}{2000 - 1000 \tan 12°}$$

$$= \frac{1000 + 425.11}{2000 - 212.55} = \frac{1425.11}{1787.4} = \mathbf{0.797} \qquad \textbf{Ans.}$$

Example 1.13

During the orthogonal cutting process, the following data is observed–

Depth of cut	***– 0.6 mm,***
Uncut chip length	***– 250 mm,***
Chip length	***– 100 mm,***
Rake angle	***– 16°,***
Horizontal cutting force	***– 2500 N,***
Vertical cutting force	***– 250 N.***

Determine; (i) Chip thickness ratio,
(ii) Shear angle,
(iii) Friction angle, and
(iv) Resulting cutting force.

Solution: $l_c = 100$ mm, $l = 250$ mm, $\alpha = 16°$ $F_H = (F_t) = 2500$ N, $F_v = (F_f) = 250$ N, $\alpha = 16°$

(i) $r = \dfrac{t}{t_c} = \dfrac{V_c}{V} = \dfrac{l_c}{l}$,

$$r = \frac{l_c}{l} = \frac{100}{250} = \mathbf{0.4} \qquad \textbf{Ans.}$$

(ii) $$\tan\phi = \frac{r\cos\alpha}{1 - r\sin\alpha} = \frac{0.4 \times \cos 16°}{1 - 0.4 \sin 16°} = \frac{0.3845}{0.8897} = 0.432$$

$$\phi = \tan^{-1}(0.4321) = \mathbf{23.37°} \qquad \textbf{Ans.}$$

(iii) $$\mu = \tan\beta = \frac{F_f + F_t \tan\alpha}{F_t - F_f \tan\alpha} = \frac{250 + 2500 \times 0.2867}{2500 - 250 \times 0.2867} = \frac{966.86}{2428.32} = 0.398$$

$$\beta = \tan^{-1}(0.398) = \mathbf{21.71°} \qquad \textbf{Ans.}$$

(iv) Resultant $R = \sqrt{F_H{}^2 + F_V{}^2} = \sqrt{(2500)^2 + (250)^2}$

$$= \sqrt{6312500} = \mathbf{2512.4\ N} \qquad \textbf{Ans.}$$

Example 1.14

During the orthogonal cutting operation following data is observed–

Rake angle **– 12°,**
Chip thickness **– 0.75 mm,**
Depth of cut **– 0.45 mm,**
Width of cut **– 3.0 mm,**
Tangential force **– 1500 N,**
Feed force **– 750 N.**

Then calculate–

(i) coefficient of friction between the workpiece and tool interface.
(ii) ultimate shear stress.

Solution: $\alpha = 12°$, $t_c = 0.75$ mm,

t (Depth of cut) = 0.45

$F_f = 750$ N, $F_t = 1500$ N

$$r = \frac{t}{t_c} = \frac{0.45}{0.75} = 0.6,$$

$$\alpha = 12°$$

(i)
$$\mu = \tan\beta = \frac{F_f + F_t \tan\alpha}{F_t - F_f \tan\alpha}$$

$$= \frac{750 + 1500 \tan 12°}{1500 -- 750 \times \tan 12°} = \frac{1068.8}{1340.58} = \mathbf{0.797} \quad \textbf{Ans.}$$

(ii)
$$\tan\phi = \frac{r\cos\alpha}{1 - r\sin\alpha} = \frac{0.6 \times 0.978}{1 - 0.6 \times 0.2079} = \frac{0.5868}{0.8752} = 0.670$$

$$\phi = \tan^{-1}(0.670) = 33.84°$$

but
$$F_s = F_t \cos\phi - F_f \sin\phi$$
$$= 1500 \times 0.8306 - 750 \times 0.5568 = 828.2 \text{ N}$$
$$A = \text{width} \times \text{depth of cut}$$
$$= 3.0 \times 0.45 = 1.35$$

but
$$\tau = \frac{F_s \sin\phi}{A} = \frac{828.2 \times \sin 33.84°}{1.35} = \mathbf{341.6\ N/mm^2} \quad \textbf{Ans.}$$

Example 1.15

During the orthogonal cutting operation following data is observed:

Rake angle **= 12°,**
Length of continuous chip **= 100 mm,**
Cutting speed **= 25 m/min,**
Feed rate **= 0.18 mm/reve,**
Diameter of M.S. tube **= 80 mm,**

Tangential force $= \mathbf{4000\ N,}$

Feed Force $= \mathbf{2400\ N,}$

then find

(i) coefficient of friction

(ii) shear plane angle

(iii) velocity of chip along tool face

(iv) chip thickness.

Solution: Given, $\alpha = 12°$, $l_c = 100$ mm,

$F_t = 4000$ N, $F_f = 2800$ N,

$V = 25$ m/min, $f = 0.18$ mm/rev.

(i) $$\mu = \tan\beta = \frac{F_f + F_t \tan\alpha}{F_t - F_f \tan\alpha}$$

$$= \frac{2400 + 4000 \times \tan 12°}{4000 - 2400 \times \tan 12°} = \frac{3250.2}{3489.86} = \mathbf{0.931} \quad \textbf{Ans.}$$

(ii) $$r = \frac{t}{t_c} = \frac{l_c}{l} = \frac{V_c}{V},$$

$$r = \frac{l_c}{l} = \frac{100}{\pi \times 80} = 0.398 = 0.4$$

$$\tan\phi = \frac{r\cos\alpha}{1 - r\sin\alpha} = \frac{0.4 \times \cos 12°}{1 - 0.4 \times \sin 12°} = \frac{0.3912}{0.9168} = 0.4266$$

$$\phi = \tan^{-1}(0.4266) = \mathbf{23.10°} \quad \textbf{Ans.}$$

(iii) $$r = \frac{V_c}{V}$$

$$0.4 = \frac{V_c}{25} \quad \text{or} \quad \boldsymbol{V_c = 10\ m/min} \quad \textbf{Ans.}$$

(iv) $$r = \frac{t}{t_c} = \frac{\text{feed}}{t_c}, \quad 0.4 = \frac{0.18}{t_c}$$

or $$t_c = \frac{0.18}{0.4} = \mathbf{0.45\ mm} \quad \textbf{Ans.}$$

Example 1.16

During the orthogonal cutting of mild steel, the following data is observed:

friction between tool and work piece interface – 0.40,

rake angle = 12°,

uncut thickness = 0.55 mm,

cutting speed = 225 m/min,
width of cut = 2.75 mm,
shear stress = 450 N/mm²,

find:

(i) shear angle,
(ii) cutting and thrust component of forces.

Solution: Given $\alpha = 12°, \mu = 0.40,$

$$t = 0.55, \quad \upsilon = 225 \text{ m/min}, \ b = 2.75 \text{ mm}$$

$$\mu = \tan\beta,$$

$$0.4 = \tan\beta \quad \text{or} \quad \beta = \tan^{-1}(0.4) = 21.80$$

According to Merchant theorem

$$\phi = \frac{\pi}{4} - \frac{(\beta - \alpha)}{2}$$

$$\phi = \frac{\pi}{4} - \frac{(21.8 - 12)}{2} = 45 - 4.900 = \mathbf{40.01°} \quad \textbf{Ans.}$$

(ii) $\tau = \dfrac{F_s \times \sin\phi}{A}$; $\quad 450 = \dfrac{F_s \times \sin 40.01°}{2.75 \times 0.55}$;

$$F_s = \frac{680.625}{0.6429} = 1058.6 \text{ N}$$

but $F_s = R\cos(\phi + \beta - \alpha)$

or $$R = \frac{F_s}{\cos(\phi + \beta - \alpha)} = \frac{1058.6}{\cos 28.41} = 1203.5 \text{ N}$$

Now, $F_t = R\cos(\beta - \alpha) = 1203.5 \times \cos(21.8 - 12°) = \mathbf{1186\ N}$ **Ans.**

$F_f = R\sin(\beta - \alpha) = 1203.5 \times \sin(21.8 - 12°) = \mathbf{204.7\ N}$ **Ans.**

Example 1.17

During the orthogonal cutting operation of a mild steel bar, the following data was observed:

cutting speed – 35 m/min,
rake angle – 12°,
chip thickness – 3.0 mm,
coefficient of friction – 0.80,
width of cut – 20 mm/rev,
shear stress – 680 N/mm²,

then find

(i) chip thickness ratio,
(ii) shear plane angle,
(iii) shear force,
(iv) cutting force,
(v) power consumption,
(vi) friction angle.

Solution: Given $V = 35$ m/min, $\alpha = 12°$,
$t_c = 3.0$, $\mu = 0.80$, $b = 20$ mm,
$f = 20$ mm/rev, $\tau_s = 680$ N/mm^2

(i) $r = \dfrac{t}{t_c} = \dfrac{2.0}{3.0} = \mathbf{0.67}$ **Ans.**

(ii) $\tan\phi = \dfrac{r\cos\alpha}{1 - r\sin\alpha} = \dfrac{0.67 \times \cos 12°}{1 - 0.67 \times \sin 12°} = \dfrac{0.655}{0.8606} = 0.7610$

$\phi = \tan^{-1}(0.7610) = \mathbf{37.3°}$ **Ans.**

(iii) $\tau_s = \dfrac{F_s \times \sin\phi}{A}$

$680 = \dfrac{f_s \times \sin 37.3°}{20 \times 2}$

$F_s = \dfrac{680 \times 20 \times 2}{\sin 37.3°} = \mathbf{44.885\ KN}$ **Ans.**

(iv) $\mu = \tan\beta$ or $0.8 = \tan\beta$ $\beta = \tan^{-1}(0.8) = \mathbf{38.66°}$

(v) Now, $F_s = R\cos(\phi + \beta - \alpha)$

$R = \dfrac{44.885}{\cos 63.96} = 102.24$ KN

but $F_t = R\sin(\beta - \alpha) = 10.224 \times \sin 26.66°$

$= \mathbf{45.87\ KN}$ **Ans.**

(vi) Power, $P = \dfrac{F_t \times V}{60}$ kW

$= \dfrac{45.87 \times 35}{60} = \mathbf{26.75\ kW}$ **Ans.**

Example 1.18

During the orthogonal cutting operation of mild steel for the following data found:

Rake angle – 18°,
Depth of cut – 0.45 mm,
Chip thickness – 0.7 mm,
Cutting velocity – 110 m/min,
Cutting force – 400 N,
Feed force – 150 N.

Determine the following–

(i) shear plane angle,
(ii) velocity of chip along the tool face,
(iii) work done in shear,
(iv) shear strain.

Solution: Given $\alpha = 18°$, $t_c = 0.7$ mm,

$t = 0.45$, $v = 110$ m/mm,

$F_t = 400$ N, $F_f = 150$ N.

$$r = \frac{t}{t_c} = \frac{0.45}{0.70} = 0.643$$

$$\tan\phi = \frac{r\cos\alpha}{1 - r\sin\alpha} = \frac{0.643 \times \cos 18°}{1 - 0.643 \times \sin 18°}$$

$$= \frac{0.6115}{0.8013} = 0.763$$

$$\phi = \tan^{-1}(0.763) = \mathbf{37.35°}$$ **Ans.**

(ii) Velocity of chip $r = \frac{t}{t_c} = \frac{V_c}{V}$ or $r = \frac{V_c}{V}$

$$V_c = r \times V = 0.643 \times 110$$

$$= 70.7 = \mathbf{71\ m/min}$$ **Ans.**

(iii) Work done in shear $W_s = F_s \times V_s$

but $F_s = F_t\cos\phi - F_f\sin\phi$

$$= 400 \times \cos 37.35° - 150 \sin 37.35°$$

$$= 226.9 \simeq 227.$$

and $V_s = \frac{V\cos\alpha}{\cos(\phi - \alpha)} = \frac{110 \times 0.951}{0.9435}$

$$= 110.8 \simeq \mathbf{111\ m/min}$$ **Ans.**

(iv) Strain $\varepsilon = \cot\phi + \tan(\phi - \alpha)$

$$= \cot 37.35° + \tan 19.35°$$

$$= 1.310 + 0.3511 = \mathbf{1.66}$$ **Ans.**

Example 1.19

During the turning operation of mild steel rod 0.25 kW/cm³/min power required and maximum power available is 5 kW at machine spindle. Assuming a cutting speed of 40 m/min and feed rate is 0.40 mm/rev. Determine.

(i) maximum metal removal rate,
(ii) depth of cut,
(iii) cutting force,
(iv) normal pressure on the chip.

Solution: Given – Power during turning = 0.25 kW/cm^3/min.

Maximum power available at spindle = 4 kW

Cutting velocity, $V = 40$ m/min,

feed rate $f = 0.40$ mm/rev.

(i) Maximum metal removal rate

$$(MRR)_{max} = \frac{\text{Max. available Power at machine spindle}}{\text{power required/cm}^3\text{/min}}$$

$$= \frac{4}{0.25} = \mathbf{16\ cm^3/min} \qquad \textbf{Ans.}$$

(ii) Depth of cut

$$(MRR)_{max} = V \times t \times f$$

$$16 = 40 \times t \times 0.40 \times 10,$$

$$t = \frac{16}{40 \times 0.40 \times 10} = 0.1\ \text{cm} \simeq \mathbf{1.0\ mm} \qquad \textbf{Ans.}$$

(iii) Cutting force F_t

$$P = \frac{F_t \times V}{1000 \times 60}\ \text{kW}$$

$$4 = \frac{F_t \times 40}{1000 \times 60}, \quad F_t = \mathbf{6000\ N} \qquad \textbf{Ans.}$$

(iv) Normal pressure on chip

$$P = \frac{F_t}{\text{Chip area}} = \frac{6000}{1 \times 0.40} = 15000\ \text{N/mm}^2 \qquad \textbf{Ans.}$$

Example 1.20

While machining a 20 mm diameter thin pipe on a lathe, the length of chip for one revolution of pipe was measured and found to be 40 mm long. Tool had a rake angle of 10°. determine the share angle.

[UPTU 2011-12]

Solution: Given $\alpha = 10°$, $l_c = 40$ mm, $d = 20$ mm

$$l = \pi \times d = \pi \times 20 = 62.80$$

$$r = \frac{l_c}{l} = \frac{40}{62.8} = 0.6369$$

$$\tan\phi = \frac{r\cos\alpha}{1 - r\sin\alpha} = \frac{0.6369 \times \cos 10°}{1 - 0.6369 \times \sin 10°}$$

$$= \frac{0.6369 \times 0.9848}{1 - 0.6369 \times 0.1736} = \frac{0.6272}{0.8894} = 0.705$$

$$\phi = \tan^{-1}(0.705) = 35.19° \qquad \textbf{Ans.}$$

PROBLEMS

Q.1. Cutting and Thrust component of the machining during orthogonal cutting of an alloy with a rake angle 18° cutting force and feed force found to be 400 N and 225 N respectively–

(i) Estimate the coefficient of friction between the tool and chip interface.

(ii) If the rake angle is reduced to 0°, keeping all other parameters the same and if the coefficient of friction also remains unchanged, estimate the new value of F_f and F_t using Lee and Shaffer's solution. *[UPTU– 2011]*

Q.2. The following data was obtained from an orthogonal cutting test:

Rake angle = 18°,
depth of cut = 5 mm,
feed rate = 0.25 mm/rev,
chip length before cutting = 30 mm,
chip length after cutting = 15 mm,
vertical cutting force = 1200 N,
horizontal cutting force = 700 N,

Using Merchant's' analysis, **determine–**

(i) shear angle plane
(ii) friction force and friction angle
(iii) resultant of forces.
(iv) energy consumed during the process.

Q.3. In an orthogonal cutting of a steel bar with carbide tip tool, the following data was obtained–

Rake angle = 12°,
width of chip = 5 mm,
uncut chip thickness = 0.20 mm,
chip thickness ratio = 0.40,
horizontal cutting force = 1300 N,
vertical cutting force = 1700 N,

then **calculate the:**

(i) Shear plane angle, and
(ii) Mean shear strength.

Q.4. In orthogonal cutting of C35 steel by High Speed Steel tool and following data was observed:

rake angle = 12°,
cutting ratio = 0.40,
width of cut = 1.5 mm,
cutting force = 800 N,
thurst force = 800 N,

then **determine–**

(i) Shear plane angle, and
(ii) Component of forces.

Q.5. Following data was observed during orthogonal cutting of EN8.

chip thickness = 0.80 mm,
uncut chip thickness = 0.30 mm,

width of chip = 3.0 mm,
rake angle = 0°,
cutting force = 800 N,
thrust force = 400 N,
and **then determine–**
(i) angle of friction
(ii) shear strength of work material,

Q.6. During the orthogonal cutting operation of mild steel the following data was observed–
rake angle = 12°,
chip thickness ratio = 0.45,
horizontal component of cutting forces = 1200 N,
vertical component of cutting force = 1800 N,
then calculate,
(i) coefficient of friction, and
(ii) various component of cutting forces.

Q.7. During the orthogonal cutting operation of EN8 by carbide tool the following data was observed–
rake angle = 10°,
chip thickness = 0.6 mm,
uncut chip thickness = 0.25 mm,
cutting speed = 0.8 m/sec,
cutting force = 1000 N,
thrust force = 700 N,
then determine–
(i) friction angle
(ii) shear stress
(iii) energy consumed in friction and shear

Q.8. During the orthogonal cutting of mild steel, the following data was observed–
rake angle = 18°,
depth of cut = 5 mm,
feed rate = 0.30 mm/rev,
cutting speed = 0.65 m/sec,
chip length before cutting = 30 mm,
chip length after cutting = 10 mm,
horizontal cutting force = 650 N,
vertical cutting force = 1000 N,
using Merchant's analysis **determine–**
(i) shear plane angle
(ii) friction force and angle
(iii) energy consumed.

Q.9. A metal is being cut orthogonally with a tool having zero rake angle. Show that the rate of heat generation in the shear plane can be expressed as $F_c \times V \times (1 - \mu r)$. Where

F_c = principal cutting force, V = cutting speed,
μ = friction coefficient and r = cutting ratio. *[UPTU–2002-03]*

[Hint: Heat generation = $F_s \times V_s$

where $F_s = \dfrac{F_c \cos(\phi + \beta - \alpha)}{\cos(\beta - \alpha)}$ and $V_s = \dfrac{V \cos \alpha}{\cos(\phi - \alpha)}$]

Q.10. During the turning of Mild Steel by a single point cutting tool of specification 0° – 10° – 5° – 6° – 10° – 60° – 1 mm (ORS) at feed rate of 0.2 mm/rev and depth of cut of 2 mm, a chip thickness of 0.4 mm was obtained. If for mild steel the shear strength τ_s = 74 N/mm^2. Find the three (3) component of cutting forces. Assuming coefficient of friction at tool chip interface as 0.78. *[UPTU–2002-03]*

Q. 11. A tool shape with 8° rake (back) and 45° side cutting edge angle has to be used under orthogonal cutting condition tangential component of cutting force is 480 N and component of cutting force is 240 N (thrust force) cutting velocity is 100 m/min, determine–

(i) required side rake angle, so that cutting is orthogenal,

(ii) Kinetic coefficient of friction.

(iii) Energy consumed in friction per unit volume of material removal, if chip reduction coefficient is 2.75. *[UPTU–2002-03]*

[Ans. (i) 14° to 16°, (ii) 0.69, (iii) 0.8 kJ]

Q. 12. Show that in orthogonal cutting with zero rake angle, ratio of shear strength τ_s and specific cutting energy P_c is given by

$$\frac{\tau_s}{P_c} = (1 - \mu \cdot r)\left[\frac{r}{1 + r^2}\right]$$

[Hint: $\tau_s = \dfrac{F_s \times \sin\phi}{w \times t}$

and $F_s = \dfrac{F_c \cos(\phi + \beta - \alpha)}{\cos(\beta - \alpha)}$ and $P_c = \dfrac{F_c}{w \times t}$ and $\alpha = 0$]

OBJECTIVE TYPE QUESTIONS

1. In an orthogonal machining operation, the chip thickness and the uncut chip thickness are equal to 0.45 mm, if the tool rake angle is 0° the share plane angle is

(a) 18° (b) 45°
(c) 30° (d) 60° *[GATE 1998]*

2. In an orthogonal cutting test, the cutting forces and thrust forces were observed to be 1000 N and 500 N respectively. If the rake angle of tool is 0°, then μ is

(a) 2 (b) $\dfrac{1}{\sqrt{2}}$

(c) $\sqrt{2}$ (d) $\dfrac{1}{2}$ *[GATE 2000]*

3. In a tool life test doubling the cutting speed reduces the tool life to $\frac{1}{8}$th of the original the Taylor's total life index is

(a) $\frac{1}{2}$ (b) $\frac{1}{3}$

(c) $\frac{1}{4}$ (d) $\frac{1}{8}$ ***[IES 2000]***

4. What is the approximate percentage change in life of a tool with 0° rake angel used in orthogonal cutting when its clearance angle is change from 10° to 7°?

(a) 30% increases (b) 30% decreases

(c) 70% increases (d) 70% decreases ***[IES 1999]***

5. Crater wear starts at some distance form the tool tip because

(a) tool temperature is maximum at that region

(b) cutting field cannot penetrate that region

(c) tool strength is minimum at that region

(d) stress on rake face is maximum at that region ***[IES-2000]***

6. In ASA system if tool nomenclature is 8°–6°–°5°–5°–10°–15°– 2 mm then side rake angle will be

(a) 5° (b) 10°

(c) 15° (d) 6° ***[IES-1993]***

7. If α is the rake angle of the cutting tool, ϕ is the shear angle and *V* is the cutting velocity, then the velocity of chip sliding along the shear plane is given by

(a) $\dfrac{V\cos\alpha}{\sin(\phi-\alpha)}$ (b) $\dfrac{V\sin\alpha}{\sin(\phi-\alpha)}$

(c) $\dfrac{V\sin\alpha}{\cos(\phi-\alpha)}$ (d) $\dfrac{V\cos\alpha}{\cos(\phi-\alpha)}$ ***[IES-2001]***

8. For cutting of brass with single point cutting tool on a lathe machine, tool should have

(a) negative rake angle (b) Zero side relief angle

(c) Positive rake angle (d) zero rake angle ***[IES-2001]***

9. Power consumption in metal cutting is mainly due to

(a) friction at the metal tool interface

(b) tangential component of the force

(c) Longitudenal component of the forces

(d) normal component of forces ***[IES-2001]***

10. In single point cutting operation (Turning) with a cemented carbide and steel combination having a Taylor's exponent of 0.25, if the cutting speed is halved, then tool life will be

(a) sixteen time (b) half

(c) two times (d) eight time ***[IES-1999]***

11. Angle between the face and the flank of a single point cutting tool is known as

(a) Point angle (b) rake angle

(c) tip angle (d) clearance angle ***[IES-1995]***

12. Single point thread cutting tool should ideally have

(a) Positive rake (b) zero rake

(c) normal rake (d) negative rake ***[IES-1995]***

13. Primary tool force used in calculating the tool power consumption in machining is the
(a) axial force (b) frictional force
(c) Tangential force (d) radial force ***[IES-1993]***

14. In economics of machining with one of the following cost remains constants
(a) machining cost/price
(b) Tool cost/piece
(c) Tool changing cost/piece
(d) tool handling cost/piece ***[IES-2000]***

15. Which one of the following hardest cutting tool material next only to diamond?
(a) silicon (b) cemented carbide
(c) ceramics (d) cubic boran nitrides CBN ***[IES-2002]***

16. Creater wear on tools always starts at some distance form the tool tip because at that point
(a) cutting fluid does not penetrate
(b) normal stress on rake face is maximum
(c) temperatures is maximum
(d) tool strength is minimum ***[IES-2002]***

17. Math List I with List II and select the correct answer using the codes given below the list

List I	***List II***
A. Plan approach angle	1. Tool face
B. Rake angle	2. Tool flank
C. Clearances angle	3. Tool face and flank
D. Wedge angle	4. Cutting edge
	5. Tool nose

[IES-2004]

(a) A B C D
1 4 2 5
(b) A B C D
1 4 3 5
(c) A B C D
4 1 3 2
(d) A B C D
4 1 2 3

18. Rake angle of cutting tool is 15°, shear angle 45° and cutting velocity 35 m/min. What is the velocity of chip along the tool face.
(a) 23.5 m/min (b) 28.5 m/min
(c) 27.3 m/min (d) 25.3 m/min ***[IES-2001]***

[Hint: $V_c = \dfrac{\sin\phi}{\cos(\phi - \infty)} \times V_c = \dfrac{\sin 45°}{\cos(45 - 15)} \times 35$

$= 28.5 \text{ m/min}$ **Ans.]**

19. A medium carbon steel workpiece is turned on a lathe at 50 m/min cutting speed 0.8 mm/sec feed and 15 mm depth of cut. What is the metal removal rate
(a) 1000 mm^3/min (b) 60,000 mm^3/min
(c) 20,000 mm^3/min
(d) not possible to calculate by given data ***[IES-2004]***

[Hint: MRR = Cutting speed × feed × depth of cut
$= 50 \times 1000 \times 0.8 \times 1.5 = 60{,}000 \text{ mm}^3\text{/min}$]

20. In a machining operation chip thickness ratio is 0.3 and the back rake angle of the tool is 10°. What is the value of shear strain

(a) 3.34 (b) 0.31
(c) 0.13 (d) 3.00 **[IES-2004]**
(MNNIT PhD.2011)

[Hint: Shear strain = $\cot \phi + \tan (\phi - \alpha)$

$r = 0.3$ and $\alpha = 10°$.

$$\tan \phi = \frac{r \cos \alpha}{1 - r \sin \alpha}$$

$$\tan \phi = \frac{0.3 \times \cos 10°}{1 - 0.3 \times \sin 10°} = \frac{0.2954}{0.9479} = 0.312, \quad \phi = 17.31°$$

$$\varepsilon = \cot 17.31° + \tan(17.31 - 10°)$$

$$= 3.2086 + 0.1283$$

$$= 3.34 \quad \textbf{Ans.}$$

21. Which one of the following is the correct expression for the merchant's machinability constant?
(a) $\phi + \gamma - \alpha$ (b) $2\phi - \gamma + \alpha$
(c) $2\phi + \gamma - \alpha$ (d) $2\phi - \gamma - \alpha$ **[IES-2005]**

22. Which of the following is a single point cutting tool?
(a) grinding wheel (b) hacksaw blade
(c) parting tool (d) milling cutter **[IES-2006]**

23. Which of the following value of index n is associated with carbide tool when Taylor's tool life equation?

$VT^n = C$ is applied

(a) 0.45 – 0.6 (b) 0.1 – 0.15
(c) 0.65 – 0.9 (d) 0.2 – 0.4 **[IES-2006]**

24. Dynamometer is a device which is used to determine the
(a) deflection of the cutting tool
(b) chip thickness ratio
(c) wear of the cutting tool
(d) forces during metal cutting **[IES-1993]**

25. During the orthogonal cutting of mild steel with a 10° rake angle the chip thickness ratio is 0.4. Shear angle evaluated from data is
(a) 22.94° (b) 6.53°
(c) 20.22° (d) 50.0° **[GATE 2010]**

[Hint: $r = 0.4, \quad \alpha = 10° \quad \tan \phi = \frac{r \cos \alpha}{1 - r \sin \alpha} = \frac{0.4 \cos 10°}{1 - .4 \times \sin 10°}$

$$= \frac{0.3939}{0.9305} = 0.4233$$

$$\phi = 22.94°$$

26. Consider the following tool material:
1. Carbide, 2. Cement, 3. Ceramic, 4. Boranzon, correct sequence of there tool material in increasing order of their ability to retain their hardness is
(a) 2, 1, 4, 3 (b) 1, 2, 3, 4
(c) 1, 2, 4, 3 (d) 2, 1, 3, 4 **[IES-1994]**

27. Tool geometry of single point cutting tool is specified by the following element: 1. back rake angle, 2-side cutting edge angle, 3-side rake angle, 4-end cutting edge angle, 5-side relief angle, 6-end belief angle, 7-nose radius, the correct sequence of these tool elements used for correctly specifying the tool geometry is

(a) 1, 2, 6, 3, 5, 4, 7
(b) 1, 2, 6, 5, 3, 4, 7
(c) 1, 2, 3, 6, 5, 4, 7
(d) 1, 2, 5, 6, 3, 4, 7 ***[IES-1994]***

28. Creater wear is predominant is

(a) ceramic tools
(b) carbon tool steel
(c) tungsten carbide tools
(d) high speed steel tools ***[IES-1995]***

29. Tool life of a cutting tool is affected mostly by

(a) no using lubricant and coolent
(b) cutting speed
(c) tool geometry
(d) micro structure of material being cut
(e) depth of cut of feed. ***[IES-2010]***

30. Continuous chips are produced by cutting of

(a) mild steel
(b) cast iron
(c) high carbon steel
(d) none of the above.

31. Discontinuous chips are produced during the machining of

(a) cast iron
(b) mild steel
(c) aluminium
(d) none of the above.

32. Type of chips produced during machining of cast iron

(a) continuous chips
(b) continuous with built up edge
(c) discontinuous chip
(d) none of the above.

33. Machining is a metal cutting (removal) process, performed by the

(a) cutting tool
(b) machine tool
(c) successful manufacturing
(d) none of the above.

34. Basic conditions for metal cutting are

(a) Tool always harder than the work material
(b) Tool and work have relative motion with respect to each other
(c) Tool shape is capable of producing maximum cutting effect
(d) All of the above.

35. High Speed Steel (18 – 4 – 1) contain the following elements in the ratio of 18 : 4 : 1 ***[JMI-2006]***

(a) V_d : W : Cr
(b) W : Cr : V_d
(c) Cr: W : V_d
(d) Cr : V_d : W

36. In machining operation metal is removed by

(a) pushing the metal with tool
(b) distortion of metal
(c) tearing chips
(d) cutting the metal across a zone
(e) shearing the metal across a zone.

37. A negative rake angle on a single point carbide tipped tools enabled their application, without any chance of failure for machining

(a) Extra hard surface
(b) In Intermittently
(c) hardend steel parts
(d) all of the above.

38. An orthogonal cutting operation is being carried out under the following conditions: cutting speed = 2 m/sec, depth of cut = 0.5 mm, chip thickness

= 0.6 mm, then the chip velocity is

(a) 2 m/sec (b) 1 m/sec
(c) 2.4 m/sec (d) 1.66 m/sec **[UPSC-2003]**

Hint: $r = \frac{t_1}{t_2} = \frac{\text{Depth of cut}}{\text{chip thickness}} = \frac{0.5}{0.6}$ and $r = \frac{v_{chip}}{v_c}$ or $\frac{0.5}{0.6} = \frac{v_{chip}}{v_c}$

$v_{chip} = 0.833 \times 0.833 \times 2 = 1.67$ m/sec **Ans.**

39. In an orthogonal cutting the cutting speed is 60 m/min, depth of cut is 2.4 mm and the cutting ratio is found to be 0.75 which of the following are then correct 1. chip velocity will be 45 m/min, 2. chip velocity, will be 80 m/min, 3. chip thickness will be 1.8 mm, 4. chip thickness will be 3.2 mm, select the correct answer using the code given below

(a) 1 and 3 (b) 1 and 4
(c) 2 and 3 (d) 1, 2 and 4

Hint: $r = \frac{\text{Depth of cut}}{\text{chip thickness}} = \frac{t_1}{t_2}$

$$0.75 = \frac{2.4}{t_2}$$

$t_2 = \frac{2.4}{0.75} = 3.2$ and further chip velocity $r = \frac{v_{chip}}{v}$

$$\frac{2.4}{3.2} = \frac{v_{chip}}{60},$$

$$v_{chip} = \frac{2.4 \times 60}{3.2} = 45 \text{ m/sec}$$

40. Detail pertaining to an orthogonal metal cutting process are given below

Chip thickness ratio – 0.4
Undeformed thickness – 0.6 mm
Rake angle – 1.0°
Cutting speed – 2.5 m/sec
Mean thickness of primary shear zone – 25 m

The shear strain rate in s^{-1} during the process is

(a) 0.1781×10^5 (b) 0.7754×10^5
(c) 1.0104×10^5 (d) 4.397×10^5 **[GATE-2012]**

41. Flanks wear occurs mainly on

(a) cutting edge
(b) nose paths and nose tips
(c) face of cutting tool at a short distance from the cutting edge
(d) nose part front relief face and side relief face of the cutting tool **[UPSC-2007]**

42. Flank wear occuss on

(a) cutting edge (b) relief face of the tool
(c) nose of tool (d) none of above **[UPSC-2010]**

43. Notch wear at the outside edge of the depth of cut is due to

(a) oxidation (b) chipping
(c) abrasive action of the work hard and chip material
(d) slip-stick action of the chip **[UPSC-1996]**

44. Angle of inclination of the rake face with respect to the tool base measured in a plane perpendicular to the base and parallel to the width of tool is called as

(a) side rake angle
(b) back rake angle
(c) end cutting edge angle
(d) side cutting edge angle
[UPSC-2003]

45. With increase in side cutting edge angle in an orthogonal single point metal operation 1. the tangential force increase 2. the longitudinal face drops 3. the radial force increases, which of there statement are correct
(a) 1 and 3
(b) 2 and 3
(c) 1 and 2
(d) 1, 2 and 3 ***[UPSC-2010]***

46. Which of the following cutting tool bits are made by power metallurgy process
(a) HSS tool bits
(b) carbon tool steel
(c) stellite tool
(d) ceramic tool ***[UPSC-2010]***

47. In taylors tool life equation VT^n = constant, what is the value of n for ceramic tools
(a) 0.15 to 0.25
(b) 0.4 to 0.55
(c) 0.6 to 0.75
(d) 0.8 to 0.9 ***[UPSC-2008]***

48. Relationship between the shear angle ϕ, the friction angle β and the cutting rake angle α is
(a) $2\beta + \phi - \alpha = c$
(b) $2\phi + \beta - \alpha = c$
(c) $2\alpha + \beta - \phi = c$
(d) $\beta + \phi - \alpha = c$ ***[UPSC-2010]***

49. In a machining process, the percentage of heat carried away by the chips is typically
(a) 5%
(b) 25%
(c) 50%
(d) 75% ***[UPSC-2002]***

50. Which of the following are the machinability criteria,
1. tool life
2. cutting forces
3. surface finish,
Select the correct answer using the code given below
(a) 1, 2 and 3
(b) 1 and 3
(c) only 2
(d) 2 and 3 ***[UPSC-2009]***

ANSWERS

1.	(b)	**2.**	(d)	**3.**	(c)	**4.**	(d)	**5.**	(d)	**6.**	(d)	**7.**	(d)
8.	(b)	**9.**	(b)	**10.**	(c)	**11.**	(d)	**12.**	(c)	**13.**	(c)	**14.**	(d)
15.	(c)	**16.**	(b)	**17.**	(b)	**18.**	(b)	**19.**	(a)	**20.**	(a)	**21.**	(c)
22.	(c)	**23.**	(d)	**24.**	(a)	**25.**	(b)	**26.**	(b)	**27.**	(c)	**28.**	(c)
29.	(b)	**30.**	(a)	**31.**	(a)	**32.**	(c)	**33.**	(a)	**34.**	(d)	**35.**	(b)
36.	(e)	**37.**	( )	**38.**	(d)	**39.**	(b)	**40.**	(d)	**41.**	(d)	**42.**	(b)
43.	(c)	**44.**	(a)	**45.**	(d)	**46.**	(d)	**47.**	(c)	**48.**	(b)	**49.**	(d)
50.	(a)												

QUESTIONS

1. What is machining and how it is differ from casting and forging?
2. What is the mechanism of metal cutting? Explain the same in detail.
3. How chip formation take place in metal cutting operation? ***[UPTU– 2007-08]***
4. How many types of chips are produced during the machining operations?
5. What do you understand by the machine tool? What are the essential elements? What is a machine shop?
6. How many factors influence the machining operation?
7. What is the essential criterion for a cutting tool to give maximum production with minimum maintenance and trouble? ***[UPTU– 2010-11]***

8. Discuss the importance of shear angle from standpoint of metal cutting performance. What factors affect its value? *[UPTU– 2010-11]*
9. What is machinability? Explain the factors that affect the machinability of a material? *[UPTU– 2010-11]*
10. How does the cutting process parameters affect the cutting tool wear in a single point tool? *[UPTU– 2010-11]*
11. What are the factors influecing in the selection of cutting speed and feed for a turning (machining) operation?
12. In which type of material negative rake angle is provided and Why?
13. Differentiate between the orthogonal cutting and oblique cutting operation. *[UPTU– 2002-03, 2007-08]*
14. What is the tool signature (Nomenclature)?
15. How many types of cutting tools are availables?
16. Differentiable between machinability and tool life? *[UPTU– 2002-03]*
17. Differential between continuous chips and discontinuous chip. *[UPTU– 2002-3, 2007-08]*
18. What is the meant by built up edge? Explain conditions which promote the growth of build up edge along with its consequence. *[UPTU– 2003-04, 2004-05, 2007-08]*
19. What is meant by machinability? Explain the method of representing the machinability? *[UPTU– 2003-04]*
20. How is metal removed in metal cutting? Explain the process by given any simple model to explain the metal removal process. *[UPTU– 2004-05]*
21. What are the factors that control surface finish in turning. How do you select the cutting process parameter for finish turning. *[UPTU– 2004-05]*
22. Draw a neat and labelled sketch of right hand side single point cutting tool, showing various tool angle. What are effect of various tool angle in machining? Explain what is built up edge? *[UPTU– 2004-06]*
23. Under what machining conditions, a ductile material like aluminium will give discontinuous chip. *[UPTU– 2007-08]*
24. What do you mean by 'Tool designation' or *tool signature*?
25. What are basic requirements of cutting fluids in metal machining? Explain with their applications. *[UPTU– 2004-05]*
26. What are the uses of cutting fluids? Discuss some of the cutting fluids used during machining and their method of application.
27. Explain the cutting fluids. *[UPTU– 2006-07]*
28. What is machinability? Explain the factors that affect the machinability of a material. *[UPTU– 2010-11]*
29. How do you define tool life? Explain the parameters that control the tool life of a single point cutting tool? *[UPTU– 2003-04]*
30. Explain: (i) tool life, (ii) Machinability, (iii) Cutting fluids, (iv) flank wear and creater wear. *[UPTU– 2006-07]*
31. What are the functions of a cutting fluid.
32. What are the properties should be in a cutting tool material.
33. Write short notes on HSS tool material.
34. What are the effect to various alloying elements in a tool material.
35. Define the Taylor's tool life.
36. Write the short notes on economics of machining.

❑❑❑

UNIT-II

Machine Tools

Chapter 2

Lathe Machine

2.1 LATHE MACHINE

History of lathe machine is very old. It is the most versatile and widely used machine tool, mostly all types of the operations can be performed on the lathe machine, such as **turning, drilling, grinding, boaring, thread cutting** etc. hence lathe machine is also known as the **mother** of all types of machine.

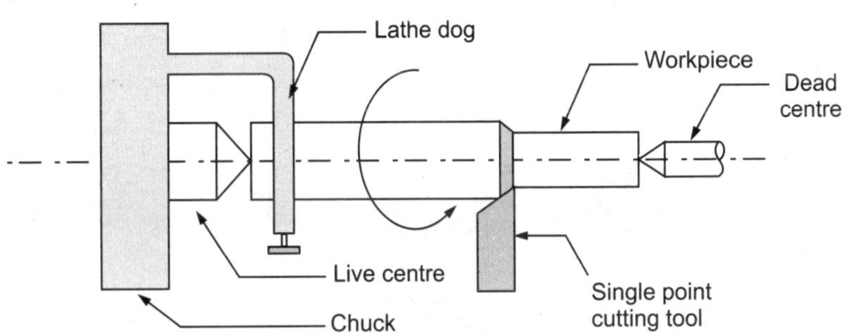

Fig. 2.1 Working principle of a lathe machine.

Commonly cylindrical jobs are produced on the lathe machine. A working principle of lathe machine is shown in Fig. 2.1 for turning (metal removal) purpose workpiece (job) is held in between the two centres, one is live centre (left hand side) and other dead centre (right hand side) and workpiece is rotating in **clockwise** direction. A single point cutting tool is used for cutting (removal) of material which is moving parallel to the axis of workpeice. Rotation to rotating the chuck power is used as live centre is attached with chuck resulting rotating the centre.

2.1.1 Selection of Lathe Machine for Metal Cutting

Various factors are considered before the metal removal from the workpiece axis (surface) on a lathe machine, are given below:

(i) Speed required

(ii) Size and shape of workpiece

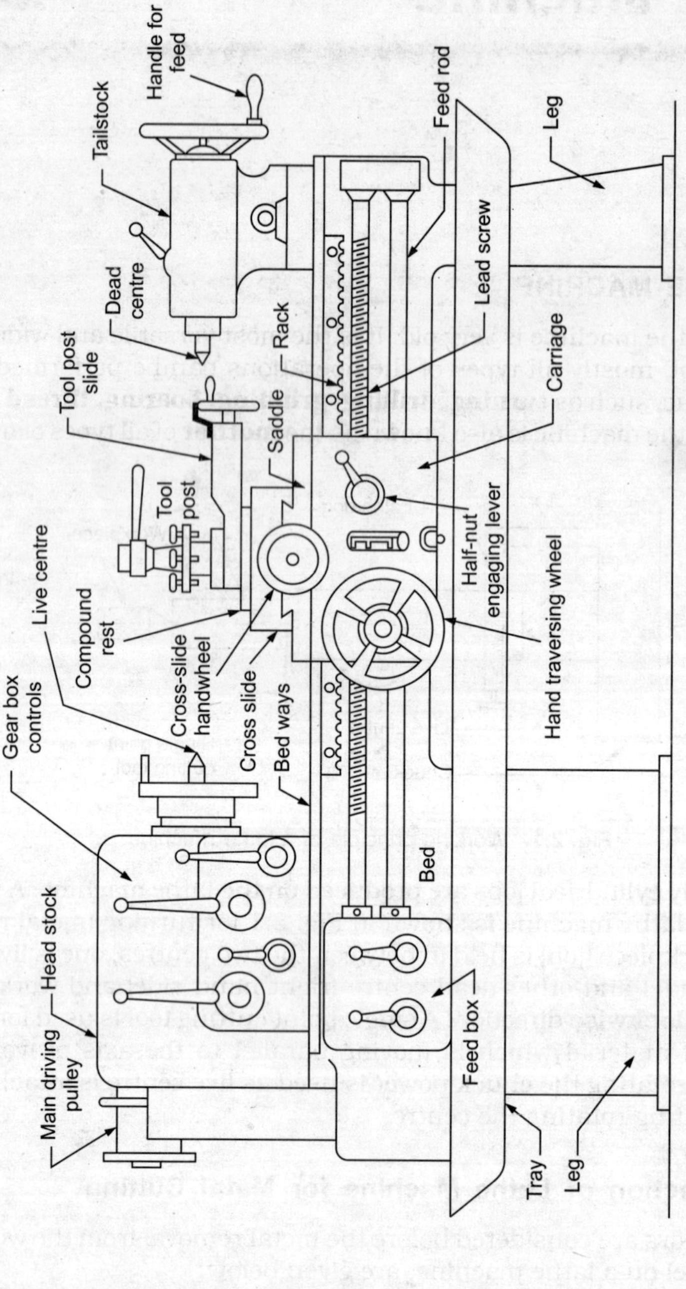

Fig. 2.2 Geared head centre lathe (Engine lathe)

(iii) Surface finish required (surface roughness).
(iv) Power required.
(v) Number of parts to be produced.
(vi) Degree of accuracy.

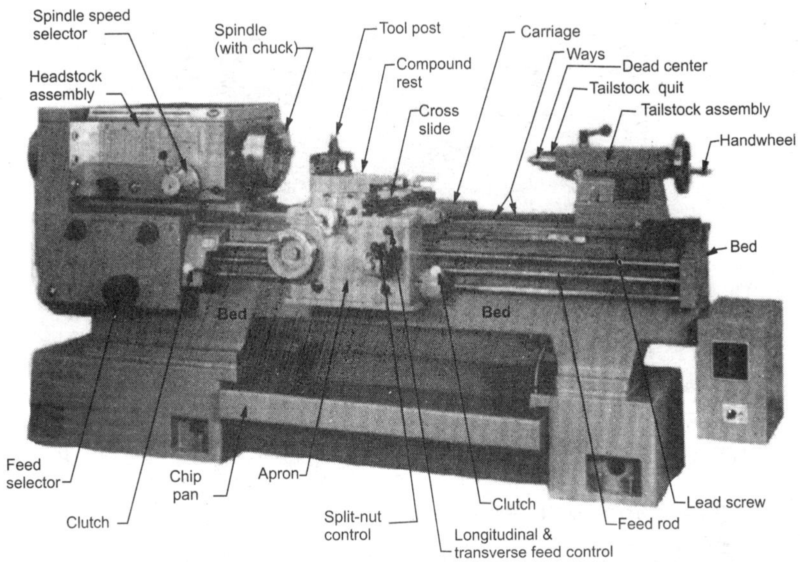

Fig. 2.3 Various parts of a lathe machine

2.1.2 Types of Lathe Machines

Several types of Lathe machines are available. They are following types:

(i) Tool room lathe machine
(ii) Centre lathe machine
(iii) Speed lathe machine
(iv) Bench lathe machine
(v) Automatic lathe machine
(vi) Special purpose lathe machine
(vii) Turret lathe machine
(viii) Capstan lathe machine
(ix) C.N.C. (Computer Numerical Control) lathe machine
(x) Heavy duty lathe machine.

2.1.2.1 Tool Room Lathe

Tool room lathe is equipt with all modern facility (all necessary accessories) for producing the accurate work. This lathe machine basically consist geared head stock with a considerable range of spindle speed. This lathe machine is equipped with feed rod, thread dial, quick change gear, lead screw, centre steady rest, draw in collect attachment, pump for coolant etc., this type of the lathe machine basically used in tool room, hence named as tool room lathe. All tool room lathes are care fully tested for accuracy and, as the name implies this type of lathe specially for producing **small die, Punch, small tools** and some other precision parts.

Fig. 2.4 Tool room lathe.

2.1.2.2 Engine Lathe or Centre Lathe

Engine lathe is also known as centre lathe. It is a powers driven has automatic feed, interchangeable speeds. It also consists a lead screw for cutting the thread. It is a most versatile of all machine tools and mostly all operation such as **turning, reaming, threading facing, screw cutting, polishing** etc. can be performed. Workpiece held in between the two centre i.e., live centre and dead centre by a chuck or a face plate.

Main components of engine lathe are **bed, head stock, Tail stock, carriage** and **feed mechanism**. A engine lathe is shown in Fig. 2·5. Entire weight of lathe machine is beared by the bed which is made out of cast iron (cast iron can absorb all vibration it self).

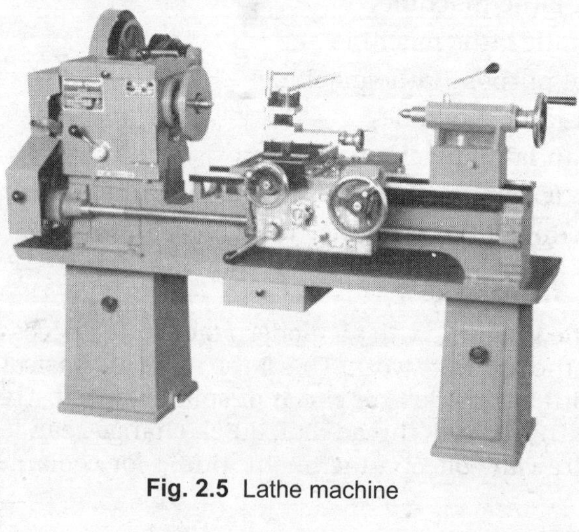

Fig. 2.5 Lathe machine

Head stock is fitted to the bed at the left end of the M/C and permanently secured. For different speed and feed, a chart is permanently fastened on the front of heat stock. Spindle is hollow through out its length and its front end is bored with a taper to fit the live center.

In engine lathe/centre lathe tail stock is fitted on the right end of the machine on bed ways. Tail stock can be adjusted to accommodate different length of work. Main components of tail stock are spindle, spindle wheel, clamp, bolt and center. Carriage is also another important part of engine lathe which consists the apron, saddle, compound rest, tool post, cross and longitudinal feed mechanism. Carriage can be moved along the ways by hand or by power.

2.1.2.3 Speed Lathe

Speed lathe is one of the simplest lathe in construction. Its main parts are head stock, tailstock etc. an adjustable slide is provided for supporting the cutting tool which is manipulated by hand. Workpiece is held in between the centre or supported by face plate and this lathe is driven by variable speed motor. This speed lathe is generally used for light turning process such as **wood turning, centering, metal spinning, polishing** etc. Speed lathe operating with high speed, for mapping a light cut by manual operated to cutting tool. Speed lathe usually used in pattern shop where wooden pieces are turned at high speed to converting in to final shape.

Fig. 2.6 Speed lathe machine

2.1.2.4 Bench Lathe

Bench lathe is same as engine lathe in construction. The basic difference between engine lathe and bench lathe is in size, means bench lathe are smaller in size therefore they are usually mounted/fixed on a heavy duty bench hence known as bench lathe. Bench lathe is adopted only for small work with maximum 250 mm swing capacity. Bench lathe is adapted for mass production with precision work. Mostly all types of attachements are present in bench lathe except one or two attachements.

Fig. 2.7 Bench lathe

2.1.2.5 Automatic Lathe

Automatic lathe are used to produced component in mass quantity with high precision accuracy. In automatic lathe tool feed automatically to the workpiece and unload workpiece after operation cycle is completed. As in this type of lathe machine loading and unloading of workpiece is completed by automatic hence known as automatic lathe. Those lathe machine which are fully automatic are provided with a mechanized feed so that a large numbers of work piece can be machined. In automatic lathe machine a very little attentation of operator is needed. These machine consists two side tool slide.

Fig. 2.8 CNC Lathe machine

2.1.2.6 Special Purpose Lathe

There are a huge number of special types of lathe machine available.

Each machine is designed to accommodate a specific work, one is gap bed lathe. In this type of lathe machine head stock can be removed to swining a large diameter of job.

Another example of special purpose lathe is a wheel lathe which is designed to perform the turning of the journals and wheel thread of rail road car, wheel and axle. Other special purpose lathe are horizontal, vertical, single multi spindle automatic.

2.1.3 Specification of a Lathe Machine

Lathe machine can be specified as given below:

1. **Swing or maximum diameter of the workpiece** that can be rotated area the bed as shown in Fig. 2.9(*A*).
2. **Distance between the two centre:** Maximum length of workpiece that can be held between the two centres as shows in the Fig. 2.9(*B*).
3. **Swing or maximum diameter over carriage:** How much diameter (maximum) workpiece can be rotated over the bed way or over the carriage as shown in Fig. 2.9(*C*).
4. **Length of bed:** Over all langth of bed is known as lenght of bad.
5. Diameter of hole through spindle.
6. Width of bed.
7. Lead screw diameter and number of thread per cm.

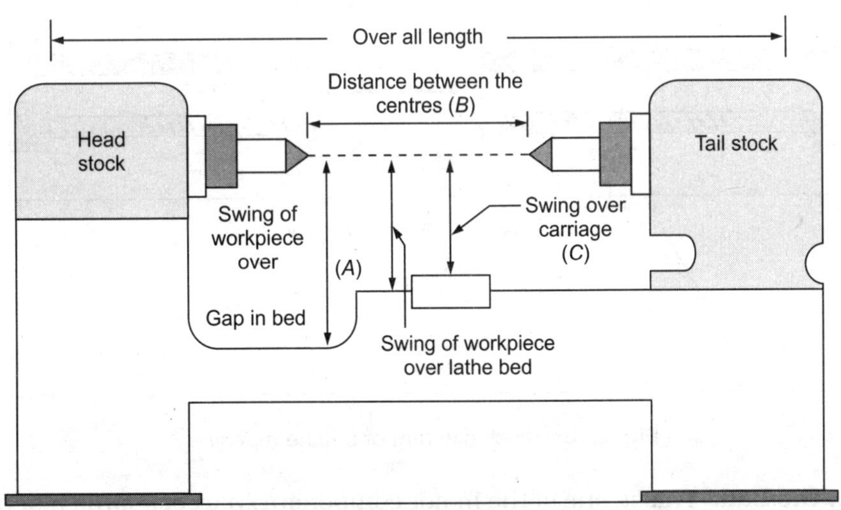

Fig. 2.9 Specification of a lathe machine

2.1.4 Common parts of Lathe Machine

A lathe machine have the following main parts:

1. Lathe bed
2. Head stock
3. Tailstock
4. Carriage
   (i) Compound slide
   (ii) Cross-slide
   (iii) Tool post
   (iv) Apron
5. Feed rod
6. Lead screw
7. Gear box
8. Lathe chuck or face plate
9. Half nut lever

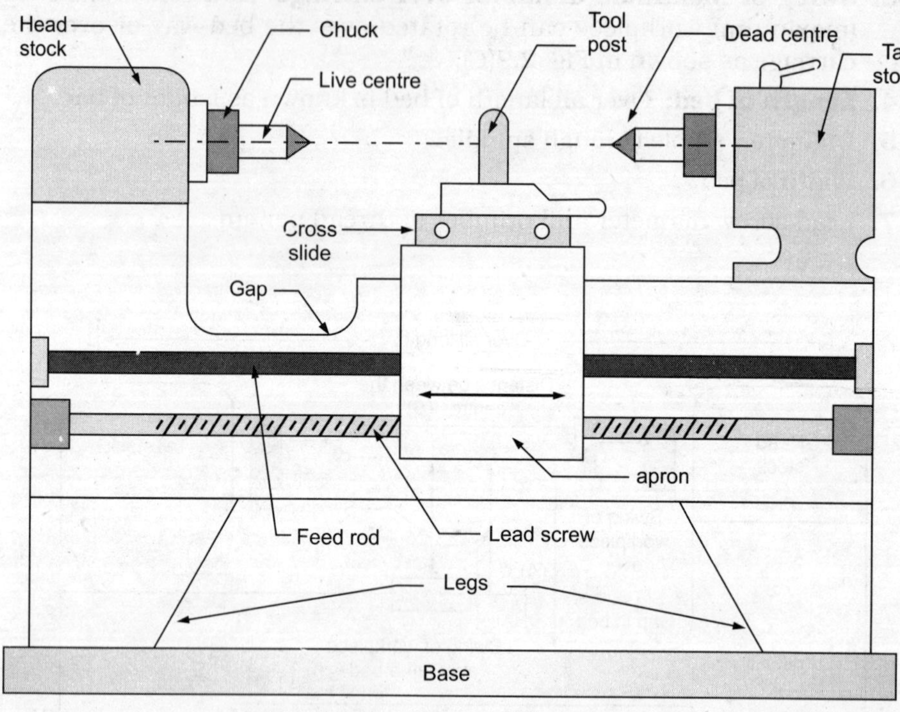

Fig. 2.10 Block diagram of a lathe machine

1. Lathe Bed: This is one of the major component/part of a lathe machine entire weight of lathe machine is beared by lathe bed and machine foundation is done on it. It is heavy riged in construction; bed is in single piece and made by cast iron, as cast iron having very good damping capacity and absorb the all vibration is self (Cost iron can be casted into any shape very easily). Bed consist the two guide ways, **flat guide** ways and **included** on the other hand flat guide ways provide larger bearing surface with corresponding reduction in wear but requires some special care for clearing.

The names given to various parts of the lathe have changed over the years and are still not completely standardised.

2.2 GUIDE WAYS

The bed of the lathe provides the foundation for the whole machine and holds the headstock, tailstock and carriage in alignment. The surfaces of the bed that are finely machined-and upon which the carriage and tailstock slide are known as "ways".

Some beds have a gap near the headstock to allow extra-large diameters to be turned. Sometimes the gap is formed by the machined ways stopping short of the headstock, sometimes by a piece of bed that can be unbolted, removed and lost.

Some very large lathes have a "sliding bed" where the upper part, on which the carriage and tailstock sit, can be slid along a separate lower part and so make the gap correspondingly larger or smaller.

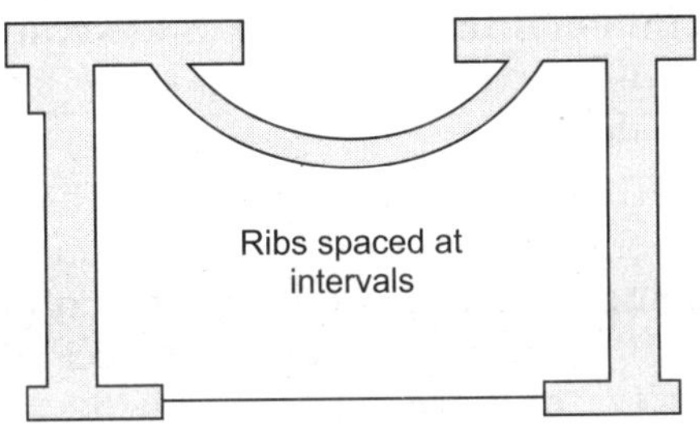

(*a*) Flat guides

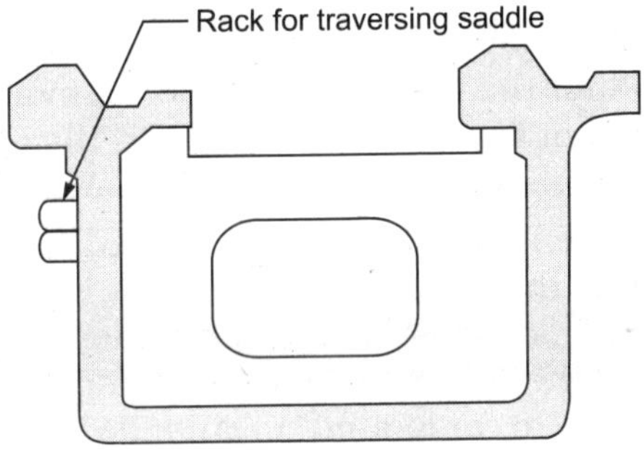

(*b*) Flat & inverted V-guide

Fig. 2.11 Lathe guide ways

2.2.1 Apron

The vertical, often flat and rectangular "plate" fastened to the front of the "Saddle" is known as the "Apron" and carries a selection of gears and controls that allow the carriage to be driven (by hand or power) up and down the bed. The mechanism inside can also engage the screw cutting feed and various powered tool feeds, should they be fitted. The leadscrew, and sometimes a power shaft as well, are often arranged to pass through the apron and provide it with a drive for the various functions. The sophistication of the apron-mounted controls, and their ease of use, is a reliable indicator of the quality of a lathe. Virtually all screw-cutting lathes have what is commonly called a "half-nut" lever that closes down one and sometimes two halves of a split nut to grasp the leadscrew and provide a drive for screw cutting.

Apron design can be roughly divided into "single-wall" and "double-wall" types. The "single-wall" apron has just one thickness of metal and, protruding from it (and unsupported on their outer ends) are studs that carry gears.

The "double-wall" apron is a much more robust structure, rather like a narrow, open-topped box with the gear-carrying studs fitted between the two walls and hence rigidly supported at both ends. This type of construction produces a very stiff structure–and one that is far less likely to deflect under heavy-duty work; another advantage is that the closed base of the "box" can be used to house an oil reservoir the lubricant ion which is either splashed around or, preferably, pumped to supply the spindles, gears and even, on some lathes, the sliding surfaces of the bed and cross slide as well.

2.2.2 Compound Slide Rest Consisting of the Cross Slide and Top Slide

Sitting on top of the "Saddle" is the "Cross Slide" that, as its name implies, moves across the bed and on top of that there is often a "Top Slide" or "Tool Slide" that is invariably arranged so that it can be swivelled and locked into a new position.

Very early lathes had a simple **T-shaped** piece of metal against which the turner "rested" his tool (all turning being done by hand) but when it became possible to move this "Rest" across the bed by a screw feed it became known, appropriately enough, as a "Slide-rest".

After the "Top Slide" became a more common fitting the term "Slide-rest" was not so frequently used and the different functions of the two slides led to their specific names being more widely adopted.

When two slides are provided (or sometimes, on watchmaker's lathes, three) the complete assembly is known as the "Compound" or "Compound Slide" or even "Compound Slide-rest". Some makers have been known to label the "Top Slide" as the "Compound Rest" or even the "Compound Slide" but as "to compound" means the 'joining of two or more not "one" this use of the term in incorrect. The top and cross slide together should be referred to as "the compound".

2.2.3 Carriage

The whole assembly of **Saddle**, **Apron**, **Top** and **Cross Slide** is known as the "Carriage". It is mounted on the guide ways. Main objection of carriage is to carry the cutting tool for feed motion purpose. Carriage can moved from left to right or vice versa between the head stock and tail stock, either by power-feed or by hand with wheel. For thread cutting purpose half nut is closed over the lead screw than traverses the carriage along the bed ways.

Main parts of carriage: Lathe machine carriage having the following main parts:

(i) Saddle
(ii) Cross slide
(iii) Compound slide or compound rest
(iv) Apron
(v) Tool holder (post).

(i) ***Saddle:*** Saddle is some what like **H-shape** and it is casted which is fitted on the bed and moves on the guide ways. On the top of the saddle there is a cross-slide and tool post saddle can be clamped during the machine process.

(ii) ***Cross slide:*** It is mounted on the carriage and has two fold purpose. Basic aim of cross slide is to carriage the compound slide and tool post. It can be moved by hand or power.

(iii) ***Compound slide or compound rest:*** It is mounted on the top of the cross slide. Compound rest consist a base having graduated **scale** and same can be swivelled around vertical axis. Compound rest can be set (fixed) at any angle for generating the taper on the periphery of work piece. Cross slide and compound rest screws are equipped with micrometer collars. Compound rest feed only by manual not power operated.

(iv) ***Apron:*** Apron is bolted to the front of saddle and hangs over the front of the bed. Apron housing consist clutch, gear and levers for operating the carriage by manual or power feed. Apron also consist friction clutch for automatic feed and a split half nut. Split half nut is used when generating the threads on the periphery of workpiece and same can be closed over the lead screw threads.

(v) ***Tool post:*** Tool post is mounted on the compound rest and slide in a **T-slot**. Tool post is a part of lathe machine which is used to securely and rigidly hold to a cutting tool during the turning operation. It can be moved left or right by the compound slide.

Tool post are various type-like American tool post (piller type), English tool post (clamp type), turret type tool post (4-way) etc.

Fig. 2.12 Lathe carrige

2.3 HEAD STOCK

The lathe Headstock used, at one time, to be called the "Fixed Headstock" or "Fixed Head", and the rotating shaft within it the "Mandrel". Today the mandrel is usually called the "Spindle", but this can cause confusion with the tailstock, where the sliding bar is known variously as the "ram", "barrel"- and "spindle".

The headstock is normally mounted rigidly to the bed (exceptions exist in some production, CNC, automatic and "Swiss-auto" types) and holds all the mechanisms, including various kinds and combinations of pulleys or gears, so that the spindle can be made to turn at different speeds.

2.3.1 Headstock Spindle

The end of the headstock spindle is usually machined so that it can carry a faceplate, chuck, drive-plate, internal or external collets or even special attachments designed for particular jobs. In turn, these attachments hold the workpiece that is going to be machined.

The "fitting" formed on the end of the spindle is normally one of five types:

1. A simple flange through which threaded studs on a faceplate or chuck (for example) can pass and be tightened into place with nuts. This is a secure method, and allows high-speed reverse, but is very inconvenient on a general-purpose lathe.
2. A threaded nose onto which fittings screw. This is perfectly acceptable for smaller lathes, but unsatisfactory on larger industrial machines where, for reasons of production economy, the spindle may need to be reversed at high speed. Reversing a screwed on chuck causes it to unscrew with potentially disastrous results.

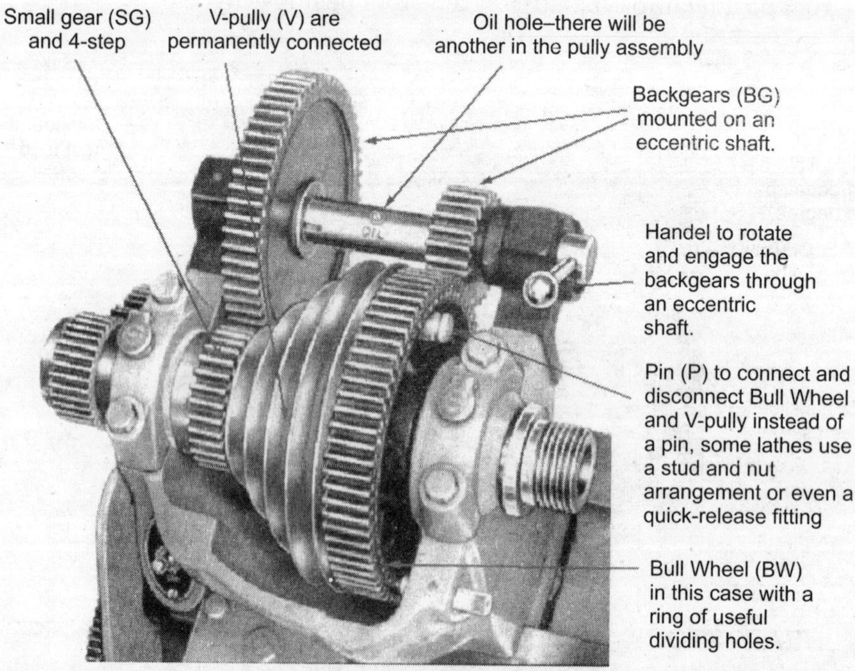

Fig. 2.13 Head stock

3. A "D1-taper Camlock" fitting-a long-used, standard system that employs three or more "studs" that are turned to lock into the back of chucks and faceplates, etc.
4. A taper-either of the simple Hardinge type or, for bigger lathes, the "taper-nose, long-key drive" an older but excellent American design where a large screwed ring was held captive on the end of the spindle and used to draw the chuck, or other fitting, onto a long, keyed taper formed on the spindle end.
5. Various fittings that became increasingly complex and apparently invented for the sake of being able to claim a National Standard (the famous not-invented-here syndrome). All these succeeded in doing was to raise manufacturing costs by preventing the interchange of spindle-nose tooling between machines and requiring firms to keep larger inventories of spares and numbers of duplicated firings.

   Some of these included: British and ISO Standard Spindle Noses-Direct Mounting; British and ISO Short Taper with Bolt or Stud Fixing; British and ISO Short Taper with Camlock Fixing; British and ISO Short Taper with Bayonet Ring Fixing and, of course, German Standard Spindle Noses. Unbelievably, there appears never to have been a French standard.

2.4 BACKGEAR

As its name implies, "backgear" is a gear mounted at the back of the headstock (although in practice it is often located in other positions) that allows the chuck to rotate slowly with greatly-increased torque (turning power). At first, the ability to run a workpiece slowly might seem unnecessary, but a large-diameter casting, fastened to the faceplate and run at 200 rpm (about the slowest speed normally available on a lathe without backgear) would have a linear speed at its outer edge beyond the turning capacity of a small lathe. By engaging backgear, and so reducing the speed but increasing the torque, even the largest faceplate-mounted jobs can be turned successfully.

Screw cutting also requires slow speeds, typically between 25 and 50 rpm especially if the operator is a beginner, or the job tricky. A bottom speed in excess of those figures (as usually found on most Far Eastern and European machines but not those built in the United Kingdom) means that screw cutting especially internally, into blind holes is, in effect, impossible. These lathes are advertised as "screwcutting" but what that means in reality is just power feed along the bed. Even if you go to the trouble of making up a pulley system to reduce the spindle speeds you will find the torque needed to turn large diameters at low speeds causes the belts to slip. The only solution is a gear-driven low speed and so a proper small lathe, with a backgear fitted, not only becomes capable of cutting threads but can also tackle heavy-duty drilling, big-hole boring and large-diameter facing in other words, it is possible to use it to the very limits of its capacity and strength.

Fig. 2.14 Back gear in lathe machine

Beginners are sometimes confused about how to engage backgear–especially if the lathe lacks a handbook-but with a little care anyone can work out how it should be done, at least on a conventional machine. On the main spindle of the lathe, the one carrying the drive pulley, will be found a large gear, generally referred to as the "Bull Wheel". The Bull Wheel is attached to the pulley by a nut and bolt, a spring-loaded pin, a pawl that presses into a gear on the pulley (or some other means) and, if this fastening is undone–by slackening the nut and pushing it towards the pulley, or by pulling the pin out. It should be found that the pulley will spin freely on the shaft. By moving the "backgears" into position-they generally slide sideways, or are mounted on an eccentric pin–the mechanism will come into operation. If the pulley will not spin on the shaft, or there seems to be no obvious way of disconnecting the Bull Wheel from the pulley, it may be that you are dealing with an "over-engineered" machine where some clever device has been introduced to make life "easy" for the operator. Sometimes there will be a screw, flush with the surface of the drive pulley and beneath this a spring-loaded pin that pushes into the back face of the Bull Wheel. Quick-action "Sliding-cam" mechanisms are occasionally used (as on the Drummond and Myford M Series lathes) where a knob on the face of the Bull Wheel has to be pushed sideways, and so ride up a ramp, which action disengages the connecting pin automatically. Some lathes, with enclosed headstocks (like later Boxford models) have a "single-lever" backgear; in this system moving the first part of the lever's movement disengages the connection whilst the next brings the backgear into mesh.

2.5 LEADSCREW

Originally termed a "master thread", or described as the "leading screw", but now always referred to as the "leadscrew", this is a long threaded rod normally found running along the front of the bed or, on some early examples running between the bed ways down the bed's centre line. Lead screw provides automatic feed and makes thread cutting possible. It is a precision-threaded shaft, driven by gears as the head stock turns. By using a train of gears to connect the lathe spindle to the leadscrew and the leadscrew to the lathe carriage the latter, together with its cutting tool, could be forced to move a set distance for every revolution of the spindle.

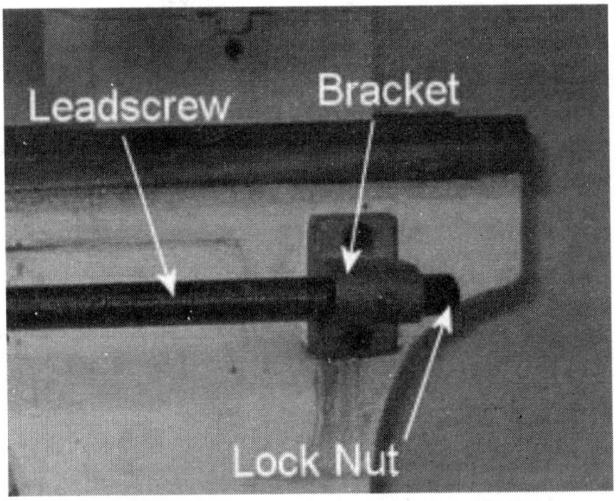

Fig. 2.15 Lead screw

2.6 TAILSTOCK

Tail stock is situated on the right hand end on the bed of a lathe machine. It consist a body which is bolted to the base and base is mounted on the bed guide ways. The tailstock supports long workpiece that would otherwise sag or flex too much to allow for accurate machining. Without a tailstock, long pieces cannot be turned straight and will invariably have a taper. Some tailstocks can be intentionally misaligned to accurately cut a taper if needed. The tailstock has a centering device pressed into a shallow, specially drilled hole in the end of the workpiece. The center can be either "live" or "dead". Live centers have a bearing, allowing the center to rotate along with the workpiece. Dead centers do not rotate and must be lubricated to prevent overheating due to friction with the workpiece. Instead of a center, a drill chuck can be mounted in the tailstock.

Fig. 2.16 Tailstock

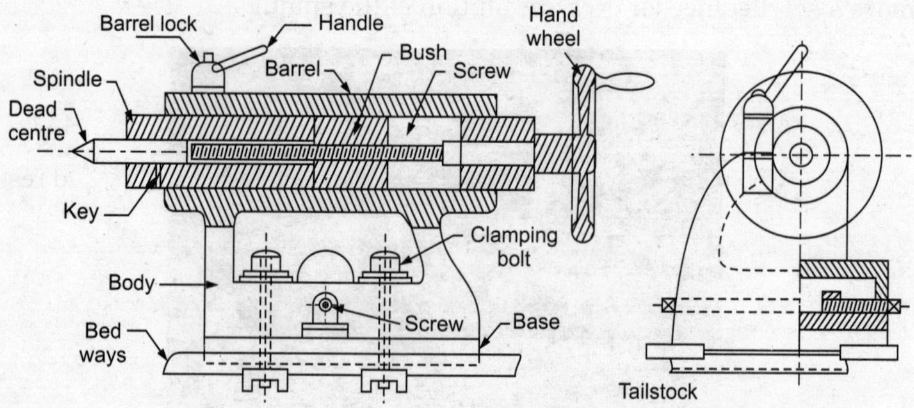

Fig. 2.17 Block diagram of a tailstock

2.7 COUNTER SHAFT

Most small electric motors in Britain run at 1425 rpm, whilst those in the USA and Europe are usually marked a little faster at 1600 to 1700 rpm or so.

If the lathe spindle was to be driven directly from one of these motors, even using a small pulley on the motor shaft, and a larger one on the lathe, it would be turning far too quickly to be useful for the great majority of jobs; hence, it is necessary to introduce some way of reducing the lathe's spindle speed and that is the job of the countershaft.

In a typical arrangement, illustrated here, the motor is fastened to an upright, hinged, cast-iron plate and fitted with a small pulley on its spindle. Because the 1500 rpm motor is driving a much larger pulley in a ratio of something like 5 :1 the speed is reduced to 300 rpm (1500 divided by 5).

On the same shaft as the very large pulley is a set of three smaller pulleys, arranged in the "reverse" order from those on the lathe. If the middle pulley on the countershaft is made to drive the identically-sized pulley on the lathe spindle that too, of course, will turn at 300 rpm. The pulleys each side of it are normally arranged to halve and double that speed–hence the creation of a speed set covering a useful 150 rpm, 300 rpm and 600 rpm.

It is a simple matter to fit both a small and a large pulleys to the motor shaft, and two correspondingly larger pulleys on the countershaft, and so double the number of available speeds to six. If a two-speed electric motor is used the range doubles again to 12 and, should the lathe designer have managed to squeeze a four-step pulley between the spindle bearings, a total of 16 would be available; with a backgearfitted the total would rise to thirty-two speeds that, typically, might start at 25 rpm and extend all the way up to over 3000 rpm.

Cross Slide

- The cross-slide is mounted to the top of the carriage to provide movement perpendicular to the length of the bed for facing cuts. An additional motion assembly, the compound rest, with an adjustable angle, is often added to the top of the cross slide for angular cuts. The cutting tools that do the actual metal removal during turning are mounted in an adjustable tool holder clamped to the compound rest.

2.8 WORK HOLDING DEVICES

A lathe chuck is a device on the driven spindle on the head (headstock) of the machine that holds the workpiece. It grips and spins it.

The chuck has "fingers" called jaws that "pinch" the workpiece between them to secure it. Chucks can be 2, 3, 4 jaw or more. And the chuck can be self-centering or of the independent type.

In an engine lathe standard practice of holding the work piece is between two centers, finds no place in a capstan or turret lathe as there is dead centre to support the work piece at the other end. Work piece is, therefore supported at the spindle end by the help of chucks and fixtures.

The usual methods of holding work piece in a capstan or turret lathe are:

1. Collet chucks
2. Push out type
3. Draw in type
4. Dead length type
5. Self centering chuck
6. Independent chuck
7. Combination chuck
8. Air operated chuck

2.8.1 Coolet Chucks

The collet chucks are used for gripping bars introduced through the head stick spindle of a capstan or turret lathe and is one of the most common method of holding work.

Different sizes of spring collets having **square, hexagonal, round** or any other shaped bore are fitted in the chuck body for holding different sizes of bar having different sections.

2.8.2 Push Out Type

To grip the work, the tapped portion of the spring collet is pushed in to the mating taper of the chuck. There is a tendency of the bar to be pushed slightly outward when the collet is pushed in to the chuck body for gripping. If the bar is fed against a stop bar fitted on the turret head, this slight outward movement of the bar ensures accurate setting of the length for machining.

2.8.3 Draw in Type

To grip the work, the tapered portion of the spring collet is pulled back in to the mating taper of the chuck which causes the split end of the collet to close in and grip the bar. The machining length of the bar in this type of chuck cannot be accurately set as the collet while closing will draw the bar slightly outward towards the spindle.

2.8.4 Dead Length Type

For accurate positioning of the bar, both the push out and draw in type collet present some error due to the movement of the bar along with the collet while gripping. This difficulty is removed by using a stationary collet on the bar. A sliding sleeve closes up on the tapered collet which is prevented from any end movement by the shoulder stop.

2.8.5 Three Jaw Chuck

The self-centering chuck will have a "scroll" of threads behind the jaws that engages them. The scroll is the screw that will close each of the jaws on the workpiece (simultaneously) with the use of a tool to tighten it. And this chuck will automatically center the workpiece in the chuck.

2.8.6 Four Jaw Chuck

Independent chucks will have jaws that must have a tool used on each of them independently to tighten them. And of course, the workpiece will have to be "fooled around with" until it is centered in the jaws.

(a) Collect chuck

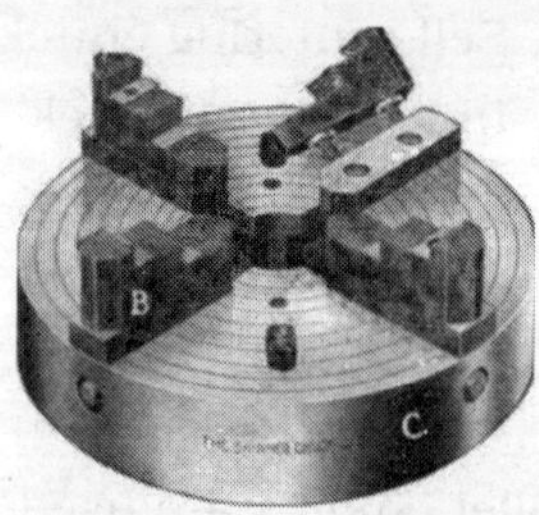

(b) Four jaw chuck

(c)

(d) Three jaw Chuck

Fig. 2.18 Various lathe machine chucks

2.9 LATHE OPERATIONS

The most common operations which can be carried out on a lathe machines are **facing, turning, drilling, reaming, boring, knurling, thread cutting** etc. These operations are discussed as follows:

2.9.1 Facing

Facing is one of prime operation during the machining of a workpiece on a lathe machine. It is essential in all most machining operation. In this operation workpiece is held in chuck and facing tool is fed the center of workpiece which is properly hold in tool post, towards the periphery with the help of cross slide.

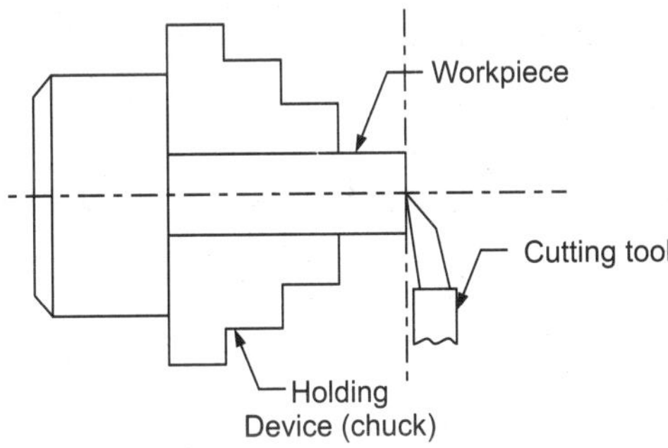

Fig. 2.19 Facing operation

2.9.2 Turning

Another most common and important operation which can be performed on the lathe machine is turning operation. Turning is defined as removal of unwanted material from the periphery of the workpiece. In turning operation workpiece is held in chuck or in between the two centers.

Turning operations are following types:

(i) Plain turning,

(ii) Step turning,

(iii) Taper turning.

(i) Turning process

(ii) Tool holders

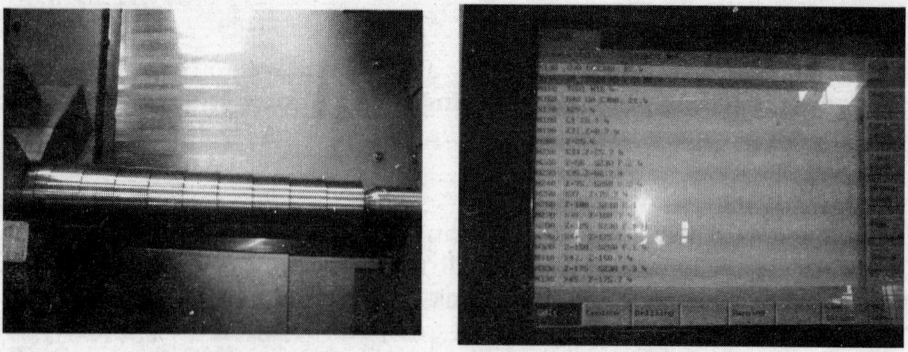

(iii) Step turning (iv) CNC programing

Fig. 2.20 Turning process by a CNC machine [30]

2.9.2.1 Plain Turning

It is a metal removal process in which the excess material is removed from the cylindrical workpiece as shown in the Fig. 2.21 in this process a long workpiece is held in between the two centers and for cutting perpose a longitudinal feed is given to the cutting tool by manually or by powered.

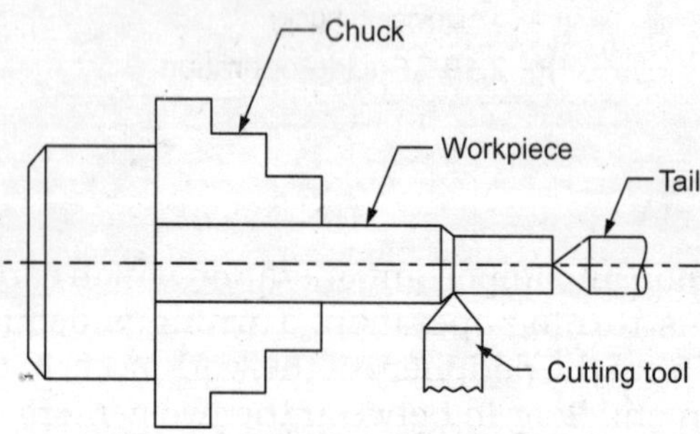

Fig. 2.21 Plain turning

2.9.2.2 Step Turning

In step turning process various steps of different diameters are produced by the single point cutting tool. This operation is carried out in the similar way as plain turning. In turning process feed, speed and depth of cut are the machining variables (Fig. 2.22).

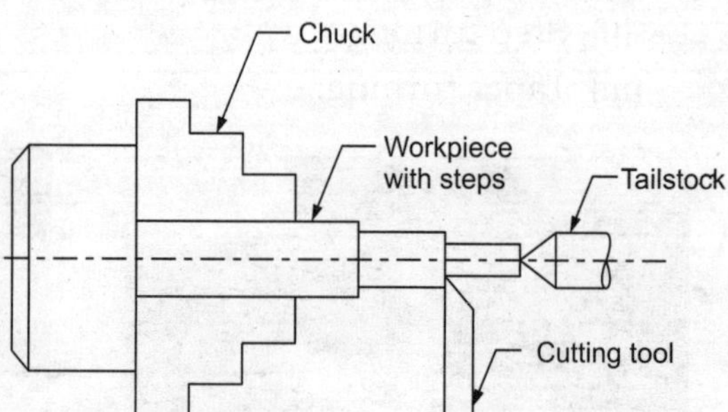

Fig. 2.22 Step turning

2.9.2.3 Taper Turning

Taper turning is a turning operation in which a conical work piece is produced. It is defined as uniform increasing or decreasing in diameter of cylindrical work measured along the length. As per B.I.S. morse tapers are standards and provided in all machine parts and tool wherever it is required.

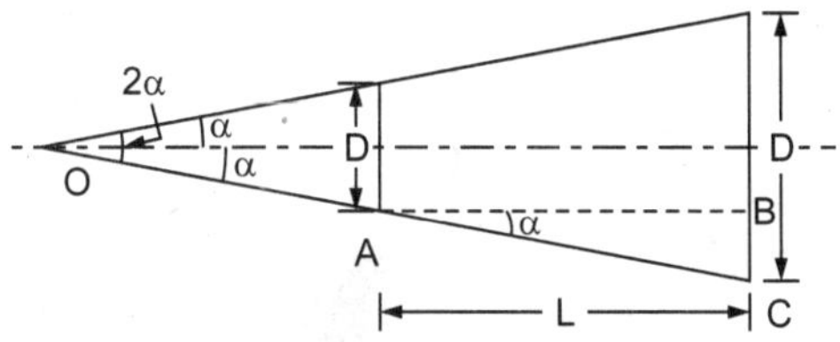

Fig. 2.23

From ΔABC

$$\tan \alpha = \frac{\frac{D}{2} - \frac{d}{2}}{L} = \left(\frac{D - d}{2L}\right)$$

$$\alpha = \tan^{-1}\left(\frac{D - d}{2L}\right)$$

where D = Major diameter of taper in mm

d = Small diameter of taper in mm

L = Length of tapered in mm

α = half of taper angle.

2.9.2.4 Taper Turning Method

Taper can be generate on a work piece by the following methods:

(i) By setting of compound rest method,
(ii) By taper turning attachment method,
(iii) By tail stock set over method,
(iv) Taper turning with a form tool
(v) Taper by C.N.C. lathe machine.

2.9.2.4.1 By setting of compound rest method

It is very common method for producing a taper for a short length. In this method centering of work piece is not affected. In this process work piece is held in the chuck firmly and rotating on the lathe axis cutting tool (single point) which is held in the tool post is fed at an angle to the axis of rotation of the work piece. Set the compound rest by swiveling it from the center line of the lathe center through an angle to half of taper angle as shown in the Fig. 2.24. Short length taper up to 45° angle be generated by this process by clamping the lathe carriage in place and after adjusting and clamping the tool. Several cuts applied for generate the taper on the work piece. Compound rest handle is used to provide the feed while cross slide is used for depth of cut.

Main **limitation** of this process is slow production rate and poor surface finish as feed is given by completely hand.

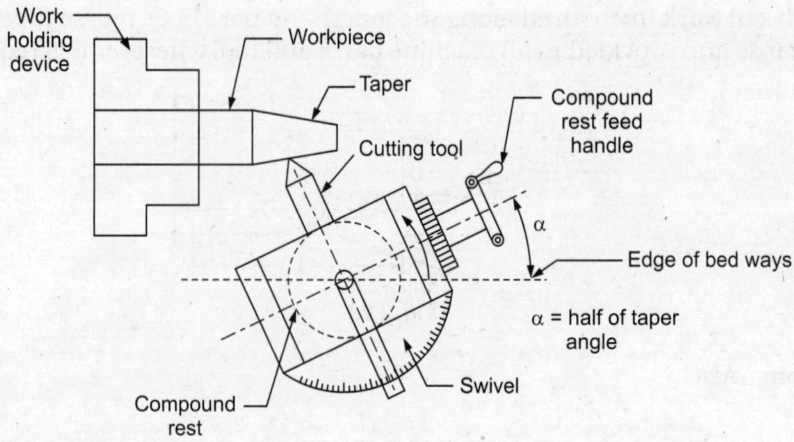

Fig. 2.24 Taper turning by swiveling of compound rest

2.9.2.4.2 By taper turning attachment

It is another method for taper generation. One of the unique feature of this method that longer taper can be easily produced by this method. Up to 20° angle taper can be easily produced by this process. Work piece is held in between the two center and rotating. Now set the link at desired angle to give the required taper on the work piece. Taper turning attachment is bolted on the back or rear side of the lathe which consist a supporting bar with a top bar having graduations in degree and arranged to swivel to the angle (Fig. 2.25).

After every cut feed to the tool is given by moving the compound rest which is positioned parallel to the cross slide.

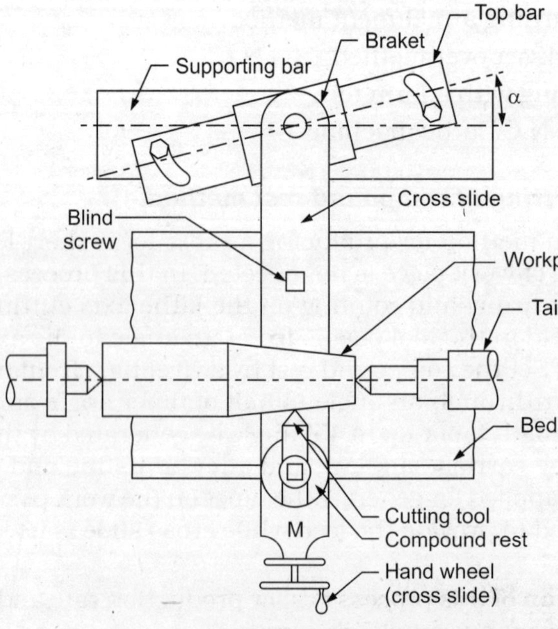

Fig. 2.25 Taper by taper turning attachment

2.9.2.4.3 By tail stock set over method

Tail stock set over method is shown in Fig. 2.26. This process is best suitable to produce long taper of small angle. In this method tail stock is set over from its center line equal to half of taper. Cutting tool move along the bed, it generate the taper on work piece.

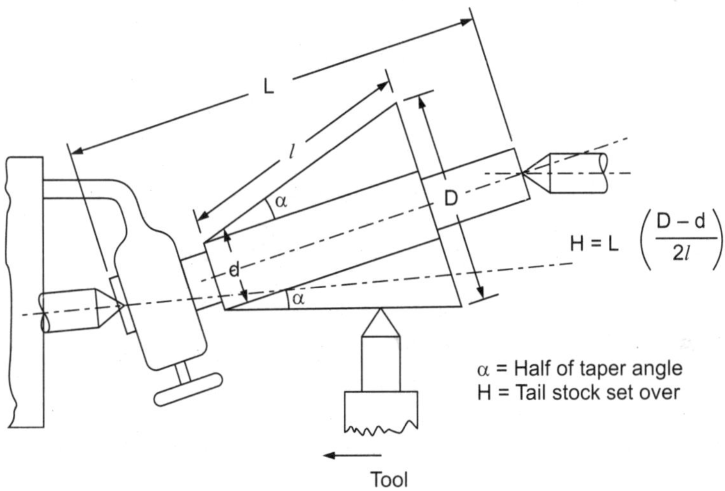

Fig. 2.26 Taper turning with tail stock set over method.

CASE I for giving taper 1 in 20 on a job 160 mm long. Find taper.

$$\text{on 160 mm length} = \frac{160}{20} = 8 \text{ mm.}$$

$$\text{tail stock set over} = \frac{8}{2} = 4 \text{ mm.}$$

CASE II for a taper of 16° on a work piece having 80° mm length here

$$\sin\frac{\alpha}{2} = \frac{\text{set over}}{\text{taper length}}$$

$$\text{set over} = \text{taper length} \times \sin\frac{\alpha}{2}$$

$$= 80 \times \sin\frac{16}{2} = 80 \times \sin 8^\circ$$

CASE III when D = major diameter in mm

d = minor diameter mm

l = tapered length mm

L = total length of job mm

$$\text{Tail stock set over (mm)} = \frac{D-d}{2}\times\frac{L}{l}$$

2.9.3 Drilling

A process through which a through or a blind hole is produced with the help at twist drill is known as drilling operation. In drilling process workpiece is held by chuck while twist drill (multi point cutting tool) is held by tailstock. Workpiece is rotated in the chuck by powered while twist drill is fed manually by rotating the hand wheel of tailstock.

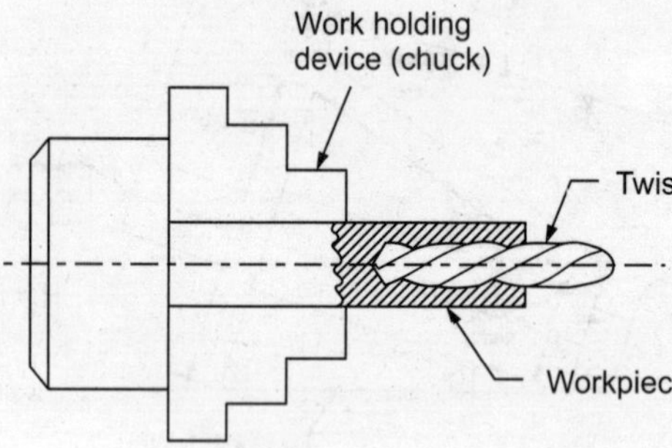

Fig. 2.27 Drilling

2.9.4 Reaming

Reaming is another important operation performed on a lathe machine. Usually reaming operation is performed after drilling operation, if drilled hole need be finished to correct size. For finish the drilled holed a tool is used which is known as reamer it is a multi point cutting tool, working principal of reaming operation is shown in Fig. 2.28 In this process reamer tool rotate inside a workpiece which having a hole and hold by a chuck while reamer tool is hold by the tailstock, reamer fed into the hole to the proper depth by means of tailstock hand wheel.

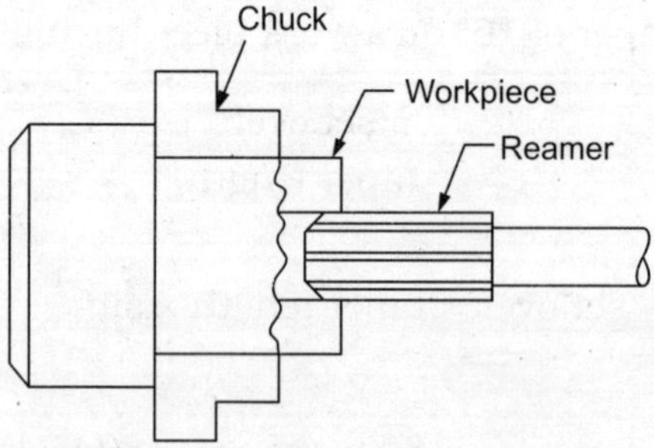

Fig. 2.28 Reaming

2.9.5 Boring

It is an operation of enlarging a hole which is produced by drilled, forging or casting process. Boring operation enlarge the hole as well as correct any eccentricity of the hole. For boring operation a tool is used which is known as bored tool. It is a single point tool but quite different with tool used in a lathe machine Fig. 2.29 shows the working principal of boring operation in which work piece is held by a rotating chuck while bored tool is held by the tool post tip of the tool is brought into contact with the surface of hole. Feed is given to tool by hand or by powered. In boring operation feed, depth of cut and speed is adjusted for a finish cut to finish the enlarged hole to size. Boring operation can not produced a hole.

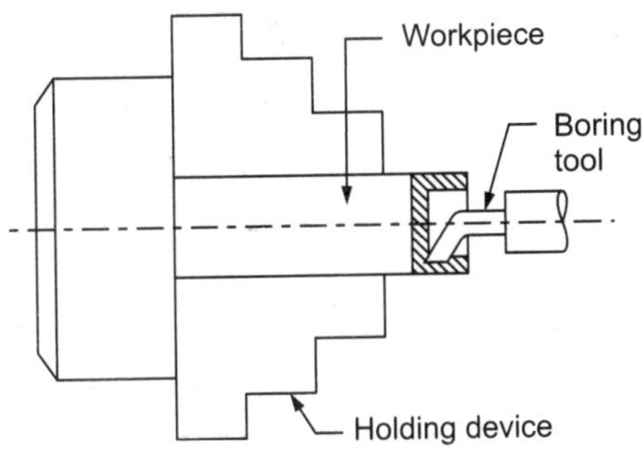

Fig. 2.29 Boring operation

2.9.6 Knurling

Basic function of knurling operation to embose a diamond shaped pattern on the periphery of the workpiece for better griping purpose. In knurling operation a tool is used to empose the pattern is known as knurling tool. It is a multi point cutting tool. One of the unique feature of knurling operation that it is a chip less process. Process is performed at slow speed. Fig. 2.30 shows the working principal of knurling operation. In this process a work piece is held between the two centers and rotating. A knurling tool is gripped in tool post and comes contact to rotating work piece. Proper depth of cut applied untill an impression is made on the work piece surface.

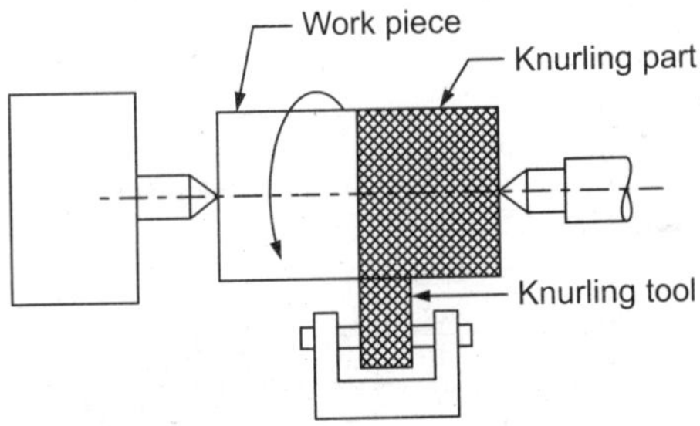

Fig. 2.30 Knurling operation

2.9.7 Thread Cutting

Thread cutting is also another important operation that can be performed on lathe machine both internal and external, left and right threads can be generated on the workpiece.

Figure 2.31 shown the threading operation in which a work piece is roating in between the two centres and a threading tool is fed longitudinally to the rotating work piece. Longitudinal feed is equal in the pitch of the thread to be cut.

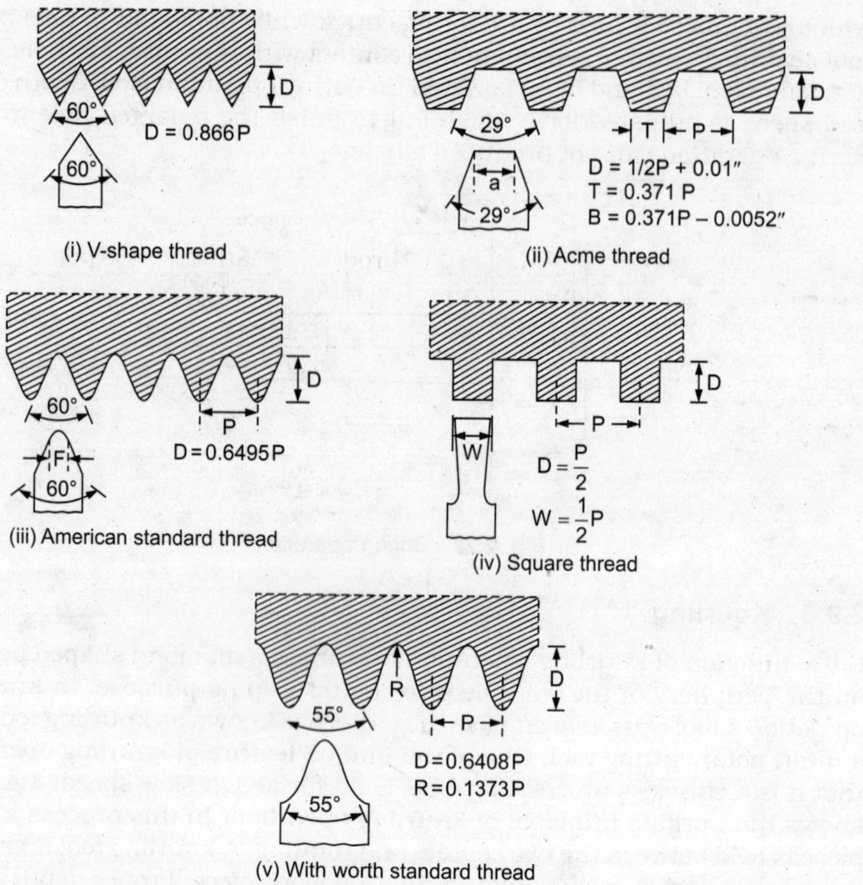

Fig. 2.31 Various type of threads

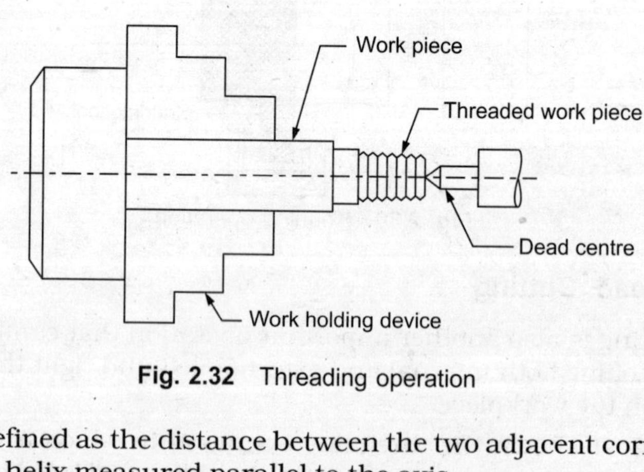

Fig. 2.32 Threading operation

Pitch is defined as the distance between the two adjacent corresponding points on the helix measured parallel to the axis.

Lead is the axial movement of the screw travelled in its one revolution.

$$\text{Lead} = \text{No of start} \times \text{pitch}$$

For thread cutting the ratio of gears between the driver gear to driven gear (lead screw) found from the following relation.

$$\frac{\text{Driver gear}}{\text{Driven gear}} = \frac{\text{Lead of threads to be cut on work piece}}{\text{Lead of thread on lead screw}}$$

2.10 CAPSTAN LATHE

INTRODUCTION: A capstan lathe is a production lathe used to manufacture any number of identical pieces in the minimum time. These lathes are development of engine lathes. The capstan lathe was first developed in the United States of America by Pratt and Whitney sometimes in 1860.

Special characteristics of a capstan lathe enable it to perform a series of operations such as **drilling, turning, boring, thread cutting, reaming, chamfering, cutting-off** and many other operations in a regular sequence to produce a large number of identical pieces in a minimum time.

Fig. 2.33 Capstan lathe machine

2.10.1 Different Parts of Capstan Lathe

- Head stock
- Cross-slide tool post
- Hexagonal turret
- Saddle for auxiliary slide
- Auxiliary slide
- Lathe head
- Feed rod
- Saddle for cross-slide.

2.10.1.1 Bed

The bed is a long box like casting provided with accurate guide ways upon which are mounted the carriage and turret saddle. The bed is designed to ensure strength, rigidity and permanency of alignment under heavy duty services. Like engine lathe precision surface finishing methods must be applied to keep it resistant to wear during service period.

2.10.1.2 Headstock

The headstock is a large casting located at the left hand end of the bed. The headstock of capstan and turret lathe may be of the following types:

1. Step cone pulley driven headstock
2. Direct electric motor driven headstock
3. All geared headstock
4. Pre-optive or pre-selective headstock.

2.10.1.3 Step Cone Pulley Driven Headstock

This is the simplest type of headstock and is fitted with small capstan lathes where the lathe is engaged in machining small and constant diameter workpieces. Three or four steps of pulleys can cater to the needs of the machine.

2.10.1.4 Direct Electric Motor Driven Headstock

In this type of headstock the spindle of the machine and the armature shaft are one and the same. Any speed variation or reversal is effected by simply controlling the motor. The machine is suitable for smaller diameter of workpieces rotated at high speeds.

2.10.1.5 All Geared Headstock

On the larger lathes, the headstocks are geared and different mechanisms are employed for speed changing by actuating levers. The speed changing may be affected without stopping the machine.

2.10.1.6 Preoptive or Pre-selective Headstock

- It is an all geared headstock with provisions for rapid stopping, starting and speed changing for different operations by simply pushing a button or pulling a lever.
- Different speeds are required for different operations. These speed variations are obtained by placing the speed changing lever in the required position.
- After the first operation is complete, a button or lever is simply actuated and the spindle starts rotating at the selected speed required for the second operation without stopping the machine.

2.10.1.7 Cross-slide and Saddle

In small size capstan lathes, hand operated cross-slide and saddle are used which are clamped on the lathe bed at the required position. The larger capstan lathes and heavy duty turret lathes equipped with usually two designs of carriage:

1. Conventional type carriage
2. Side hung type carriage

Conventional type carriage bridges the gap between the front and rear bed way sand is equipped with four station type tool post at the front, and one rear tool post at the back of the cross-slide.

The side hung type carriage is generally fitted with heavy duty turret lathes where the saddle rides on the top and bottom guide ways on the front of the lathe bed.

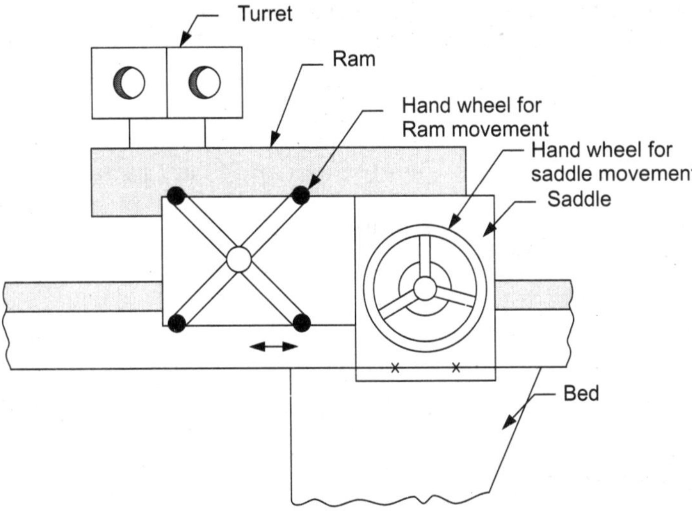

Fig. 2.34 Saddle arrangement for capstan lathe

The design facilitates swinging of larger diameter of workpieces without being interferred by the cross slide. The saddle and the cross slide may be fed longitudinally or crosswise by hand or power. The tools are mounted on the tool post and correct heights are adjusted by using rocking or packing pieces.

2.10.2 The Turret Saddle and Auxiliary Slide

In a capstan lathe, the turret saddle bridges the gap between two bed ways, and the top face is accurately machined to provide bearing surface for the auxiliary slide.

The saddle may be adjusted on the lathe bed ways and clamped at required position. In a turret lathe the turret is directly mounted on the top of the saddle and any movement of the turret is effected by the movement of the saddle.

The turret is the hexagonal tool holder intended for holding six or more tools. The centre line of the each hole is perfectly aligned with the axis of the lathe when aligned with the head stock spindle.

In addition to these holes, there are four tapped holes on each face of the turret for securing different tool holding attachments. After one operation is completed, as the turret is brought back away from the spindle nose, the turret index automatically by a mechanism incorporated on the bed and in turret saddle, so that the tool mounted on the next face is aligned with the work.

2.11 TURRET LATHE

INTRODUCTION: ***Turret lathes*** are a major departure from engine or basic lathes. These machines possess special features that adapt them to production. The "skill of the worker" has been built into these machines, making it possible for inexperienced operators to reproduce identical parts.

In contrast to this, the engine lathe requires a skilled operator and takes more time to produces parts that are dimensionally the same. The principal characteristic of this group of lathes is that the tools for consecutive operations can be set up in readiness for use in the proper sequence. Although skill is required to set and adjust the tools properly, once they are correct less skill is required to operate them and many parts can be produced.

2.11.1 Different Parts of Turret Lathe

(i) Head stock
(ii) Cross-slide tool post
(iii) Hexagonal turret
(iv) Turret saddle
(v) Feed rod
(vi) Saddle for cross-slide

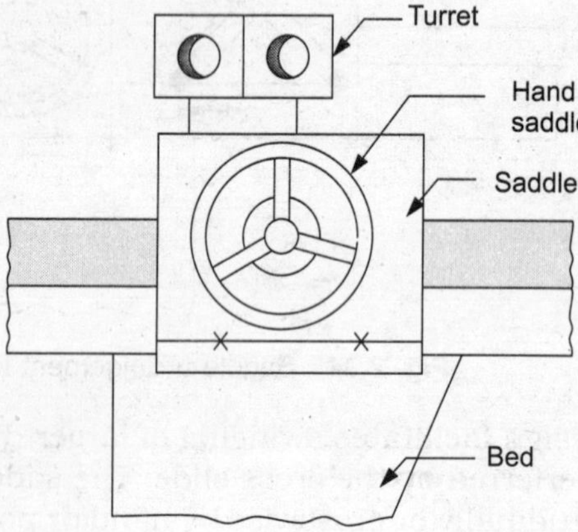

Fig. 2.35 Saddle arrangement for turret lathe

Turret lathes are constructed similarly to engine lathes. The headstock in most cases is geared, with provision for 6 to 16 spindle speeds and double these with a two-speed motor. The spindle speeds as well as forward and reverse movement are controlled by levers extending from the head. The drive motor is sometimes located in or on the motor leg below the headstock and connected to the geared-head sheave by V-belts, or a flange-mounted motor may be directly connected to the drive shaft in the head.

The cross-slide unit on which the tools are mounted for facing, forming, and cutting off is somewhat different in construction from the tool post and carriage arrangement used on lathes. It is made up of four principal parts: **cross slide**, **square turret**, **carriage**, and **apron**. These parts are discernible in the various turret lathe illustrations. Some of the cross slides are supported entirely on the front and lower front ways, permitting more swing clearance. This arrangement is frequently utilized on saddle-type machines that are to be used for large-diameter chucking jobs. The other arrangement for mounting has the cross slide riding on both upper bedways and is further supported by a lower way. This is used on machines engaged in bar work and in other processes where a large swing clearance is not necessary. An advantage of this type is the added tool post in the rear, frequently used in cutting-off operations.

On top of the cross slide is mounted a square turret capable of holding four tools. If several different tools are required, they are set up in sequence and can be quickly indexed and locked in correct working position. To allow cuts to be duplicated, the slide is provided with positive stops or feed trips. Likewise, the longitudinal position of the entire assembly may be controlled by positive stops on the left side of the apron. Cuts may be taken with square-turret tools *simultaneously* with tools mounted on the hexagonal turret.

The outstanding feature is the *turret* in place of the tailstock. This turret, mounted on either the sliding ram or the saddle, carries the tools for the various operations. The tools are mounted in proper sequence on the various faces of the turret, so that as it indexes around between operations, the proper tools are brought into position. For each tool there is a stop screw that controls the distance the tool will feed. When this distance is reached, an automatic trip lever stops further movement of the tool by disengaging the driven clutch.

The difference between the engine and turret lathes is that the turret lathe is adapted to quantity production work, whereas the engine lathe is primarily used for miscellaneous jobbing, tool room or single-operation work. The features of a turret lathe that make it a quantity production machine are these:

1. Tools may be set up in the turret in the proper sequence for the operation.
2. Each station is provided with a feed stop or feed trip so that each cut of a tool is the same as its previous cuts.
3. *Multiple cuts* can be taken from the same station at the same time.

Horizontal turret lathes are made in two general designs known as the ram and saddle. In appearance they are much alike, and both may be used for either bar or chucking work. The ***ram-type turret lathe*** shown in Fig. 2.36 (a) is so named because of the way the turret is mounted. The turret is placed on a slide or ram that moves back and forth on a saddle clamped to the lathe bed. This arrangement permits quick movement of the turret and is recommended for bar and light-duty chucking work. The saddle, though capable of adjustment, does not move during the operation of the turret. Ram-type machines do not require the built-in rigidity of chucking machines, because the bar tools can be made to support the work.

A *saddle-type lathe* Fig. 2.36 (b) is used for chucking work and has the turret mounted directly on a saddle that moves back and forth with the turret. Because chucking tools overhang and are unconnected with the work through some sort of support, greater strain on both work and tool support result. Chucking tools must have rigidity. The stock is longer, which

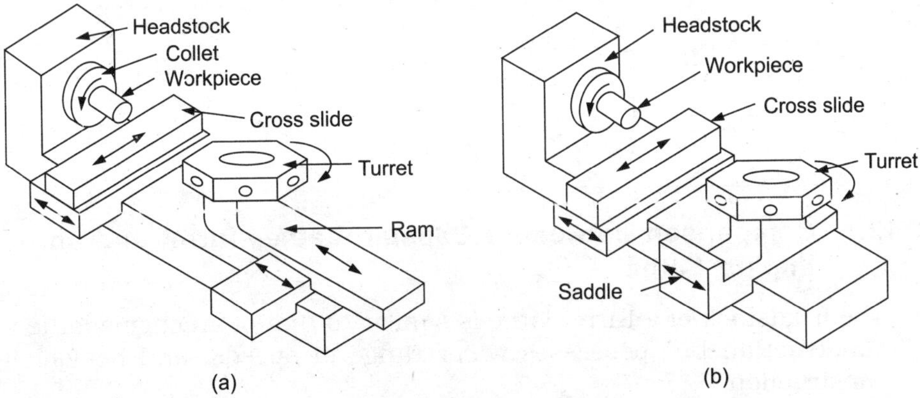

Fig. 2.36 Horizontal turret (a) Ram-type and (b) saddle-type [1]

is an advantage in long turning and boring cuts, and saddle mount assist the rigidity.

2.12 DIFFERENCE BETWEEN CAPSTAN LATHE AND TURRET LATHE

S. No.	*Capstan lathe (Ram type)*	*Turret lathe (Saddle type)*
1.	It is a light duty machine.	It is a heavy duty machine.
2.	The turret head is mounted on the ram and the ram is mounted on the saddle.	The turret head is directly mounted on the saddle and the saddle slides over the bed ways.
3.	The saddle will not be moved during machining.	The saddle is moved along with the turret head during machining.
4.	The lengthwise movement of turret is less.	The lengthwise movement of turret is more.
5.	Short workpieces only can be machined.	Long work pieces can be machined.
6.	It is easy to move the turret head as it slides over the ram.	It is difficult to move the turret head along with saddle.
7.	The turret head cannot be moved crosswise.	The turret head can be moved crosswise in some turret lathes.
8.	As the construction of lathe is not rigid, heavy cut cannot be given.	As the construction of lathe is rigid, heavy cuts are possible.
9.	It is used for machining work pieces up to 60 mm diameter.	It is used for machining work pieces up to 200 mm diameter.
10.	Collet is used to hold the work piece.	Jaw chuck is used to hold the workpiece.
11.	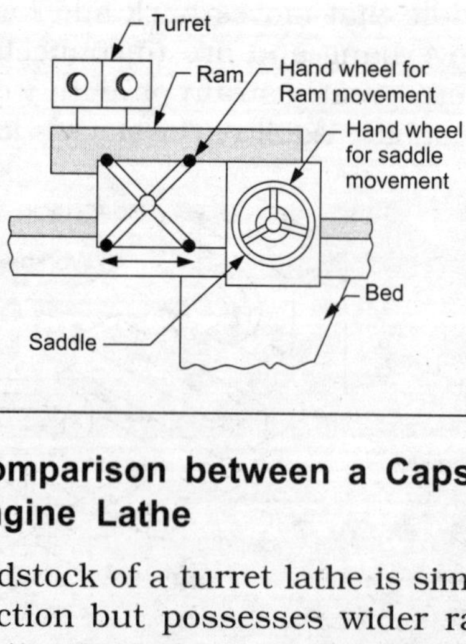	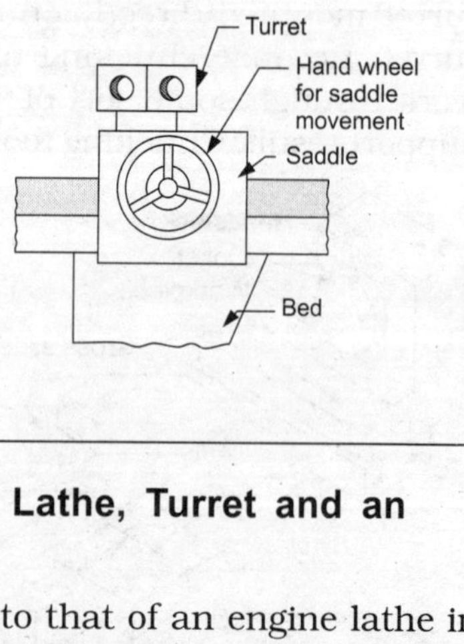

2.12.1 Comparison between a Capstan Lathe, Turret and an Engine Lathe

1. The headstock of a turret lathe is similar to that of an engine lathe in construction but possesses wider range of speeds, and heavier in construction.

2. Similar sizes of capstan and turret lathe and engine lathe, when an engine lathe will require a motor of 3 h.p. to drive its spindle and other parts, a capstan and turret lathe need power as high as 15 h.p. for high rate of production.
3. In a turret lathe, the tailstock of an engine lathe is replaced by a turret. This is a six sided block each of which may carry one or more tools. These tools may be indexed one after the other to perform different operations in a regular order. This is a decisive advantage in mass production.
4. In a turret lathe, combination of cuts can be taken. Two or more tools may be mounted on the same face of the turret, making it possible to machine more than one surface at a time. This feature reduces total operational time.
5. A semi skilled operator can operate a capstan or turret lathe after the machine has been set up by a skilled machinist. A skilled machinist may be requisitioned for setting up only for a large number of machines, where as actual production may be given by a semi skilled operator.
6. Capstan and turret lathe is fundamentally a production machines, capable of producing large numbers of identical pieces in a minimum time. Where as centre lathe is suitable for odd jobs having different shapes and sizes.
7. Capstan and turret lathes are not usually fitted with lead screws for cutting threads. A short length of lead screw called "Chasing screw " are sometimes provided for cutting threads by a chaser in a turret lathe. Where as lead screw is provided in Engine Lathe for cutting the threads.
8. The turret has essentially the same parts as the engine lathe except the turret and the complex mechanism incorporated in it for making it suitable for mass production work.

2.13 CAPSTAN AND TURRET LATHE MECHANISM

The carriage, cross-slide, turret-slide, and the saddle holding the turret may be fed into the work by hand or power.

Two main mechanisms are:

1. Turret indexing mechanism
2. Bar feeding mechanism.

2.13.1 Turret Indexing Mechanism

A simple line sketch of the mechanism is shown in Fig. 2.37.

1. The turret is mounted on the spindle, which rests on a bearing on the turret saddle.
2. The index plate, the bevel gear and an indexing ratchet are keyed to the spindle.
3. The plunger fitted with in the housing and mounted on the saddle locks the index plate by spring pressure and prevents any rotary movement of the turret as the tool feeds in to the work.

4. A pin is fitted on the plunger projects out of the housing. An actuating cam and the indexing pawl are attached to the lathe bed at the desired position.

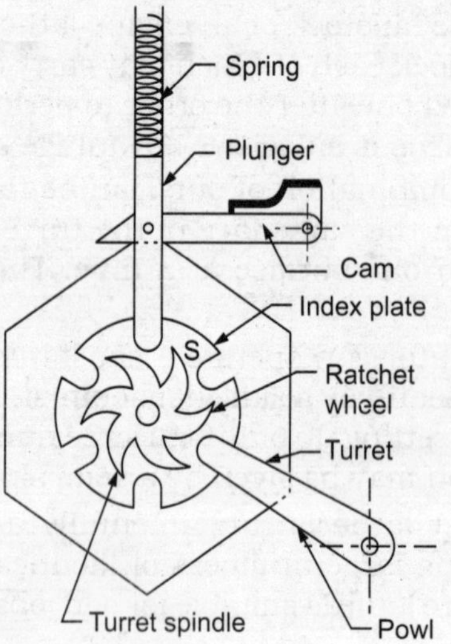

Fig. 2.37 Turret Indexing mechanism

5. Both the cam and the pawl are spring loaded. As the turret reaches the backward position, the actuating cam lifts the plunger out of the groove in the index plate due to the riding of the pin on the bevelled surface of the cam and thus unlocks the index plate.
6. The spring loaded pawl which by this time engages with a groove of the ratchet plate, causes the ratchet to rotate as the turret head moves backward.
7. When the index plate or the turret rotates through one-sixth of revolution, the pin and the plunger drops out of the cam and the plunger locks the index plate at the next groove. The turret is thus indexed by one-sixth of revolution and again locked in to the new position automatically.

2.13.2 Bar Feeding Mechanism of Capstan Lathe

The Fig. 2.38 typically shows the kinematic arrangement of feeding and clamping of bar stock in capstan lathes. The bar stock is held and tightly clamped in the push type spring collet which is pushed by a push tube with the help of a pair of bell-crank levers actuated by a taper ring as shown in figure. Bar feeding is accomplished by four elementary operations:

- unclamping of the job–by opening the collet
- bar feed by pushing it forward
- clamping of the bar by closing the collet
- free return of the bar–pushing element.

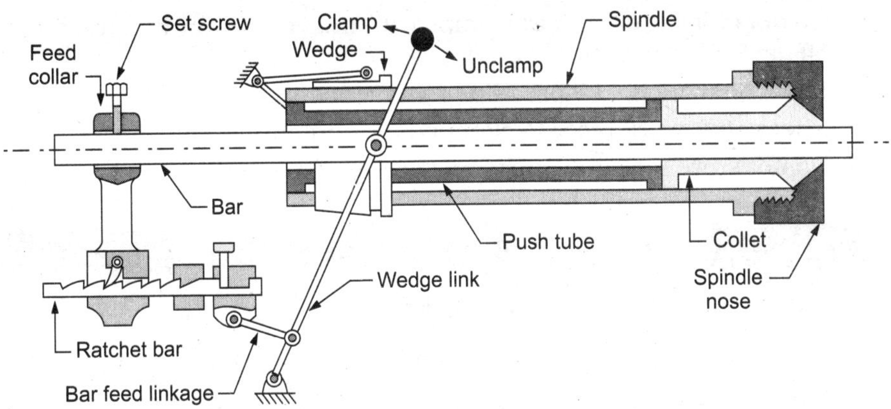

Fig. 2.38 Typical bar feeding mechanism in capstan lathe

After a job is completed and part off, the collet is opened by moving the lever manually rightward to withdraw the push force on the collet. Further moving of the lever in the same direction causes forward push of the bar with the help of the ratchet–paul system. After the projection of the bar from the collet face to the desired length controlled by a pre-set stop- stock generally held in one face of the turret or in a separate swing stop, the lever is moved leftward resulting closing of the collet by clamping of the barstock. Just before clamping of the collet, the leftward movement of the lever pushes the bar feeder (ratchet) back freely against the paul.

2.14 SAFETY GUIDELINE FOR WORKING ON LATHE MACHINE

Following are the safety guidelines for working on lath machine:

(i) Always keep safety in mind.

(ii) Do not operate the lathe machine without prior approval of operator.

(iii) Area surrounding the lathe machine should be free from oil or coolant and it should be free from any obstructions.

(iv) Do not use compressed air to blow chips from the spindle of machine surface, cabinets, control or the floor around the machine.

(v) Do not wear the neckties, wristwatches rings jewellary etc., while working on a lathe machine.

(vi) Wear the proper apron while working on the lathe machine.

(vii) Wrenches, tools and other equipment/tools should be kept off the machine spindle as well as off all moving parts of the machine.

(viii) Clean machine and its surrounding after use.

(ix) Make sure chuck wrench is removed from chuck before starting lathe.

(x) Before starting motor make sure the clutch levers are in their neutral position.

(xi) In any accident condition immediately report to the higher authority.

(xii) If work is turned between the centers, make sure that proper adjustment is made between centers and that the tailstock is locked in place.

(xiii) Do not dispose off or touch chips with finger. Use always chip dispose off device

(xiv) Stop the machine before taking any measurements.

(xiv) Cheek that work piece is clamped tight in chuck.

(xv) Do not store tools and parts on the top of machine.

OBJECTIVE TYPE QUESTIONS

1. Lathe machine bed is generally made by
(a) cast steel (b) mild steel
(c) wrought iron (d) cast iron

2. Lathe centres are provided with a standard taper is known as
(a) morse taper (b) chapman taper
(c) jarno taper (d) Seller's taper

3. How many number of jaws consist a self centeric chuck
(a) three (b) two
(c) four (d) eight ***[UPSC–2008]***

4. Which one of the following method should be used for turning external taper only
(a) tailstock offset (b) form tool
(c) taper attachement (d) compound rest ***[UPSC–2009]***

5. In a centre lathe, the spindle speed will be lowest during
(a) thread cutting (b) parting off
(c) knurling (d) taper turning

6. In turning operation a four jaw-self-centering chuck is to be refered only for
(a) turning of copper and brass casting
(b) setting up of heavy and irregular shaped work
(c) quick and efficient centering of heavy workpiece
(d) none of above

7. Which type of cutting tool is used in lathe machine
(a) single point cutting tool (b) multi point cutting tool
(c) two point cutting tool (d) six point cutting tool
(e) four point cutting tool.

8. A unit of a lathe machine which consist the lathe spindle and control levers for speed selction is known as
(a) carriage (b) head stock
(c) tail stock (d) cross slide
(e) apron

9. Embossing a diamond shaped pattern on the job by lathe machine is known as
(a) threading (b) knurling
(c) broaching (d) reamming
(e) parting off

10. Which one of the following is a lathe accessories
(a) tool post (b) lathe carriage
(c) cross slide (d) compound slide

11. In automatic machine where large numbers of components are machined from a bar, it is held in

(a) face plate (b) collect chuck

(c) magnatic chuck (d) three jaw self centric chuck

12. Purpose of tumbler gear in a lathe machine to ***[PH.D-2015 GLA]***

(a) reverse spindle rotation

(b) give derived direction of movement to the lathe carriage

(c) generate the threads

(d) none of above

13. Angle of a lathe center is

(a) 20° (b) 40°

(c) 60° (d) 80°

(e) 90°

14. A study rest in lathe machine is used to

(a) performing the knurling operation

(b) supporting the lathe bed

(c) holding very long and heavy jobs

(d) external taper turning

15. Taper of a lathe spindle is

(a) 1 : 10 (b) 1 : 15

(c) 1 : 20 (d) 1 : 25

16. Which is the most accurate of the following chuck

(a) magnatic chuck (b) three jaw chuck

(c) collect chuck (d) independing chuck

17. During the taper turning operation, compond rest is swivelled through an angle given by

(a) $\tan \phi = \frac{D-d}{2l}$ (b) $\tan \phi = \frac{D-d}{l}$

(c) $\tan 2\phi = \frac{D-d}{2l}$ (d) $\tan 2\phi = \frac{D-d}{l}$

where D = large diameter, d = small diameter, l = length of tapered piece.

18. Cutting power consumptions in turning can be significantly reduced by

(a) increasing the clearance angle

(b) increasing the rake angle

(c) increasing the nose radius

(d) increasing the cutting angle ***[GATE–2009]***

19. In turning of a cylinderical rods. it is necessary to keep the transverse force to minimum mainly to

(a) reduce vibration and chatter (b) improve cutting efficiency

(c) improve the surface finish (d) increases the producivity***[UPSC–1998]***

20. To obtained good surface finish in turning operation, one should be use a sharp tool with, feed and speed of a rotation of job

(a) maximum minimum (b) maximum, maximum

(c) minimum, maximum (d) minimum, minimum

21. In turning of a solid bar, the travel of the cutting tool in the direction of feed motion is 1000 mm, rotational speed of the workpiece is 500 r.p.m and feed rate is 0.2 mm/revolution this is corresponding to a machining time of

(a) 5 mm (b) 10 min
(c) 20 min (d) 100 min

22. In a lathe machine, the material is removed by
(a) feelding the job against a cutter having reciprocating motion
(b) rotating the work piece against a cutting tool
(c) feeding the job against a rotating cutting tool
(d) rotating the job against a rotating cutting tool

23. The most important parameter on which basies a lathe machine is specified
(a) bed length (b) maximum speed of spindle
(c) distance between the two centers
(d) maximum diameter of the job which can swing

24. In a lathe machine, function of a manderal is
(a) for knurling operation
(b) taper turning operation
(c) checking the roundness of a job
(d) for holding and rotating a hollow piece of job

25. Lath carriage consist
(a) ACME thread (b) square thread
(c) buttres thread (d) V-thread

26. A mechanism that lock the lathe carriage to the lead screw for thread cutting is known as
(a) half nut (b) full nut
(c) morse tapper (d) saddle

27. In a lathe gear box for changing spindle speeds the provision for speeds is made in
(a) arithmatic progression (b) geometrical progression
(c) harmonic progression (d) random fashion

28. Which is not true in the content of turret and capstan lathe as compared to engine lathe
(a) used for producing large number of identical pieces
(b) possibility of wide range of speeds
(c) skilled operator is required for operation
(d) need of greater horse power motor for driving spindle

29. Which of the following is not a part of capstan lathe
(a) tool post (b) chuck
(c) tail stock (d) spindle

30. Turret lathe upon which long turning cuts can be made by moving the saddle along the bed ways of machine are known as
(a) ram type turret (b) drum type turret
(c) saddle type turret (d) universal lathe

ANSWERS

1.	(d)	**2.**	(a)	**3.**	(a)	**4.**	(a)	**5.**	(d)	**6.**	(b)	**7.**	(a)
8.	(b)	**9.**	(b)	**10.**	(b)	**11.**	(b)	**12.**	(b)	**13.**	(c)	**14.**	(c)
15.	(c)	**16.**	(c)	**17.**	(a)	**18.**	(b)	**19.**	(c)	**20.**	(c)	**21.**	(b)
22.	(b)	**23.**	(d)	**24.**	(d)	**25.**	(b)	**26.**	(a)	**27.**	(b)	**28.**	(c)
29.	(c)	**30.**	(a)										

QUESTIONS

1. Define the lathe machine with the help of a block diagram.
2. Why lathe machine is called as the mother of all machine?
3. What types of operation can be performed on a lathe machine?
4. How a lathe machine can be specified?
5. Classify the lathe machine.
6. Define the various parts of a lathe machine with the help of the figures.
7. How many types of lathe machine you know?
8. Differentiate between the capstan lathe machine and turret lathe machine. ***[UPTU–2006-07]***
9. How does a turret lathe differ from a capstan lathe? What special tooling is associated with the turret lathe? ***[UPTU–2002-03]***
10. What are the differences between an automobile lathe and a capstan lathe? Give an example of a component suitable for a capstan lathe with dimensions. ***[UPTU–2004-05]***
11. What are the types of lathe you are familiar with? State their application. ***[UPTU–2004-05]***
12. List the different types of lathe available giving salient feature of each. What are the use of (i) lead screw, (ii) feed rod, (iii) Tail stock, (iv) half nut, (v) Compound slide in lathe. ***[UPTU–2010-11]***
13. Draw the tool layout for manufacturing hexagonal bolt on turret lathe in mass. Assume the dimensions. ***[UPTU–2007-08]***
14. Describe the following operations briefly:
    (i) Step turning, (ii) Taper turning, (iii) Drilling, (iv) Reaming, (v) Thread cutting, (vi) Kurling, (vii) Boring, (h) Porting.
15. Explain the tailstock with a neat sketch.
16. Define the following with their importance:
    (i) Feed, (ii) Speed, (iii) Depth of cut.
17. Write the short notes on the guide ways with the help of neat sketch.
18. Describe the method of taper turning by compound rest swiveling method and taper turning attachment.

❑❑❑

Chapter 3

Shaper, Slotter, Planner Machine

3.1 SHAPER MACHINE

Shaper is a machine tool device which is used to remove (cut) the unwanted material from a workpiece (block). Unique feature of shaper that it cut (remove) the material only in forward stroke while backward stroke is idle and the time of returning of backward stroke is less as compared to forward stroke. Shaper basically used to generate the various shapes out of block like **dove tial**, ***v*-block, guide gib** etc.

In shaper a single point cutting is used to remove the material. In shaper, tool has a reciprocating motion while workpiece (block) stationary.

Classification of Shaper: Shaper can be classified on the following basics:

1. **According to the ram driving mechanism:**
   - (i) Geared shaper
   - (ii) Hydraulic shaper
   - (iii) Crank shaper.
2. **According to the type of cutting stroke:**
   - (i) Push cut shaper
   - (ii) Draw out shaper.

3.1.1 Principal Parts of a Shaper

- (i) Base
- (ii) Clapper box
- (iii) Table
- (iv) Column
- (v) Ram
- (vi) cross rail
- (vii) Tool head.

Fig. 3.1 Shaper machine

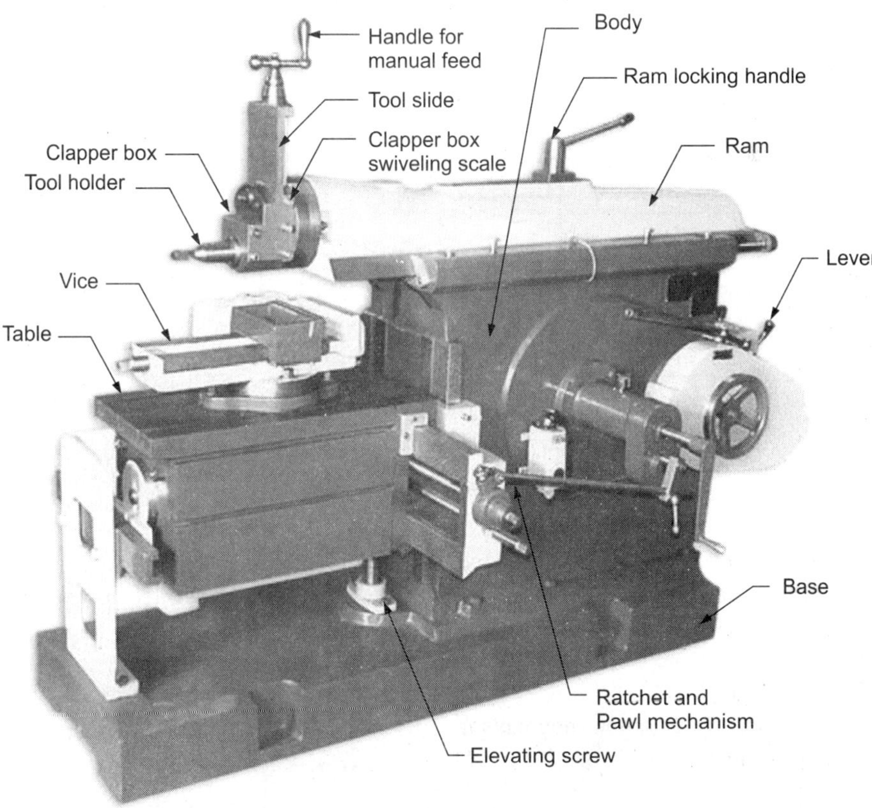

Fig. 3.2 Various parts of a Shaper machine.

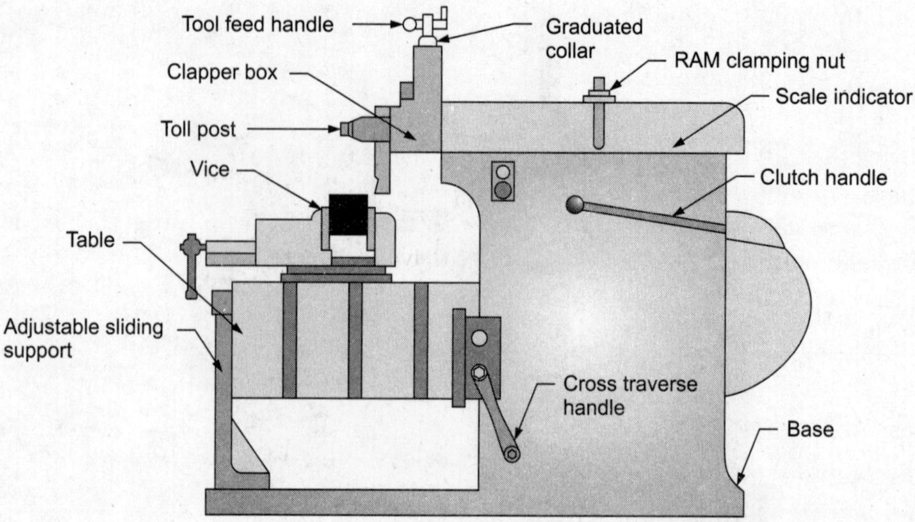

Fig. 3.3 Block diagram of a shaper machine

The drive for the main and feeding motion is effected by means of a three-phase motor which is mounted in or on the machine frame.

The main gearing consists of a gear train. It is a toothed gearing which in most cases is switchable in 4, 6 or 8 steps, thus providing various speeds of the ram.

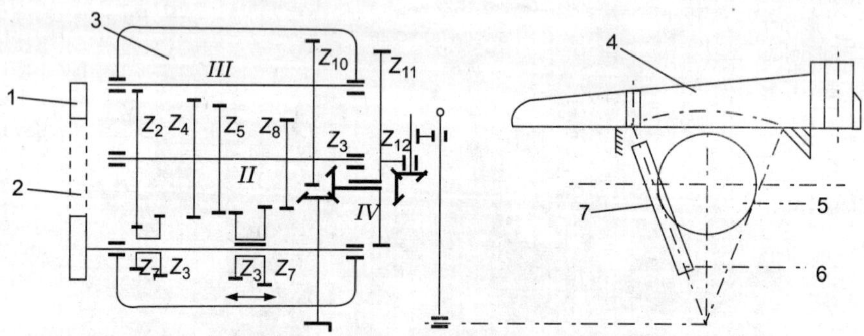

1. electric motor
2. vee-belt drive
3. toothed gearing with toothed gears z_1-z_{12}, z_{12} simultaneously being crank gear/I–III shafts
4. ram
5. crank gear
6. oscillating slider crank mechanism
7. sliding block

Fig. 3.4 Main gearing of a mechanical shaping machine

The toothed gearing is followed by the heart of the shaping machine, namely the transmission for transforming the rotary motion into the straight-lined primary motion of the ram. The feed gear produces the automatic transverse motion of the table with the workpiece set up on it (feed motion). It is guided from the main gearing by transmission elements to the table spindle. The transmission elements are arranged on the operating side of the machine. A clutch can be actuated by means of a control lever as a result of which engagement or disengagement of the drive is effected.

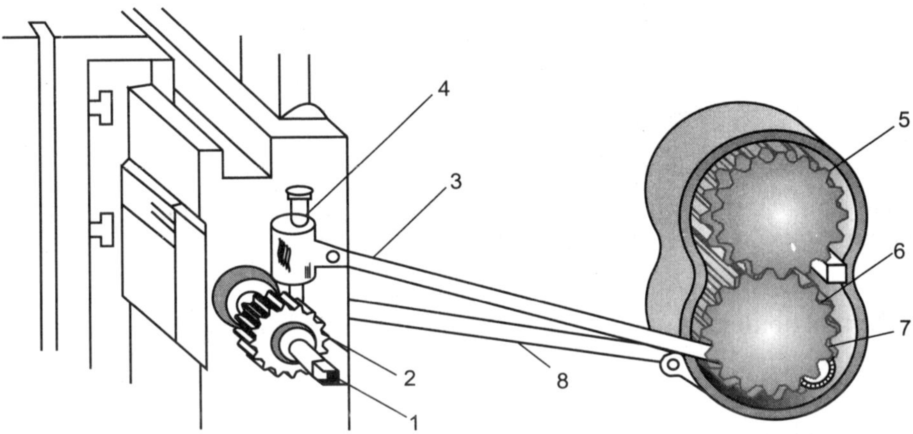

1. Table spindle
2. Ratchet wheel
3. Connecting rod (push rod)
4. Retaining pawl (catcher)
5. Toothed gear z_1
6. Toothed gear z_2
7. Eccentric pin
8. Link

Fig. 3.5 Feed gear

Shaping is a technique with straight-lined motion (cutting and feed motion). The feed motion takes place in steps.

Parameters of a shaping operation:

- tool
- workpiece
- cutting motion
- feed motion

The workpiece clamping fixture (compound slide) consists of saddle and machine table (see Fig. 3.4, parts 6 and 5).

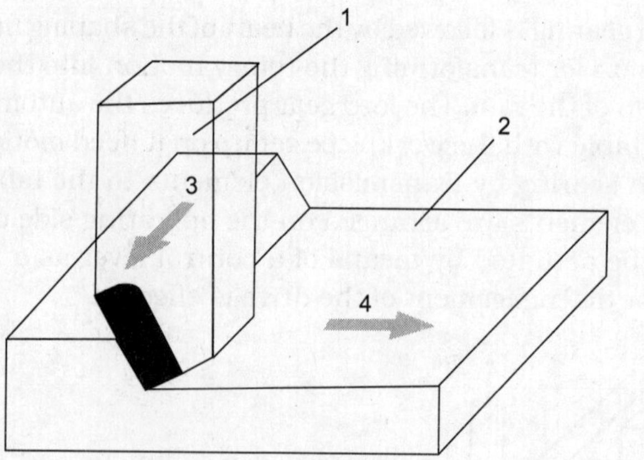

Fig. 3.6 Conditions of motion during shaping

The machine table is additionally supported.

The ram carries the ram head consisting of swivel-head plate, tool slide, clapper box with tool block and tool holder (tool post).

1. clapper
2. tool holder
3. tool base
4. clapper axis
5. clapper box
6. tool slide
7. crank with actuating screw
8. dividing ring
9. swivel-head plate
10. ram

During the working stroke the tool block rests on the clapper box (as a result of the cutting pressure), during the return stroke it is lifted. In this way the tool tip is protected. With older shaping machines the tool is dragging

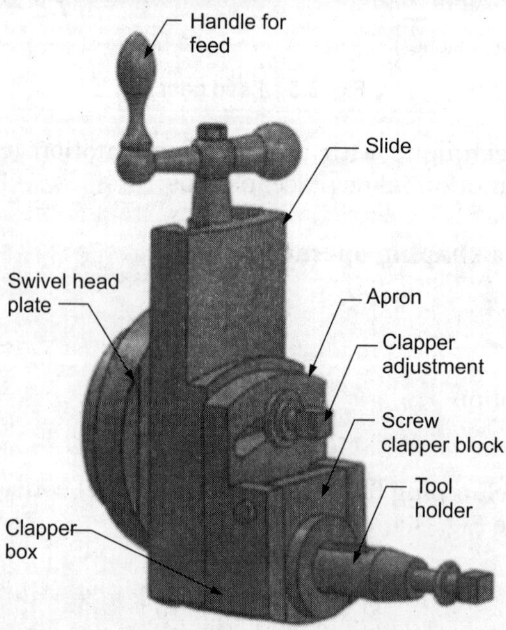

Fig. 3.7 Ram head with tool slide

over the workpiece during the return stroke. The working position is reached by the dead weight, and thus, by falling back. Modern machines have an automatic tool lifter.

For shaping oblique surfaces the tool slide can be swivelled on the swivel-head plate. In order to maintain the mobility of the tool block, it is also possible to set the clapper box on the tool slide at an angle, i.e., to swivel it in the circular slot.

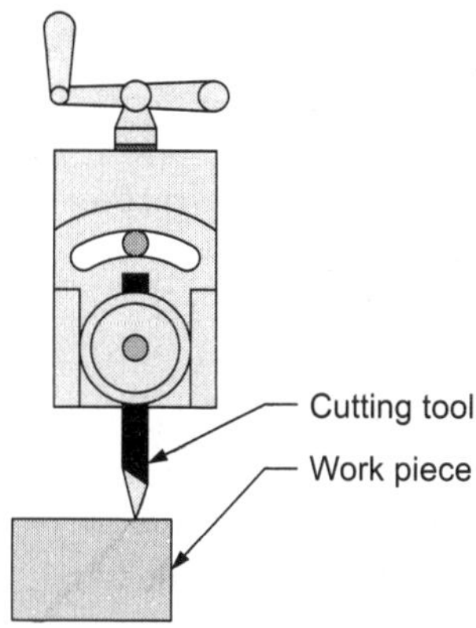

Fig. 3.8 Setting of the tool box when oblique surfaces are to be shaped

3.1.2 Additional Components

Additional components are mounted for special operations. Recessing and shaping with formed tool can be carried out with automatic infeed, if adjustable stops or starting cams are mounted. For machining circular–arc shaped parts special ram heads have been developed. The decision in favour of one or the other kind of machining depends on the shape or configuration of the workpiece to be manufactured.

3.2 SHAPER DRIVE MECHANISMS

A shaper drive mechanism changes the rotary motion of the power source (Electric motor) into the reciprocating motion of the ram.

Metal cutting is carried out during the forward stroke of the ram only; the return stroke of the ram does no cutting and hence is called idle stroke. Since return stroke does no cutting the drive system incorporates a quick return mechanism so that the ram moves faster during return stroke in order to minimize the idle time. The two common mechanism used for this purpose are,

(*i*) Crank and slotted link quick return mechanism
(*ii*) Whitworth quick return mechanism

3.2.1 Crank and Slotted Link Mechanism

The crank and slotted link quick return mechanism is shown in Fig. 3.9. Slotted link mechanism is very common in mechanical shapers. It is a simple and compact mechanism. It converts the rotary motion of the electric motor and gear box into the reciprocating motion of the ram. The slotted link mechanism gives the ram a higher velocity during the return non-cutting stroke that during its forward cutting stroke. Thereby the time wasted during the return stroke is reduced.

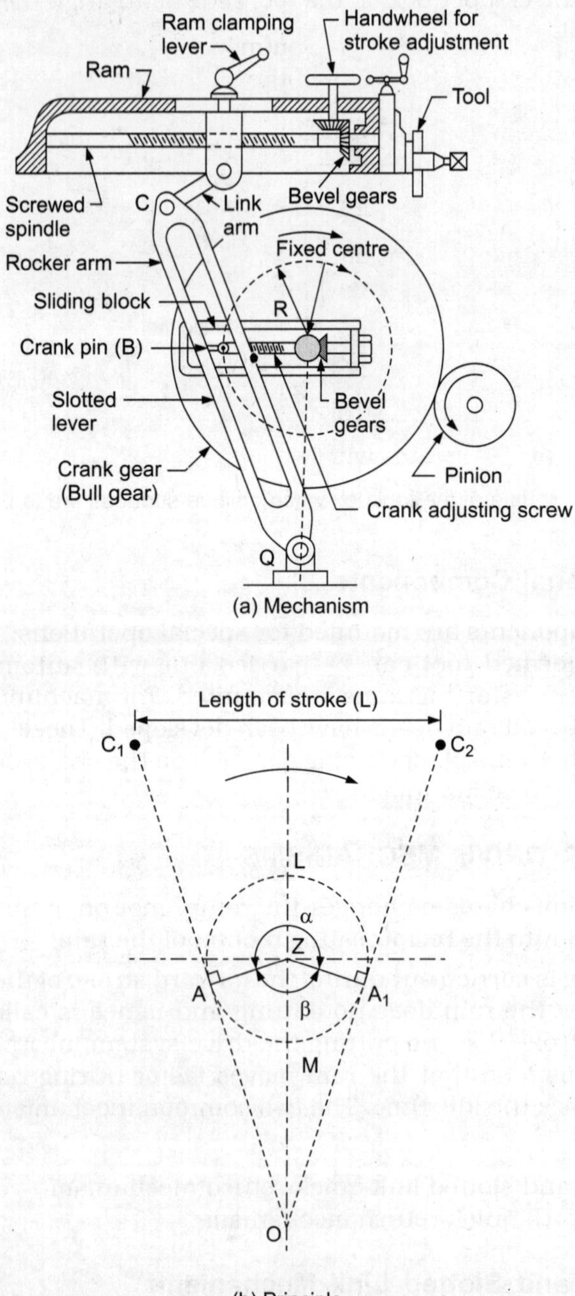

Fig. 3.9 Crank and slotted link quick return mechanism

The bull gear [(Fig. 3.9(*a*)] is driven by a pinion which is connected to the motor shaft through a multi speed gear box. The bull wheel is provided with a slot [(Fig. 3.9(*b*)]. The crank pin 'A' is secured into this slot; simultaneously it can slide in the slotted crank 'B'. When the bull wheel rotates the crank pin 'A' also rotates and side by side slides through the slot in the slotted crank 'B'. This makes the slotted crank to oscillate about its one end C. This oscillating motion of the slotted crank makes the ram to reciprocate. The intermediate link *D* is necessary to accommodate the rise and fall of the crank.

The position of the crank pin 'A' in the slot in the bull wheel decides the length of the stroke of the shaper. The further it is away from the centre of bull wheel, the longer is the stroke. The cutting stroke of the ram is completed while the crank pin moves from A to A_1 and the slotted crank goes from left to right. Similarly during the return stroke crank pin moves from A_1 to A and the link changes its position from right to left.

The time taken by cutting and idle strokes of the ram is proportional to the $\angle AZA_1$ and A_1ZA respectively.

$$\frac{\text{Cutting stroke}}{\text{Return stroke time}} = \frac{\angle ALA_1}{\angle AMA_2}$$

Since the crank pin 'A' rotates with uniform velocity and $\angle AMA_2$ is smaller it is obvious that the idle stroke is quicker than the forward cutting stroke. Hence the crank and slotted link mechanism is known as quick return mechanism.

3.2.2 Whitworth Quick Return Mechanism

Whitworth quick return mechanism is shown in Fig. 3.10. Crank BC revolves at a uniform speed. During cutting stroke the point *C* travels from *Y* to *X* through Z. The ram is returned at high speed as the crank rotates from *X* to Y though *T*.

Then, time for cutting stroke/ Time for return stroke = 360–6/6

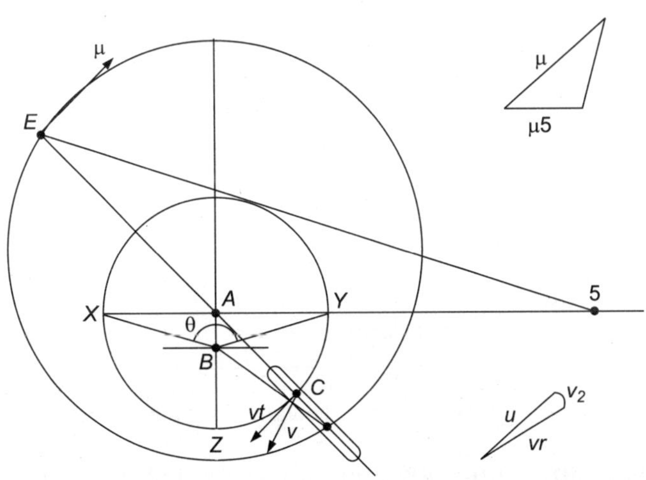

Fig. 3.10 Withworth quick return mechanism

Since 0 is smaller than 360°, the time for cutting is more than the idle stroke time. Hence it is known as quick return motion. The stroke length can be changed by varying the radius AE. Since the change in stroke length alters the cutting speed it requires a change of gear to get desired cutting speed.

3.3 COMPONENTS SHAPES PRODUCED BY SHAPER

Various type of shapes can be generated by the shaper machine.

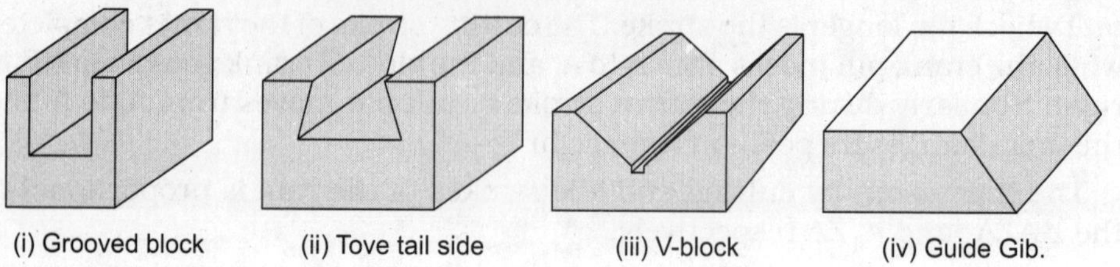

Fig. 3.11

3.3.1 Slotter Machine

Slotter is also another machine tool which is usually used to cut the key ways in a work piece. In effect slotter is a vertical shaper Fig. 3.12 shows a block diagram of a slotter machine. Basic difference between shaper and slotter is direction of cutting action of the tool. In shaper tool moves (cut) in the horizontal position and in slotter tool moves (cut) in the vertical position.

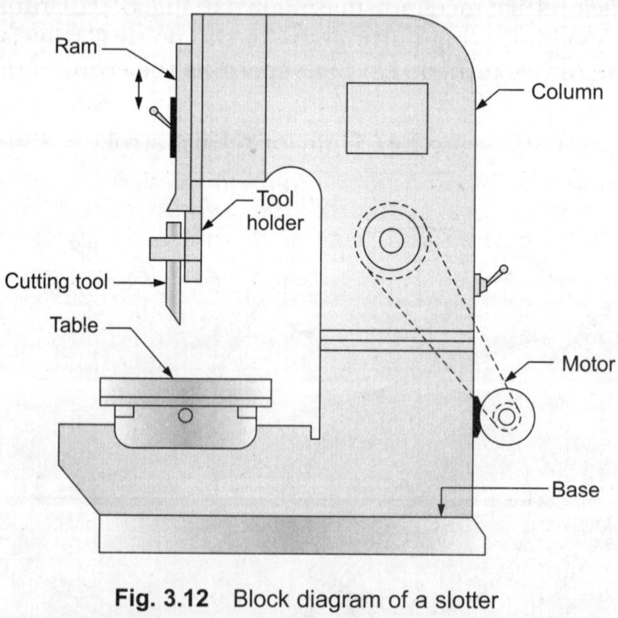

Fig. 3.12 Block diagram of a slotter

In slotter machine during operation workpiece is held stationary and ram of the slotter can be indined upto 10° to either side of the tool travel position.

3.3.2 Main Parts of a Slotter

Slotter tool consist the following parts:

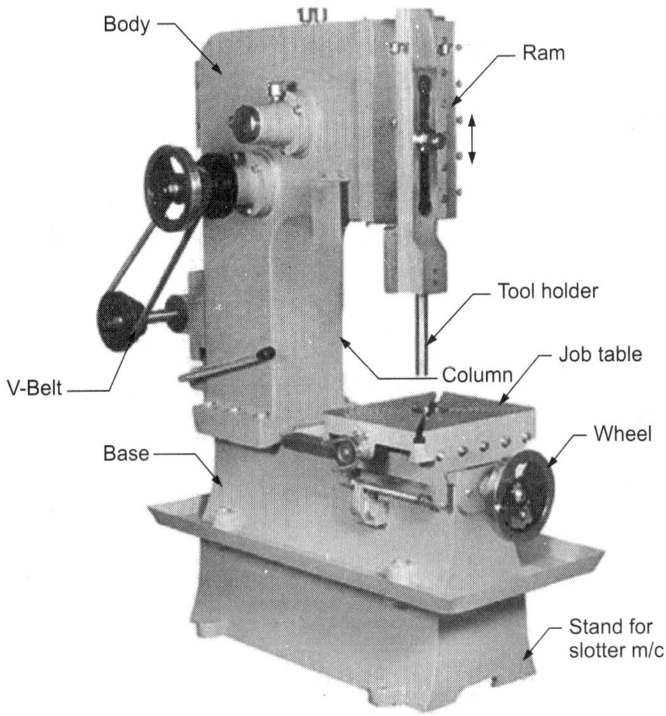

Fig. 3.13 Slotter machine

(i) Table
(ii) Ram
(iii) Housing
(iv) Column
(v) Base or bed
(vi) Tool holder.

(i) ***Table:*** Table is a one of the main part of a slotter and usually it is in circular shape (Disc type), used to hold the job in longitudinal and cross-wires direction. Table can be adjustable about its center. There is a T-slots provision in the table to clamp the work piece. Some time a work holding device, **three jaw chuck, vice, clamp** etc., is mounted on the table to easy holding the work piece.

(ii) ***Ram:*** Ram is another important part of a slotter. It is fixed with ram by the guide ways and it moves in vertical direction up and down. This ram is also same as the ram of a shaper machine, it moves on the guideways.

(iii) ***Column:*** It is a vertical member of a slotter machine line and made out of cast iron and it have the driving mechanism. On the front face of column there is a provision of guideways for reciprocating motion (up and down) of ram.

(iv) ***Base/Bed:*** Base of a slotter is always heavy casted as entire weight of the slotter machine is beared by the base. Bed is always made by the cast iron as it absorb all vibration it self and it supports column, ram, table and mechanism of driving.

(vii) ***Tool holder:*** Tool holder is a device which is used to hold the cutting tool and it is attached with the ram.

3.3.3 Specification of a Slotter Machine

Specification of a slotter machine can be given on the following points:

(i) Maximum length of the ram stroke in mm
(ii) Maximum diameter of the table (work table) in mm
(iii) Maximum power of motor in H.P.
(iv) Maximum table travel i.e., longitudinal and cross in mm
(v) Over all size of the slotter i.e., length × width × height in mm.
(vi) Material removal every stroke.

3.3.4 Types of Slotter

Slotter are the following types.

(i) General production slotter
(ii) Key seater slotter
(iii) Puncher slotter
(iv) Precision tool room slotter.

(i) ***General production slotter machine:*** General production purpose slotters are very common in used. General production slotter are medium in size neither they are bigger nor very smaller.

(ii) ***Key seater slotter:*** It is a special purpose slotter, designed to cut keyways in cams, gear hub, wheel etc.

(iii) ***Puncher slotter:*** This type of slotter are usually used to remove the heavy unwanted material from the periphery of the heavy work piece produced by stamping or forging process. It can remove the material from 1.5 to 2.5 kg/minutes.

(iv) ***Precision tool room slotter:*** As the name depitch this type of slotter used in tool room for precise and accurate machining. This slotter is fitted with various sophisticated arrangement for accurate size machining. Precision tool room slotter operates at very high speed as compared to general production slotter.

3.3.5 Drive Mechanism of a Slotter

Following drives are used in a slotter:

(i) Slotted disc mechanism
(ii) Hydraulic drive
(iii) Variable speed reversible motor drive
(iv) Slotted link drive.

(i) ***Slotted disc mechanism:*** A block diagram of slotted disc mechanism is shown in the Fig. 3.14 gear and pinion is used to rotate the slotted disc by on electric power rotation of the slotted disc is converted into the vertical reciprocating motion of the ram with help of connecting rod. Length of stroke of ram varies the eccentricity of pin.

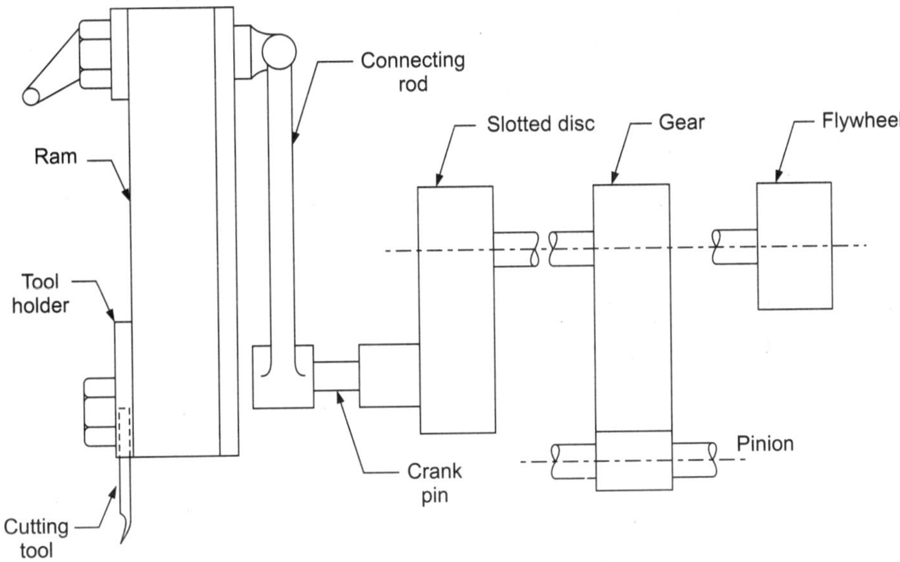

Fig. 3.14 Slotted disc mechanism

(ii) ***Hydraulic drive:*** Hydraulic drive is used in a modes slotter machine. The mechanism of hydraulic drive slotter is quite same of mechanism used in hydraulic horizontal shaper.

(iii) ***Variable speed reversible motor drive:*** This mechanism consist a reversible motor with multi-speed gear box to obtained different speed of ram.

(iv) ***Slotted link drive:***

3.3.6 Work Holding Devices

Holding of a workpiece is important aspect for holding purpose different types of devices are used:

(i) Vise,
(ii) Clamps,
(iii) T-Bolt,
(iv) Special fixture,
(v) Parallel strips.

3.3.7 Slotter Cutting Tools

For machining purpose by a slotter, cutting tools may be in the form of:

(i) solid tool steel forged bar,
(ii) In the form of inserts/bits held by a tool holder.

A single point cutting tool is used in slotter for machining purpose but it is quite different with lathe tool. Slotter cutting tool remove the material parallel with its shank rather than cutting on a line perpendicular to its shanks as in the ease of shaper machine/planer machine.

Slotter tools have front rake and side clearance but slotter tool do not consist any side rake. Range for different angle for a slotter tool is given below.

Angle	*Range*
(i) Clearance angle	4–5°
(ii) Rake angle	10–12°

Various shapes of slotter tools are given below:

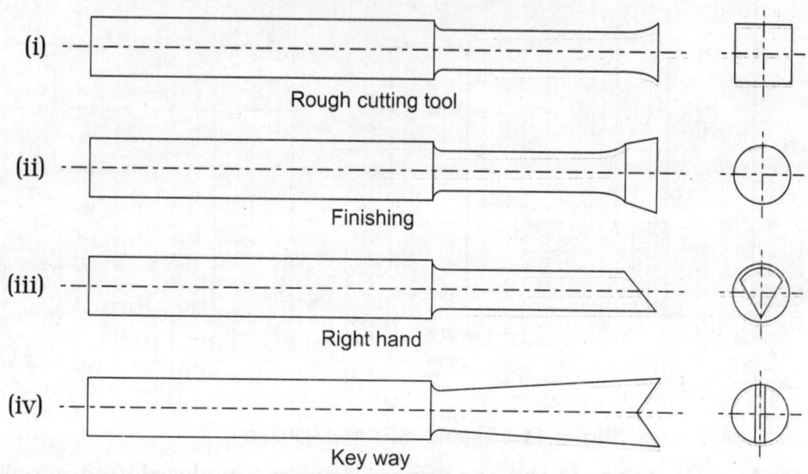

Fig. 3.15 Slotter cutting tools

3.4 PLANER

A planer is a type of metal working machine tool that uses linear relative motion between the workpiece and a single-point cutting tool to machine a

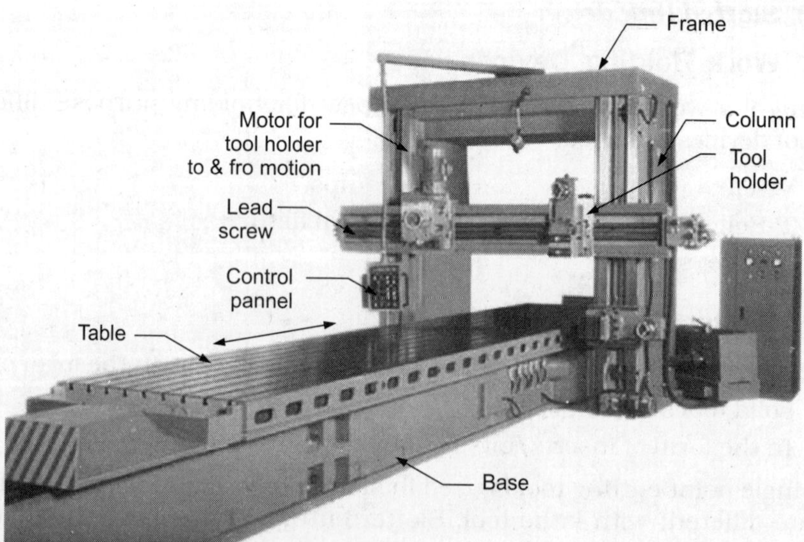

Fig. 3.16 Planer machine

linear tool path. Its cut is analogous to that of a lathe, except that it is (archetypally) linear instead of helical. (Adding axes of motion can yield helical tool paths; see "Helical planning" below.) A planer is analogous to a shaper, but larger, and with the entire workpiece moving on a table beneath the cutter, instead of the cutter riding a ram that moves above a stationary workpiece. The table is moved back and forth on the bed beneath the cutting head either by mechanical means, such as a rack and pinion drive or a leads crew, or by a hydraulic cylinder.

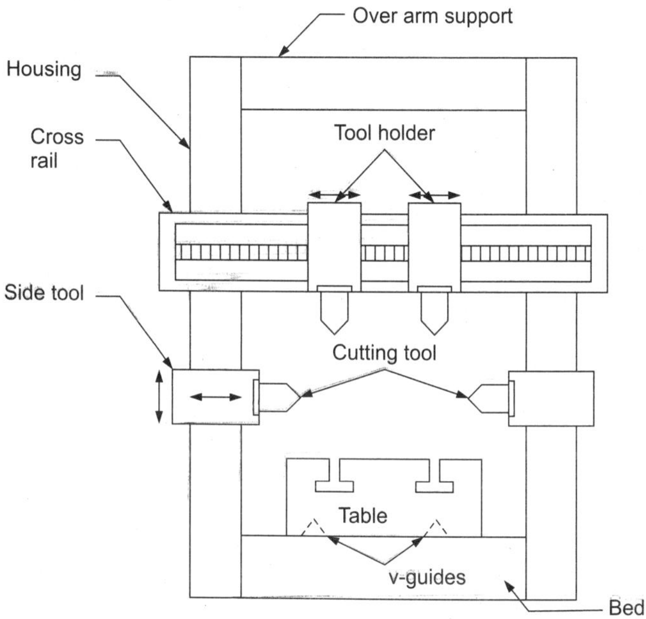

Fig. 3.17 Block diagram of a planer

3.4.1 Classification of Planer

Planers are generally divided into 5 types:

- Double housing planer
- Open side planer
- Edge type planer
- Divide table planer
- Pit type planer

3.4.1.1 Double Housing Planer

- It is the most common type of planer. It consists of mainly a massive bed on which the work table reciprocates, and two vertical columns or housing, one on each side of the bed. Each column carries a tool head that can be slide up and down on the column. A cross rail fitted between the two columns may carry one or two tool heads that can slide horizontally on the cross rail. All the tool heads can be clamped in position, and can be used collectively or individually depending on the requirements.

3.4.1.2 Open Side Planer

- Open side planer consists of only one vertical column or housing on which the cross rail is mounted. The column and the cross rail carry single and double tool heads respectively. This type of machine permits machining of wide workpieces.

3.4.1.3 Edge Type Planer

- This type of machine is used for machining the edges of heavy workpieces. The workpiece is clamped on the bed and the side mounted carriage supporting the cutting tool is reciprocated along the edge of the workpiece. Cutting can take place during both directions of carriage travel.

3.4.1.4 Divide Type Planer

- It is also called tandem planer and consists of two work tables, which may be reciprocated together or separately. It is quite well known that, mounting and setting of workpieces on the worktable consumes more time thereby restricting the machine for continuous mass production applications. In such cases, divided table planer can be used, where in, one worktable can be used for setting up a new workpiece, while the second worktable carrying workpiece is being machined. The two tables can be joined together to hold large workpieces.

3.4.1.5 Pit Type Planer

- A pit type planer differs from other planer in the sense that, the table and the workpiece resting on it remain stationary and the tool reciprocates across the work surface. This type of machine is preferred for very large work, where the weight of the workpiece and the tool required table would make reciprocating movement difficult. The job is either mounted on a stationary table, or on the floor inside a pit, and hence the name pit type planer. One or two tool heads can be mounted on the cross rail and two side tool posts the housings, if required. The entire unit travels along the horizontal ways and fro and, thus the tool moves past the work surface during operation.

3.4.2 Construction

The main parts of the double housing planer machine are bed and table, housings, cross rail, tool heads, driving and feed mechanism.

3.4.2.1 Bed and Table

The bed is a long heavy base and table made of cast iron. Its top surface is flat and machined accurately. The flat top surface has slots in which the workpiece can be securely clamped. The workpiece needs rigid fixing so that it does not shift out of its position. The standard clamping devices used on planer machine are heavy duty vice, T-holders and clamps, angle plate, planer jack, step blocks and stop. The table movement may be actuated by a variable speed drive through a rack and pinion arrangement, or a hydraulic system.

3.4.2.2 Housings

The housings are the rigid and upright column like castings. These are located near the centre on each side of the base.

3.4.2.3 Cross Rail

The cross rail is a horizontal member supported on the machined ways of the upright columns. Guide ways are provided on vertical face of each column and that enables up and vertical movement of the cross rail. The vertical movement of the cross rail allows to accommodate workpiece of different heights. Since the cross rail is supported at both the ends, this type of planer machine is rigid in construction.

3.4.2.4 Tool Heads

Generally two tool heads are mounted in the horizontal cross rail and one on each of the vertical housing. Tool heads may be swiveled so that angular cuts can be made.

3.4.2.5 Driving and Feed Mechanism

The tool heads may be fed either by hand or by power in crosswise or vertical direction. The motor drive is usually at one side of the planer near the centre and drive mechanism is located under the table.

The size of the planer is specified by the maximum length of the stroke, and also by the size of the largest rectangular solid that can be machined on it.

Application of a Modern Planer

- Modern planers are used by smaller tool and die shops within larger production facilities to maintain and repair large stamping dies and plastic injection moulds. Additional uses include any other task where an abnormally large (usually in the range of 4'×8' or more) block of metal must be squared when a horizontal grinder or floor mill is unavailable, too expensive, or impractical to use. While not as precise as grinding.

  A planer can remove a tremendous amount of material in one pass with high accuracy.

3.5 DIFFERENCE BETWEEN SHAPER AND PLANER

Shaper and planer both are the machine tools and used to remove the materials from block/workpiece.

Shaper	Planer
1. In shaper machine workpiece is stationary while tool in motion.	**1.** In planer workpiece is in motion while cutting tool is stationary.
2. Shaper machine is used only for light workpiece (block).	**2.** Planer is preferred for heavy workpiece (usually long piece).
3. Shaper machine is used for light cut (low metal removal).	**3.** Planer is used for heavy cut (large metal removal).
4. Shaper machine occupied less space as compared planer machine.	**4.** Planer machine occupied more space as compared to shaper machine.
5. In shaper machine only one cutting tool can be install at a time.	**5.** In planer machine at a time more than one cutting tool can be install.
6. Cost of shaper machine is not so high as compared to planer.	**6.** Cost of planer machine is **very high** as compared to shaper machine.
7. Workpiece setting takes less time as compared to planer. Less skillness is required.	**7.** Workpiece setting take more time and more skills as compared to shaper machine is required.
8. Shaper machines are simple, light in weight and cheaper.	**8.** Planer machines are heavier in weight complicated and costly.

OBJECTIVE TYPE QUESTIONS

1. In shaper machine
(a) tool is stationary while workpiece is in motion
(b) workpiece is stationary while tool is in motion
(c) both are in motion
(d) none of above

2. Shaper is specified by
(a) quick return ratio
(b) maximum travel of cutting tool
(c) surface area that can be machined by minute
(d) total weight of machine

3. In a mechanical shaper, length of stroke is increased by
(a) increasing the length of arm
(b) decreasing the length of the slot in the slotted lever
(c) increasing the centre distance of bull gear and crank pin
(d) decreasing the centre distance of bull gear and crank pin *[UPSC-1995]*

4. A shaper employs following for quick return motion
(a) hydraulic mechanism (b) crank and slotted link mechanism
(c) whitworth motion (d) leonard mechanism

5. Which one is not a part of shaper
(a) compound slide (b) ram
(c) clapper box (d) tool head

6. Usual ratio of forward and return stroke in a shaper
(a) 2 : 1 (b) 4 : 3
(c) 3 : 2 (d) 5 : 3

7. Feeding of a job in a shaper is done by
   (a) dividing head (b) ram movement
   (c) table moment (d) inderning centres

ANSWERS

1. (b) **2.** (b) **3.** (c) **4.** (c) **5.** (a) **6.** (c) **7.** (c)

QUESTIONS

1. Define the shaper machine in detail.
2. Define and draw a net sketch of quick return mechanism of a shaper.
3. Define the planer in detail.
4. Differentiate between shaper and planer. ***[UPTU-2006-07]***
5. What are the main difference between a shaper and planer? Discuss the difference drive mechanism used in shaper with the help of suitable diagram. ***[UPTU-2005-06]***
6. How does shaper differ from planer in term of their operations and type of workpiece. ***[UPTU-2002-03]***
7. What are the main differences between a shaper and a planer? Which are the different drive mechanism used in shaper? Discuss any one in brief. ***[UPTU-2002-03]***
8. Explain slotting operation with a neat sketch.
9. Define the slotter machine with a neat sketch.
10. Draw a block diagram of a shaper machine.
11. Write main parts of a shaper machine.
12. Sketch the different cutting tools used for a shaper machine.
13. Explain the different parts of a planer machine.
14. Classify the planers.
15. Explain open and cross belt drive mechanism for a double housing planner with neat sketch. ***[UPTU-2012-13]***

❑❑❑

Chapter 4

Milling

4.1 MILLING

Milling is the process of machining **flat, curved,** or **irregular surfaces** by feeding the workpiece against a rotating cutter containing a number of cutting edges. Usual milling machine consists a motor driven spindle, which mounts and revolves the milling cutter, and a reciprocating adjustable work table, which mounts and feeds the workpiece.

Milling machine is a machine tool which produce accurately one or more machined surface on a piece of material. In milling operation workpiece is held on a vice while one or more rotary milling cutter comes contact with workpiece. Milling is a most versatile machine tool in existence. Its adaptability is especially valuable for mass production as well as job shop production.

Classification of Milling Machine

Milling machines broadly classified as follows:

1. Column and knee type milling machine
   (a) Horizontal milling machine
   (b) Vertical milling machine
   (c) Universal milling machine

2. Fixed bed type milling machine
   (a) Simplex milling machine
   (b) Duplex milling machine
   (c) Triplex milling machine

3. Planer milling machine

4. Special type milling machine
   (a) Rotary table milling machine
   (b) Planetary milling machine
   (c) Continuous milling machine
   (d) Pantograph milling machine

Most milling machines have self-contained electric drive motors, coolant systems, variable spindle speeds, and power-operated table feeds.

Types of Milling Machines

4.1.1 Knee-Type

Knee-type mills are characterized by a vertically adjustable worktable resting on a saddle which is supported by a knee. The knee is a massive casting that rides vertically on the milling machine column and can be clamped rigidly to the column in a position where the milling head and milling machine spindle are properly adjusted vertically for operation.

Fig. 4.1 Plain knee-and-column type milling machine. [29]

The plain vertical machines are characterized by a spindle located vertically, parallel to the column face, and mounted in a sliding head that can be fed up and down by hand or power. Modern vertical milling machines are designed so the entire head can also swivel to permit working on angular surfaces.

The turret and swivel head assembly is designed for making precision cuts and can be swung 360° on its base. Angular cuts to the horizontal plane may be made with precision by setting the head at any required angle within a 180° arc.

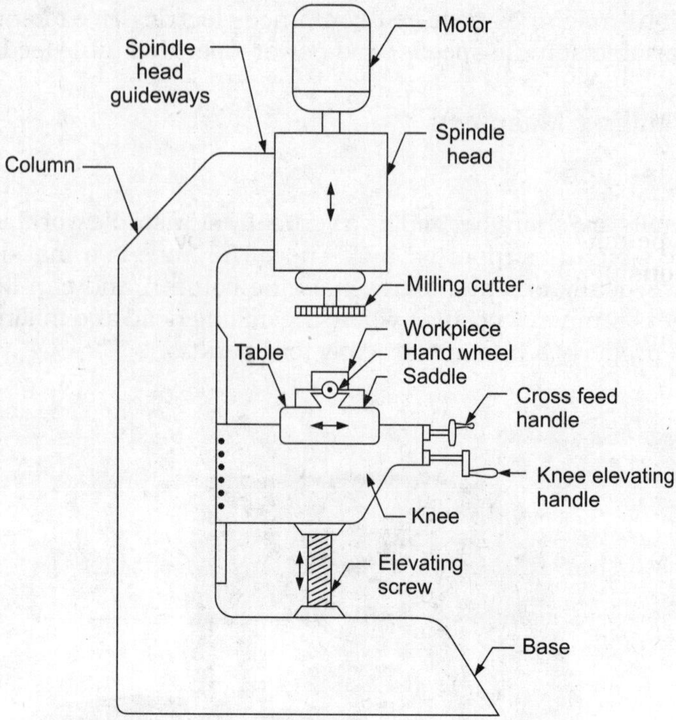

Fig. 4.2 Block diagram of a vertical milling machine

The plain horizontal milling machine's column contains the drive motor and gearing and a fixed position horizontal milling machine spindle. An adjustable overhead arm containing one or more arbor supports projects forward from the top of the column. The arm and arbor supports are used to stabilize long arbors. Supports can be moved along the overhead arm to support the arbor where support is desired depending on the position of the milling cutter or cutters.

The milling machine's knee rides up or down the column on a rigid track. A heavy, vertical positioning screw beneath past the milling cutter. The milling machine is excellent for forming flat surfaces, cutting dovetails and key ways, forming and fluting milling cutters and reamers, cutting gears, and so forth. Many special operations can be performed with the attachments available for milling machine use. The knee is used for raising and lowering. The saddle rests upon the knee and supports the worktable. The saddle moves in and out on a dovetail to control cross feed of the worktable. The worktable traverses to the right or left upon the saddle for feeding the workpiece past the milling cutter. The table may be manually controlled or power fed.

4.1.2 Universal Horizontal Milling Machine

The basic difference between a universal horizontal milling machine and a plain horizontal milling machine is the addition of a table swivel housing between the table and the saddle of the universal machine. This permits the table to swing up to 45° in either direction for angular and helical milling operations. The universal machine can be fitted with various attachments such as the indexing fixture, rotary table, slotting and rack cutting attachments, and various special fixtures.

4.1.3 Ram-Type Milling Machine

The ram-type milling machine is characterized by a spindle mounted to a movable housing on the column to permit positioning the milling cutter forward or rearward in a horizontal plane. Two popular ram-type milling machines are the universal milling machine and the swivel cutter head ram-type milling machine.

4.1.4 Universal Ram-Type Milling Machine

The universal ram-type milling machine is similar to the universal horizontal milling machine, the difference being, as its name implies, the spindle is mounted on a ram or movable housing.

4.1.5 Swivel Cutter Head Ram-Type Milling Machine

The cutter head containing the milling machine spindle is attached to the ram. The cutter head can be swiveled from a vertical spindle position to a horizontal spindle position or can be fixed at any desired angular position between vertical and horizontal. The saddle and knee are hand driven for vertical and cross feed adjustment while the worktable can be either hand or power driven at the operator's choice.

Basic milling machine configurations are shown in Fig. 4.3.

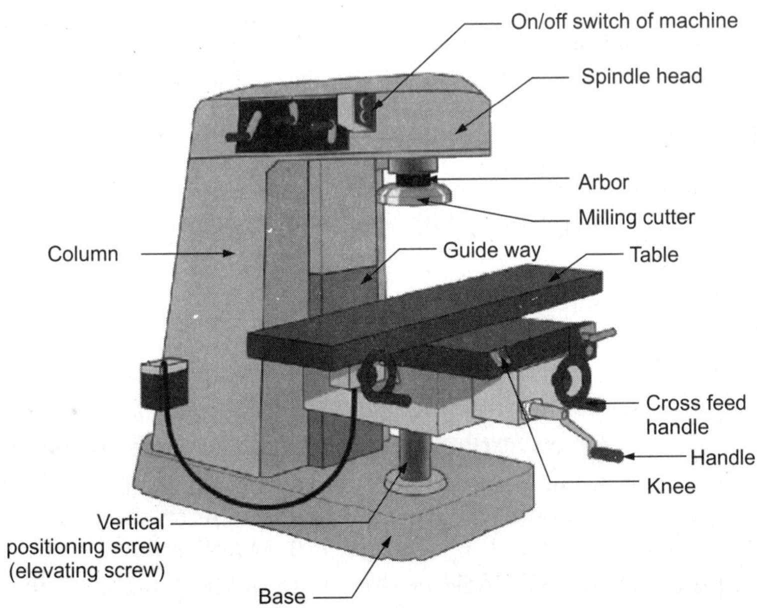

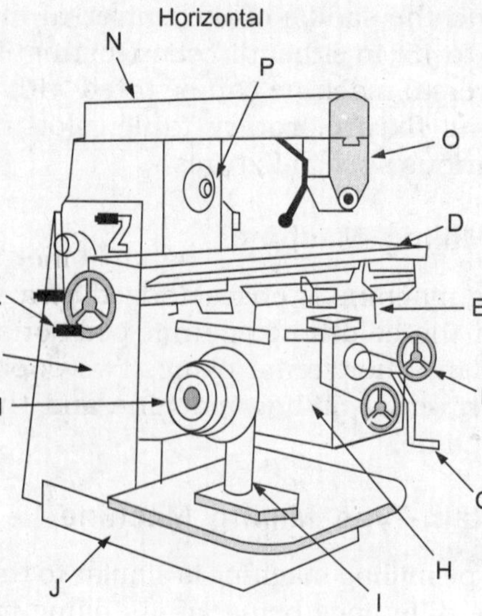

A RAM
B Vertical Head
C QuickD Table
E Crossfeed Handle
G Vertical feed Crank
K Knee
I Vertical positioning screw
J Base
K Column
L Table Handwheel
M Table transmission
N RAM type overarm
O Arbor support
P Spindle

Fig. 4.3 Milling machines (Horizontal types)

4.2 TOOLS FOR MILLING MACHINE

4.2.1 Milling Cutters

Milling cutters are usually made of high-speed steel and are available in a great variety of shapes and sizes for various purposes and it is very Important to know the names of the most common classifications of cutters, their uses, and, in a general way, the sizes best suited to the work at hand.

4.2.2 Milling Cutter Nomenclature

Figure 4.4 shows two views of a common milling cutter with its parts and angles identified. These parts and angles in some form are common to all cutter types.

- The pitch refers to the angular distance between like or adjacent teeth.
- The pitch is determined by the number of teeth. The tooth face is the forward facing surface of the tooth that forms the cutting edge.
- The cutting edge is the angle on each tooth that performs the cutting.
- The land is the narrow surface behind the cutting edge on each tooth.
- The rake angle is the angle formed between the face of the tooth and the centerline of the cutter. The rake angle defines the cutting edge and provides a path for chips that are cut from the workpiece.
- The primary clearance angle is the angle of the land of each tooth measured from a line tangent to the centerline of the cutter at the

cutting edge. This angle prevents each tooth from rubbing against the workpiece after it makes its cut.

- This angle defines the land of each tooth and provides additional clearance for passage of cutting oil and chips.
- The hole diameter determines the size of the arbor necessary to mount the milling cutter.
- Plain milling cutters that are more than 3/4 inch in width are usually made with spiral or helical teeth. A plain spiral-tooth milling cutter produces a better and smoother finish and requires less power to operate. A plain helical-tooth milling cutter is especially desirable when milling an uneven surface or one with holes in it.

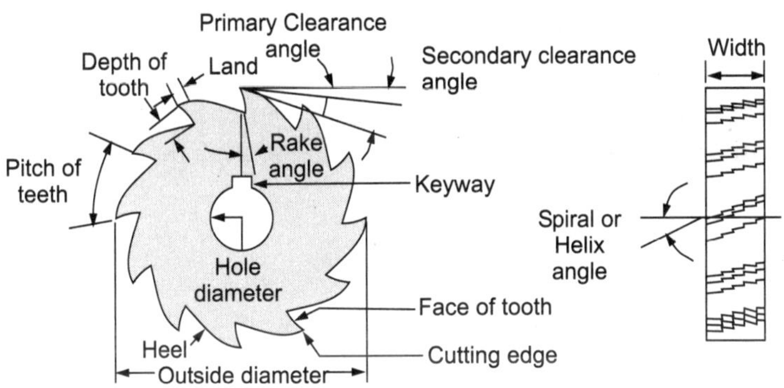

Fig. 4.4 Milling cutter nomenclature.

4.2.3 Types of Teeth

The teeth of milling cutters may be made for right-hand or left-hand rotation, and with either right-hand or left-hand helix. Determine the hand of the cutter by looking at the face of the cutter when mounted on the spindle. A right-hand cutter must rotate counterclockwise; a left-hand cutter must rotate clockwise. The right-hand helix is shown by the flutes leading to the right; a left-hand helix is shown by the flutes leading to the left. The direction of the helix does not affect the cutting ability of the cutter, but take care to see that the direction of rotation is correct for the hand of the cutter (Fig. 4.5).

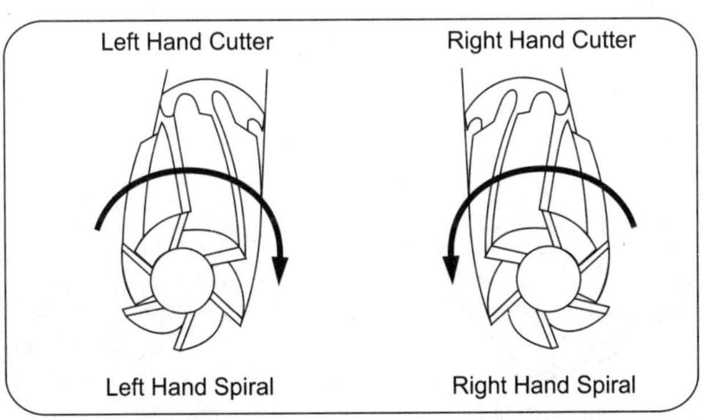

Fig. 4.5 Left and right cutters

Saw Teeth

Saw teeth similar to those shown in Fig. 4.5 are either straight or helical in the smaller sizes of plain milling cutters, metal slitting saw milling cutters, and end milling cutters. The cutting edge is usually given about 5° primary clearance. Sometimes the teeth are provided with off-set nicks which break ups chips and make coarser feeds possible.

4.3 CLASSIFICATION OF MILLING CUTTERS

1. According to the constructional all feature of cutter:
   (i) Solid cutter
   (ii) Tipped cutter
   (iii) inserted teeth cutter

2. According to the purpose of application of cutter:
   (i) Special milling cutter
   (ii) Standard milling cutter

3. According to the rotation of cutter:
   (i) Right hand rotation cutter
   (ii) Left hand rotation cutter

4. According to the method of cutter mounting:
   (i) Arbor type cutter
   (ii) Facing type cutter
   (iii) Shank type cutter

5. According to the relief characteristics of cutter teeth:
   (i) Profile relived cutter
   (ii) Form releaved cutter

4.3.1 Helical Milling Cutters

The helical milling cutter is similar, to the plain milling cutter, but the teeth have a helix angle of 45° to 60°. The steep helix produces a shearing action that results in smooth, vibration-free cuts. They are available for arbor mounting, or with an integral shank with or without a pilot. This type of helical cutter is particularly useful for milling elongated slots and for light cuts on soft metal. See Fig. 4.6.

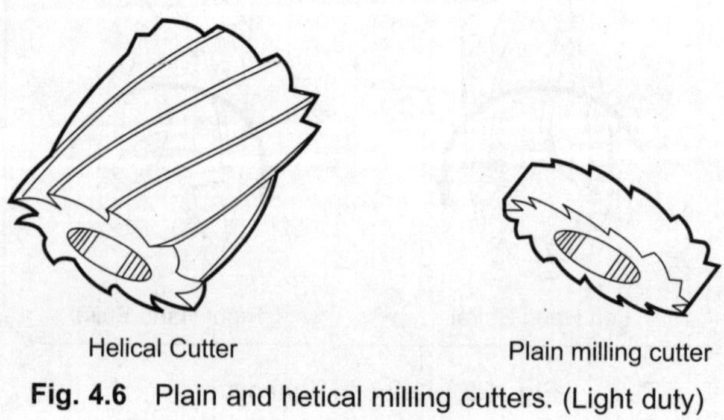

Fig. 4.6 Plain and hetical milling cutters. (Light duty)

4.3.2 Metal Slitting Saw Milling Cutter

The metal slitting saw milling cutter is essentially a very thin plain milling cutter. It is ground slightly thinner toward the center to provide side clearance. These cutters are used for cutoff operations and for milling deep, narrow slots, and are made in widths from 1/32 to 3/16 inch.

Metal-slitting saw

4.3.3 Side Milling Cutters

Side milling cutters are essentially plain milling cutters with the addition of teeth on one or both sides. A plain side milling cutter has teeth on both sides and on the periphery. When teeth are added to one side only, the cutter is called a half-side milling cutter and is identified as being either a right-hand or left-hand cutter. Side milling cutters are generally used for slotting and straddle millings.

Fig. 4.7 Side milling cutters, straddle milling both sides of reverse shifter bracket for an automotive transmission. [29]

Interlocking tooth side milling cutters and staggered tooth side milling cutters are used for cutting relatively wide slots with accuracy. Interlocking tooth side milling cutters can be repeatedly sharpened without changing the width of the slot they will machine.

Interlocking mills

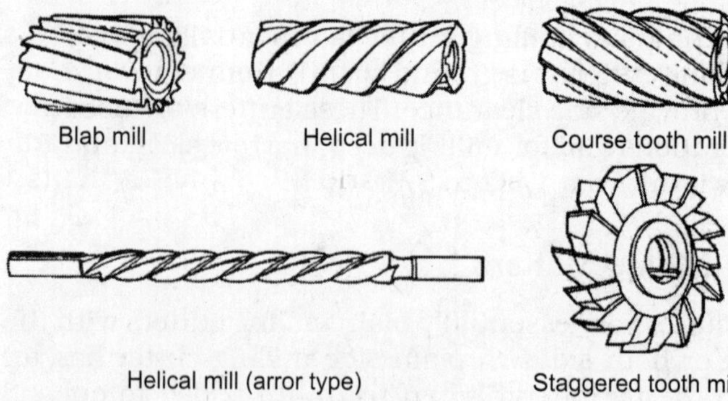

Fig. 4.8 Various milling cutters.

After sharpening, a washer is placed between the two cutters to compensate for the ground off metal. The staggered tooth cutter is the most washer is placed between the two cutters to compensate for efficient type for milling slots where the depth exceeds the width.

4.3.4 End Milling Cutters

The end milling cutter, also called an end mill, has teeth on the end as well as the periphery. The smaller end milling cutters have shanks for chuck mounting or direct spindle mounting. End milling cutters may have straight or spiral flutes. Spiral flute end milling cutters are classified as left-hand or right-hand cutters depending on the direction of rotation of the flutes. If they are small cutters, they may have either a straight or tapered shank.

The most common end milling cutter is the spiral flute cutter containing four flutes. Two-flute end milling cutters, sometimes referred to as two-lip end mill cutters, are used for milling slots and key ways where no drilled hole is provided for starting the cut. These cutters drill their own starting holes. Straight flute end milling cutters are generally used for milling both soft or tough materials, while spiral flute cutters are used mostly for cutting steel.

Large end milling cutters (normally over 2 inches in diameter) (Fig. 4.6) are called shell end mills and are recessed on the face to receive a screw or nut for mounting on a separate shank or mounting on an arbor, like plain milling cutters. The teeth are usually helical and the cutter is used particularly for face milling operations requiring the facing of two surfaces at right angles to each other.

4.3.5 T-Slot Milling Cutter

The T-slot milling cutter is used to machine T-slot grooves in worktables, fixtures, and other holding devices. The cutter has a plain or side milling cutter mounted to the end of a narrow shank. The throat of the T-slot is first milled with a side or end milling cutter and the headspace is then milled with the T-slot milling cutter.

4.3.6 Woodruff Key slot Milling Cutters

The Woodruff key slot milling cutter is made in straight, tapered-shank, and arbor-mounted types. See Fig. 4.9. The most common cutters of this type, under 1 1/2 inches in diameter, are provided with a shank. They have teeth on the periphery and slightly concave sides to provide clearance. These cutters are used for milling semi cylindrical key ways in shafts.

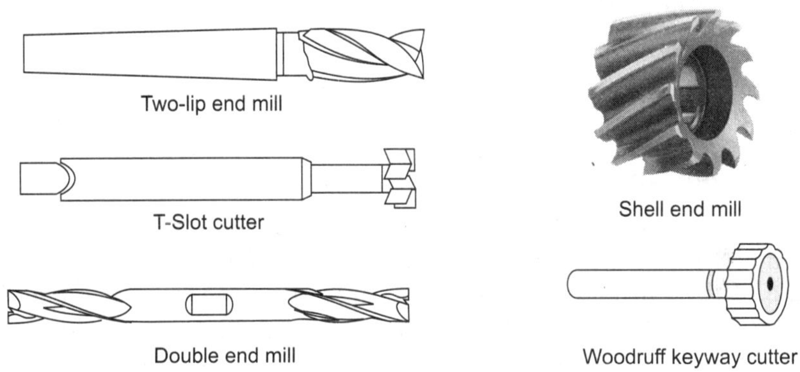

Fig. 4.9 End mill, T-slot, and Woodruff key way cutters.

4.3.7 Angle Milling Cutters

The angle milling cutter has peripheral teeth which are neither parallel nor perpendicular to the cutter axis. See Fig. 4.10. Common operations performed with angle cutters are cutting V-notches and serration's. Angle cutters may be single-angle milling cutters or double-angle milling cutters. The single-angle cutter contains side-cutting teeth on the flat side of the cutter. The angle of the cutter edge is usually 30°, 45°, or 60°, both right and left. Double-angle cutters have included angles of 45°, 60° and 90°.

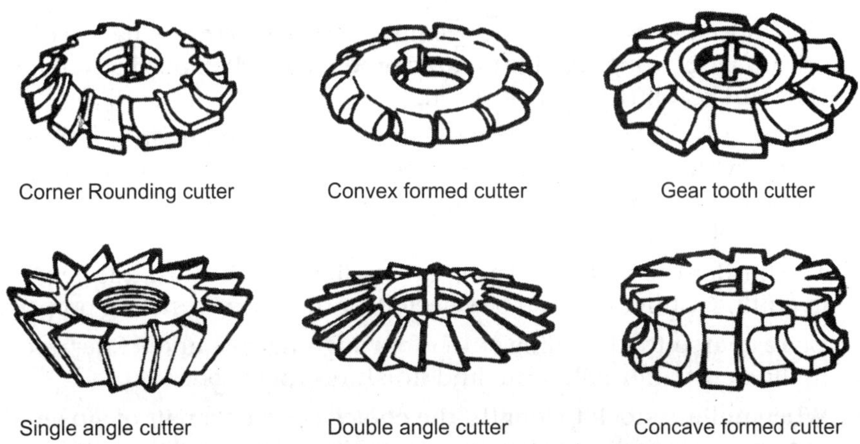

Fig. 4.10 Angle, concave, convex, corner and gear cutters.

Gear Hob

The gear hob is a formed tooth milling cutter with helical teeth arranged like the thread on a screw. These teeth- are fluted to produce the required cutting edges. Hobs are generally used for such work as finishing spur gears, spiral gears, and worm gears. They may also be used to cut ratchets and spline shafts.

4.3.8 Concave and Convex Milling Cutters

Concave and convex milling cutters are formed tooth cutters shaped to produce concave and convex contours of 1/2 circle or less. The size of the cutter is specified by the diameter of the circular form the cutter produces.

4.3.9 Corner Rounding Milling Cutter

The corner-rounding milling cutter is a formed tooth cutter used for milling rounded corners on workpieces up to and including one-quarter of a circle. The size of the cutter is specified by the radius of the circular form the cutter produces, such as concave and convex cutters generally used for such work as finishing spur gears, spiral gears, and worm wheels. They may also be used to cut ratchets and spline shafts.

4.3.10 Special Shaped-Formed Milling Cutter

Formed milling cutters have the advantage of being adaptable to any specific shape for special operations. The cutter is made specially for each specific job. In the field, a fly cutter is formed by grinding a single point lathe cutter bit for mounting in a bar, holder, or fly cutter arbor. The cutter can be sharpened many times without destroying its shape.

4.4 SELECTION OF MILLING CUTTERS

Consider the following when choosing milling cutters:

- High-speed steel, stellite, and cemented carbide cutters have a distinct advantage of being capable of rapid production when used on a machine that can reach the proper speed.
- 45° angular cuts may either be made with a 45° single-angle milling cutter while the workpiece is held in a swivel vise, or with an end milling cutter while the workpiece is set at the required angle in a universal vise.
- The harder the material, the greater will be the heat that is generated in cutting. Cutters should be selected for their heat-resisting properties.
- Use a coarse-tooth milling cutter for roughing cuts and a finer-toothed milling cutter for light cuts and finishing operations.
- When milling stock to length, the choice of using a pair of side milling cutters to straddle the workpiece, a single-side milling cutter, or an end milling cutter will depend upon the number of pieces to be cut.

- Some operations can be done with more than one type of cutter such as in milling the square end on a shaft or reamer shank. In this case, one or two side milling cutters, a fly cutter, or an end milling cutter may be used. However, for the majority of operations, cutters are specially designed and named for the operation they are to accomplish.
- The milling cutter should be small enough in diameter so that the pressure of the cut will not cause the workpiece to be sprung or displaced while being milled.

4.5 SIZE OF MILLING CUTTER

- In selecting a milling cutter for a particular job, choose one large enough to span the entire work surface so the job can be done with a single pass. If this cannot be done, remember that a small diameter cutter will pass over a surface in a shorter time than a large diameter cutter which is fed at the same speed. This fact is illustrated in Fig. 4.11.

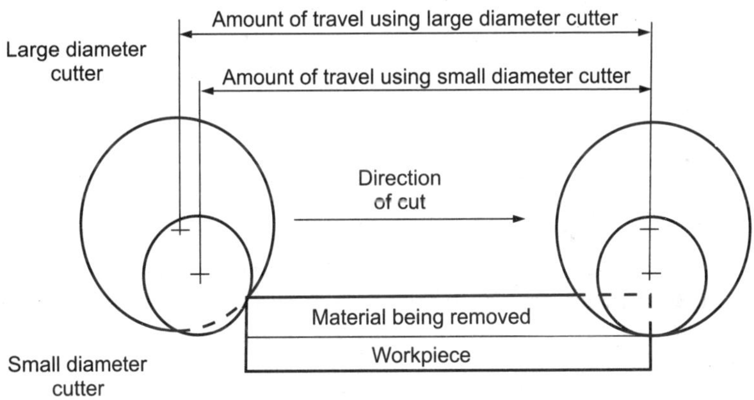

Fig. 4.11 Effect of milling cutter diameter on workplace travel.

4.5.1 Arbors

Milling machine arbors are made in various lengths and in standard diameters of 7/8,1, 1¼, and 1 ½ inch. The shank is made to fit the taper hole in the spindle while the other end is threaded.

NOTE:

The threaded end may have left or right-handed threads.

The milling machine spindle may be self-holding or self-releasing. The self-holding taper is held in the spindle by the high wedging force. The spindle taper in most milling machines is self-releasing; tooling must be held in place by a draw bolt extending through the center of the spindle.

Arbors are supplied with one of three tapers to fit the milling machine spindle the Standard Milling Machine taper, the Brown and Sharpe taper, and the Brown and Sharpe taper with tang (Fig. 4.12).

The Standard Milling Machine Taper is used on most machines of recent manufacture. See Fig. 4.10. These tapers are identified by the number 30, 40, 50, or 60. Number 50 is the most commonly used size on all modern machines.

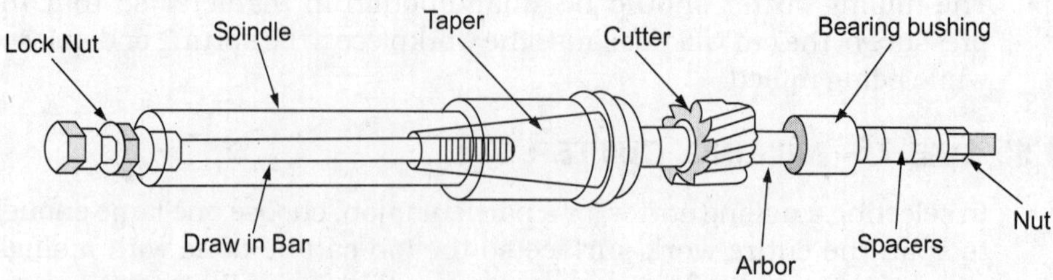

Fig. 4.12 Standard milling machine arbor.

The Brown and Sharpe taper is found mostly on older machines. Adapters or collets are used to adapt these tapers to fit machines whose spindles have Standard Milling Machine tapers.

The Brown and Sharpe taper with tang is used on some older machines. The tang engages a slot in the spindle to assist in driving the arbor.

4.5.2 Standard Milling Machine Arbor

The standard milling machine arbor has a tapered, cylindrical shaft with a standard milling taper on the driving end and a threaded portion on the opposite end to receive the arbor nut. One or more milling cutters may be placed on the straight cylindrical portion of the arbor and held in position by sleeves and the arbor nut. The standard milling machine arbor is usually splined and keys are used to lock each cutter to the arbor shaft. These arbors are supplied in three styles, various lengths and, standard diameters.

The most common way to fasten the arbor in the milling machine spindle is to use a draw bar. The bar threads into the taper shank of the arbor to draw the taper into the spindle and hold it in place. Arbors secured in this manner are removed by backing out the draw bar and tapping the end of the bar to loosen the taper.

The end of the arbor opposite the taper is supported by the arbor supports of the milling machine. One or more supports reused depending on the length of the arbor and the degree of rigidity required. The end may be supported by a lathe center bearing against the arbor nut or by a bearing surface of the arbor fitting inside a bushing of the arbor support.

The arbor may also be firmly supported as it turns in the arbor support bearing suspended from the over-arm (Fig. 4.13).

Typical milling arbors are illustrated in Fig. 4.14. Listed below are several types of style arbors.

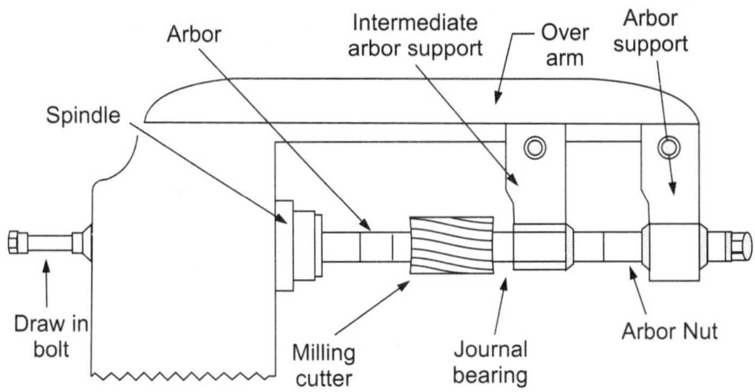

Fig. 4.13 Arbor installation.

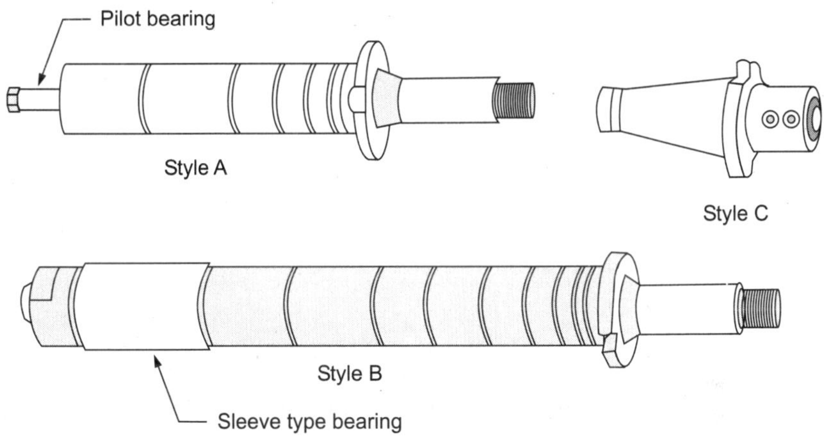

Fig. 4.14 Typical milling arbors.

4.5.2.1 Style A

Has a cylindrical pilot on the end that runs in a bronze bearing in the arbor support. This style is mostly used on small milling machines or when maximum arbor support clearance is required.

4.5.2.2 Style B

Is characterized by one or more bearing collars that can be positioned to any part of the arbor. This allows the bearing support to be positioned close to the cutter, to-obtain rigid setups in heavy duty milling operations.

4.5.2.3 Style C

Arbors are used to mount the smaller size milling cutters, such as end mills that cannot be bolted directly on the spindle nose. Use the shortest arbor possible for the work.

4.5.3 Screw Arbor

Screw arbors are used to hold small cutters that have threaded holes. See Fig. 4.15. These arbors have a taper next to the threaded portion to provide alignment and support for tools that require a nut to hold them against a taper surface. A right-hand threaded arbor must be used for right-hand cutters while a left-hand threaded arbor is used to mount left-hand cutters.

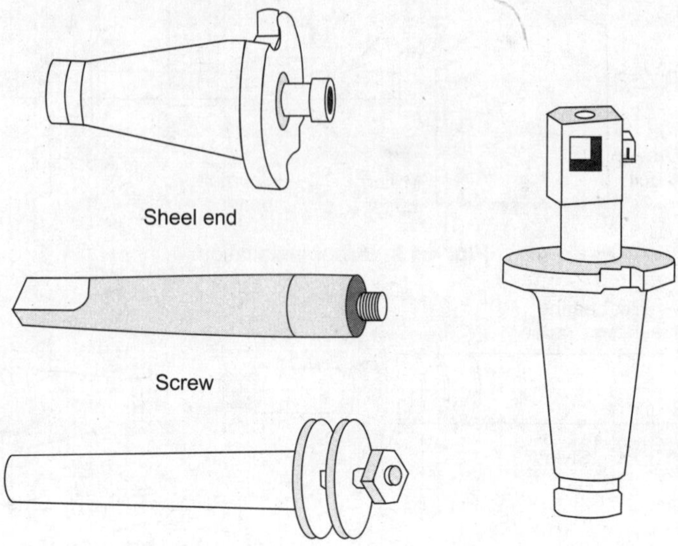

Fig. 4.15 Arbor variations.

Screw arbors are used to hold small cutters that have threaded holes. These arbors have a taper next to the that require a nut to hold them against a taper surface. A right-hand threaded arbor must be used for right-hand cutters while a left-hand threaded arbor is used to mount left-hand cutters.

The slitting saw milling cutter arbor (Fig. 4.15) is a short arbor having two flanges between which the milling cutter is secured by tightening a clamping nut. This arbor is used to hold metal slitting saw milling cutters used for slotting, slitting, and sawing operations.

The shell end milling cutter arbor has a bore in the end in which shell end milling cutters fit and are locked in place by means of a cap screw.

The fly cutter arbor is used to support a single-edge lathe, shaper, or planer cutter bit for boring and gear cutting operations on the milling machine.

4.6 COLLETS, SPINDLE ADAPTERS, AND QUICK-CHANGE TOOLING

Description

Milling cutters that contain their own straight or tapered threaded portion to provide alignment and support for tools shanks are mounted to the milling machine spindle with collets, spindle adapters, and quick-change tooling which adapts the cutter shank to the spindle.

4.6.1 Collets

A collet is a form of a sleeve bushing for reducing the size of the hole in the milling machine spindle so that small shank tools can be fitted into large spindle recesses (Fig. 4.16). They are made in several forms, similar to drilling machine sockets and sleeves, except that their tapers are not alike.

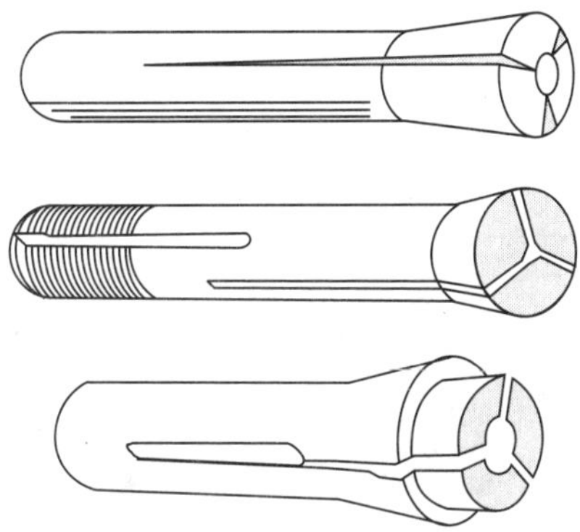

Fig. 4.16 Collet and opting collets.

4.6.2 Spindle Adapters

A spindle adapter is a form of a collet having a standardized spindle end. They are available in a wide variety of sizes to accept cutters that cannot be mounted on arbors. They are made with either the Morse taper shank or the Brown and Sharpe taper with tang having a standard spindle end (Fig. 4.17).

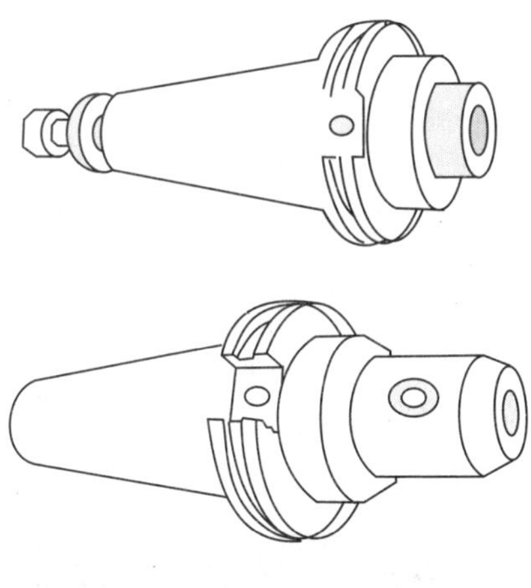

Fig. 4.17 Milling machine adaptors.

4.6.3 Chuck Adapter

A chuck adapter as shown in Fig. 4.18 and used to attach chucks to milling machines having a standard spindle end. The collet holder is sometimes referred to as a collet chuck. Various forms of chucks can be fitted to milling machines spindles for holding drills, reamers, and small cutters for special operations.

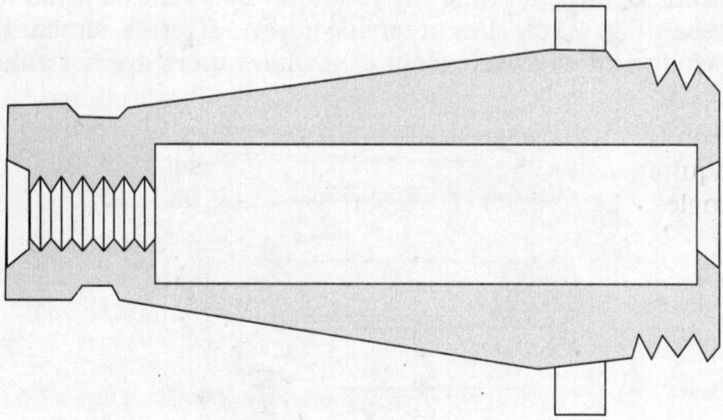

Fig. 4.18 Chuck adaptor.

4.6.4 Quick-Change Tooling

The quick-change adapter mounted on the spindle nose is used to speed up tool changing. Tool changing with this system allows you to set up a number of milling operations such as drilling, end milling, and boring without changing the setup of the part being machined. The tool holders are mounted and removed from a master holder mounted to the machine spindle by means of a clamping ring (Fig. 4.19).

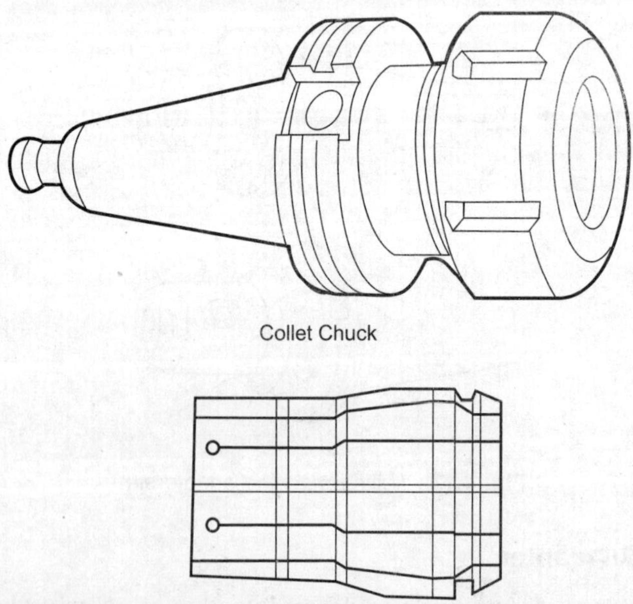

Fig. 4.19 Quick-change adaptor and tool holder.

4.7 VISES

Either a plain or swivel-type vise is furnished with each milling machine. The plain vise, similar to the machine table vise, is used for milling straight workpieces and is bolted to the milling machine table either at right angles or parallel to the machine arbor. The swivel vise can be rotated and contains a scale graduated in degrees at its base to facilitate milling workpieces at any angle on a horizontal plane. The universal vise, which may be obtained as extra equipment, is designed so that it can be set at both horizontal and vertical angles. This type of vise may be used for flat and angular milling.

Adjustable Angle Plate

The adjustable angle plate is a workpiece holding device, similar to the universal vise in operation. Workpieces are mounted to the angle plate with T-bolts and clamps in the same manner used to fasten workpieces to the worktable of the milling machine. The angle plate can be adjusted to any angle so that bevels and tapers can be cut without using a special milling cutter or an adjustable cutter head.

4.8 INDEXING FIXTURE

The index fixture consists of an index head, also called a dividing head, and footstock which is similar to the tailstock of a lathe. The index head and footstock attach to the worktable of the milling machine by T-slot bolts. An index plate containing graduations is used to control the rotation of the index head spindle. The plate is fixed to the index head, and an index crank, connected to the index head spindle by a worm gear and shaft. Workpieces are held between centers by the index head spindle and footstock. Workpieces may also be held in a chuck mounted to the index head spindle or may be fitted directly into the taper spindle recess of some indexing fixtures. There are many variations of the indexing fixture. Universal index head is the name applied to an index head designed to permit power drive of the spindle so that helixes may be cut on the milling machine. Gear cutting attachment is another name applied to an indexing fixture; in this case, one that is primarily intended for cutting gears on the milling machine.

4.8.1 High-Speed Milling Attachment

The rate of spindle speed of the milling machine may be increased from 11/2 to 6 times by using the high-speed milling attachment. This attachment is essential when using cutters and twist drills which must be driven at a high rate of speed in order to obtain an efficient surface speed. The attachment is clamped to the column of the machine and is driven by a set of gears from the milling machine spindle.

4.8.2 Vertical Spindle Attachment

This attachment converts the horizontal spindle of a horizontal milling machine to a vertical spindle. It is clamped to the column and driven from the horizontal spindle. It incorporates provisions for setting the head at any angle, from the vertical to the horizontal, in a plane at right angles to the machine spindle. End milling and face milling are more easily accomplished with this attachment, because the cutter and the surface being cut are in plain view.

4.8.3 Universal Milling Attachment

This device is similar to the vertical spindle attachment but is more versatile. The cutter head can be swiveled to any angle in any plane, whereas the vertical spindle attachment only rotates in one place from horizontal to vertical.

4.8.4 Rotary Table or Circular Milling Attachment

This attachment consists of a circular worktable containing T-slots for mounting workpieces. The circular table revolves on a base attached to the milling machine worktable. The attachment can be either hand or power driven, being connected to the table drive shaft if power driven. It may be used for milling circles, angular indexing, arcs, segments, circular slots, grooves, and radii, as well as for slotting internal and external gears. The table of the attachment is divided in degrees (Fig. 4.20).

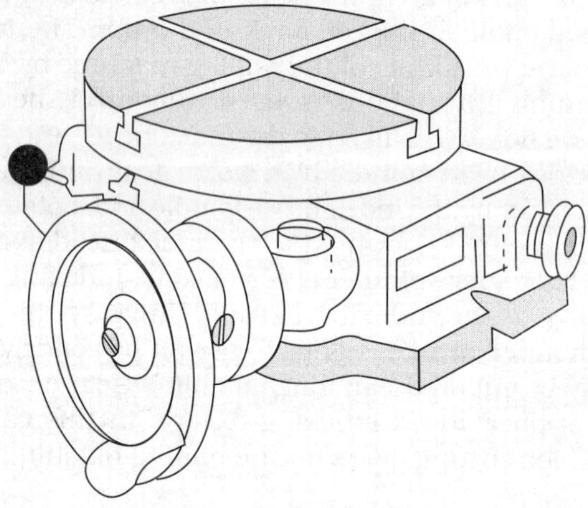

Fig. 4.20 Rotary table (circular milling attachment)

4.8.5 Offset Boring Head

Boring, an operation that is too often restricted to a lathe, can be done easily on a milling machine. The offset boring head is an attachment that fits to the milling machine spindle and permits most drilled holes to have a better surface finish and greater diameter accuracy.

4.8.6 Offset Boring Head and Tools

Note that the boring bar can be adjusted at a right angle to the spindle axis. This feature makes it possible to position the boring cutter accurately to bore holes of varying diameters.

This adjustment is more convenient than adjusting the cutter in the boring bar holder or changing the boring bar. Another advantage of the offset boring head is the fact that a graduated micrometer collar allows the tool to be moved accurately a specified amount (usually in increments of 0.001) without the use of a dial indicator or other measuring device.

NOTE:

On some boring heads, the reading on the tool slide is a direct reading. On other boring heads, the tool slide advances twice the amount shown on the micrometer dial.

4.9 MOUNTING AND INDEXING WORK

An efficient and positive method of holding workpieces to the milling machine table is important if the machine tool is to be used to its fullest advantage. The most common methods of holding are clamping a workpiece to the table, clamping a workpiece to the angle plate, clamping the workpiece in fixtures, holding a workpiece between centers, holding the workpiece in a chuck, and holding the workpiece in a vise. Regardless of the method used in holding, there are certain factors that should be observed in every case. The workpiece must not be sprung in clamping, it must be secured to prevent it from springing or moving away from the cutter, and it must be so aligned that it may be correctly machined T-slots. Milling machine worktables are provided with several T-slots which are used either for clamping and locating the workpiece itself or for mounting the various holding devices and attachments. These T-slots extend the length of the table and are parallel to its line of travel. Most milling machine attachments, such as vises and index fixtures, have keys or tongues on the underside of their bases so that they may be located correctly in relation to the T-slots.

4.9.1 Methods of Mounting Workpieces

Clamping Workpieces to the Table

When clamping a workpiece to the worktable of the milling machine, the table and the workpiece should be free from dirt and burrs. Workpieces having smooth machined surfaces may be camped directly to the table, provided the cutter does not come in contact with the table surface during milling. When clamping workpieces with unfinished surfaces in this way, the table face should be protected from damage by using a shim under the workpiece paper, plywood, and sheet metal are shim materials. Clamps should be located on both sides of the workpiece if possible to give a full bearing surface. These clamps are held by T-slot bolts inserted in the T-slots of the table. Clamp supports must be the same height as the workpiece. Never use clamp supports that are lower than the workpiece. Adjustable step blocks are extremely useful to raise the clamps, as the height of the clamp bar may be adjusted to ensure maximum clamping pressure. Clamping bolts should be placed as near to the workpiece as possible so that the full advantage of the fulcrum principle may be obtained. When it is necessary to place a clamp on an overhanging part, a support should be provided between the overhang and the table to prevent springing or possible breakage. A stop should be placed at the end of the workpiece where it will receive the thrust of the cutter when heavy cuts are being taken.

4.9.2 Clamping a Workpiece to the Angle Plate

Workpieces clamped to the angle plate may be machined with surfaces parallel, perpendicular, or at an angle to a given surface. When using this method of holding a workpiece, precautions should be taken similar to those mentioned for clamping work directly to the table. Angle plates are either adjustable or nonadjustable and are generally held in alignment by keys or tongues that fit into the table T-slots.

4.9.3 Clamping Workpieces in Fixtures

Fixtures are generally used in production work where a number of identical pieces are to be machined. The design of the fixture depends upon the shape of the piece and the operations to be performed. Fixtures are always constructed to secure maximum clamping surfaces and are built to use a minimum number of clamps or bolts in order to reduce the setup time required. Fixtures should always be provided with keys to assure positive alignment with the table T-slots.

4.9.4 Holding Workpieces between Centers

The indexing fixture is used to support workpieces which are centered on both ends. When the piece has been previously reamed or bored, it may be pressed upon a mandrel and then mounted between the centers.

Two types of mandrels may be used for mounting workpieces between centers. The solid mandrel is satisfactory for many operations, while one having a shank tapered to fit into the index head spindle is preferred in certain cases.

A jackscrew is used to prevent springing of long slender workpieces held between centers or workpieces that extend some distance from the chuck.

4.9.5 Holding Workpieces in a Chuck

Before screwing the chuck to the index head spindle, it should be cleaned and any burrs on the spindle or chuck removed. Burrs may be removed with a smooth-cut, three cornered file or scraper, while cleaning should be accomplished with a piece of spring steel wire bent and formed to fit the angle of the threads. The chuck should not be tightened on the spindle so tightly that a wrench or bar is required to remove it. Cylindrical workpieces held in the universal chuck may be checked for trueness by using a test indicator mounted upon a base resting upon the milling machine table. The indicator point should contact the circumference of small diameter workpieces, or the circumference and exposed face of large diameter pieces. While checking, the workpiece should be revolved by rotating the index head spindle.

4.9.6 Indexing

Indexing is the process of evenly dividing the circumference of a circular workpiece into equally spaced divisions, such as in cutting gear teeth, cutting splines, milling grooves in reamers and taps, and spacing holes on a circle. The index head of the indexing fixture is used for this purpose.

4.9.6.1 Index Head

The index head of the indexing fixture contains an indexing mechanism which is used to control the rotation of the index head spindle to space or divide a workpiece accurately. A simple indexing mechanism consists of a 40-tooth worm wheel fastened to the index head spindle, a single-cut worm, a crank for turning the wormshaft, and an index plate and sector. Since there are 40 teeth in the worm wheel, one turn of the index crank causes the worm, and consequently, the index head spindle to make 1/40 of a turn; so 40 turns of the index crank revolve the spindle one full turn.

4.9.6.2 Universal Dividing Head

The universal dividing head is an exceptionally accurate versatile and indexing attachment which may be employed to support the work between centres Fig. 4.21 or hold it in a chuck as shown in Fig. 4.22 . It incorporates an indexing arrangement, as well as gearing which is connected to the driving mechanism for the purpose of imparting rotary motion to the work, as required when cutting helices. The dividing head is mounted on the milling machine table, at the operator's right as he faces the machine.

Fig. 4.21 Dividing head mounted on the table of universal milling machine set up for milling helical gear teeth. [29]

Fig. 4.22 Dividing head holding the work in chuck. [27]

4.9.6.3 Index Plate

The indexing plate (Fig. 4.23) is a round plate with a series of six or more circles of equally spaced holes; the index pin on the crank can be inserted in any hole in any circle. With the interchangeable plates regularly furnished with most index heads, the spacing necessary for most gears, bolt threads, milling cutters, splines, and so forth can be obtained. The following sets of plates are standard equipment:

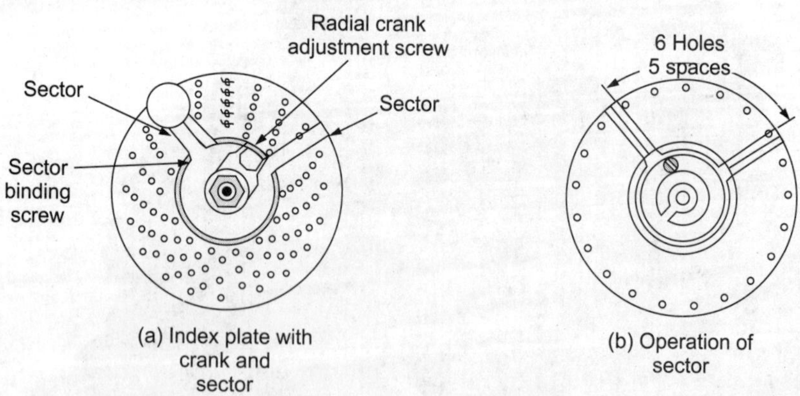

(a) Index plate with crank and sector

(b) Operation of sector

Fig. 4.23 Index plate and squtor.

Brown and Sharpe type consists of 3 plates of 6 circles each drilled as follows:

Plate 1: 15, 16, 17, 18, 19, 20 holes

Plate 2: 21, 23, 27, 29, 31, 33 holes

Plate 3: 37, 39, 41, 43, 47, 49 holes

Cincinnati type consists of one plate drilled on both sides with circles divided as follows:

First side : 24, 25, 28, 30, 34, 37, 38, 39, 41, 42, 43 holes
Second side : 46, 47, 49, 51, 53, 54, 57, 58, 59, 62, 66 holes

Sector

The sector (Fig. 4.23(a)) indicates the next hole in which the pin is to be inserted and makes it unnecessary to count holes when moving the index crank after each cut. It consists of two radial, beveled arms which can be set at any angle to each other and then moved together around the center of the index plate. Suppose that, as shown in Fig. 4.23(b), it is desired to make a series of cuts, moving the index crank 1 1/4 turns after each cut. Since the circle illustrated has 20 holes, turn the crank one full turn plus five spaces after each cut, Set the sector arms to include the desired fractional part of a turn or five spaces between the beveled edges of its arms, as shown. If the first cut is taken with the index pin against the left-hand arm, to take the next cut, move the pin once against the right-hand arm of the sector. Before taking the second cut, move the arms so that the left-hand arm is again against the pin; this moves the right-hand arm another five spaces ahead of the pin. Then take the second cut, and repeat the operation until all the cuts have been completed.

NOTE:

It is good practice always to index clockwise on the plate to eliminate backlash.

4.9.6.4 Plain Indexing

The following principles apply to basic indexing of workpieces:

Suppose it is desired to mill a project with eight equally spaced teeth. Since 40 turns of the index crank will turn the spindle one full turn, 1/8th of 40 or 5 turns of the crank after each cut will space the gear for 8 teeth. If it is desired to space equally for 10 teeth, 1/10 of 40 or 4 turns would produce the correct spacing.

The same principle applies whether or not the divisions required divide equally into 40. For example, if it is desired to index for 6 divisions, 6 divided into 40 equals $6\frac{2}{3}$ turns; similarly, to index for 14 spaces, 14 divided into 40 equals $2\frac{6}{7}$ turns. These examples may be multiplied indefinitely and from them the following rule is derived to determine the number of turns of the index crank needed to obtain one division of any number of equal divisions on the workpiece, divide 40 by the number of equal divisions desired (provided the worm wheel has 40 teeth, which is standard practice).

4.9.6.5 Direct Indexing

The construction of some index heads permits the worm to be disengaged from the worm wheel, making possible a quicker method of indexing called direct indexing. The index head is provided with a knob which, when turned through part of a revolution, operates an eccentric and disengages the worm.

Direct indexing is accomplished by an additional index plate fastened to the index head spindle. A stationary plunger in the index head fits the holes in this index plate. By moving this plate by hand to index directly, the spindle and the workpiece rotate an equal distance.

Direct index plates usually have 24 holes and offer a quick means of milling squares, hexagons, taps, and so forth. Any number of divisions which is a factor of 24 can be indexed quickly and conveniently by the direct indexing method.

4.9.6.6 Differential Indexing

Sometimes, a number of divisions is required which cannot be obtained by simple indexing with the index plates regularly supplied. To obtain these divisions, a differential index head is used. The index crank is connected to the wormshaft by a train of gears instead of a direct coupling as with simple indexing. The selection of these gears involves calculations similar to those used in calculating change gear ratio for lathe thread cutting.

4.9.6.7 Indexing in Degrees

Workpieces can be indexed in degrees as well as fractions of a turn with the usual index head. There are 360 degrees in a complete circle and one turn of the index crank revolves the spindle 1/40 or 9 degrees. Therefore, 1/9 turn of the crank rotates the spindle 1 degree. Workpieces can therefore be indexed in degrees by using a circle of holes divisible by 9.

For example, moving the crank 2 spaces on an 18-hole circle, 3 spaces on a 27-hole circle, or 4 spaces on a 36-hole circle will rotate the spindle 1 degree. Smaller crank movements further subdivide the circle: moving 1 space on an 18-hole circle turns the spindle 1/2 degree (30 minutes), 1 space on a 27-hole circle turns the spindle 1/3 degree (20 minutes), and so forth.

4.9.6.8 Indexing Operations

The following examples show how the index plate is used to obtain any desired part of a whole spindle turn by plain indexing.

- Milling a hexagon. Using the rule previously given, divide 40 by 6 which equals $6\frac{2}{3}$ turns, or six full turns plus 2/3 of a turn or any circle whose number is divisible by 3. Take the denominator which is 3 into which of the available hole circles it can be evenly divided. In this case, 3 can be divided into the available 18-hole circle exactly 6 times. Use this result 6 as a multiplier to generate the proportional fraction required.

Example: Therefore, 6 full turns of the crank plus 12 spaces on an 18-hole circle is the correct indexing for 6 divisions.

- Cutting a gear. To cut a gear of 52 teeth, using the rule again, divide 40 by 52. This means that less than one full turn is required for each division, 40/52 of a turn to be exact. Since a 52-hole circle is not available, 40/52 must be reduced to its lowest term which is 10/13. Take the denominator of the lowest term 13, and determine into which of the available hole circles it can be evenly divided. In this case, 13 can be divided into a 39-hole circle exactly 3 times. Use this result 3 as a multiplier to generate the proportional fraction required.

Example: Therefore, 30 holes on a 39-hole circle is the correct indexing for 52 divisions. When counting holes, start with the first hole ahead of the index pin.

Speed Computation

The formula for calculating spindle speed in revolutions per minute is as follows:

$$RPM = \frac{CS \times 4}{D}$$

where RPM = Spindle speed (in revolutions per minute).
CS = cutting speed of milling cutter (in SFPM)
D = diameter of milling cutter (in inches)

For example, the spindle speed for machining a piece of steel at a speed of 35 SFPM with a cutter 2 inches in diameter is calculated as follows:

$$RPM = \frac{CS \times 4}{D} = \frac{35 \times 4}{2} = \frac{140}{2} = 70 \text{ RPM}$$

Therefore, the milling machine spindle would be set for as near 70 RPM as possible.

4.10 FEEDS FOR MILLING

The rate of feed, or the speed at which the workpiece passes the cutter, determines the time required for cutting a job. In selecting the feed, there are several factors which should be considered.

Forces are exerted against the workpiece, the cutter, and their holding devices during the cutting process. The force exerted varies directly with the amount of feed and depth of cut, and in turn are dependent upon the rigidity and power of the machine. Milling machines are limited by the power they can develop to turn the cutter and the amount of vibration they can resist when using coarse feeds and deep cuts. The feed and depth of the cut also depend upon the type of milling cutter being used.

For example, deep cuts or coarse feeds should not be attempted when using a small diameter end milling cutter. Coarse cutters with strong cutting teeth can be fed at a faster rate as the chips may be washed out more easily by the cutting oil.

Coarse feeds and deep cuts should not be used on a frail workpiece if the piece is mounted in such a way that its holding device is not able to prevent springing or bending.

Experience and judgement are extremely valuable in selecting the correct milling feeds. Even though suggested rare tables are given. remember that these are suggestions only. Feeds are governed by many variable factors, such as the degree of finish required. Using a coarse feed, the metal is removed more rapidly but the appearance and accuracy of the surface produced may not reach the standard desired for the finished product. Because of this fact, finer feeds and increased speeds are used for finer, more accurate finishes, while for roughing, to use a comparatively low speed and heavy feed. More mistakes are made on overspeeding and underfeeding than on underspeeding and overfeeding.

Overspeeding may be detected by the occurrence of a squeaking, scraping sound. If vibration (referred to as chattering) occurs in the milling machine during the cutting process, the speed should be reduced and the feed increased. Too much cutter clearance, a poorly supported workpiece, or a badly worn machine gear are common causes of chattering.

Designation of Feed

The feed of the milling machine may be designated in inches per minute or millimeters per minute The milling feed is determined by multiplying the chip size (chip per tooth) desired the number of teeth on the cutter, *N* the revolutions per minute of the cutter.

Example: The formula used to find the workfeed in inches per minute.

$$IPM = CPT \times N \times RPM$$

IPM = Feed rate in inches per minute.

CPT = Chip per t

N = Number of teeth per minute of the milling cutter.

The first step is to calculate the spindle speed before the feed rate can be calculated.

$$RPM = \frac{CSD\ 4}{D} = \frac{300 \times 4}{1/2} = \frac{1,200}{0.5} = 2,400$$

The second step is to calculate the feed rate.

$$IPM = CPT \times N \times RPM$$
$$= 0.005 \times 2 \times 2,400$$
$$= 24$$

Therefore, the RPM for a 1/2-inch-diameter end mill machining aluminum revolves at 2,400 RPM and the feed rate should be 24 inches per minute.

The formula used to find workfeed in millimeters per minute is the same as the formula used to find the feed in IPM, except that mm/min is substituted for IPM.

Direction of Feed

It is usually regarded as standard practice to feed the workpiece against the milling cutter. When the workpiece is fed against the milling cutter, the teeth cut under any scale on the workpiece surface and any backlash in the feed screw is taken up by the force of the cut. See Fig. 4.24.

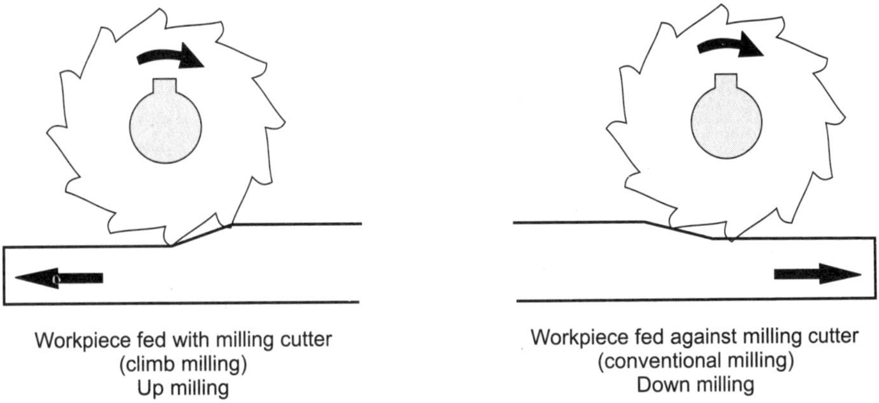

Fig. 4.24 Direction of feed.

As an exception to this recommendation, it is advisable to feed with the milling cutter when cutting off stock or when milling comparatively deep or long slots.

The direction of cutter rotation is related to the manner in which the workpiece is held. The cutter should rotate so that the piece springs away from the cutter; then there will be no tendency for the force of the cut to loosen the piece. No milling cutter should ever be rotated backward; this will break the teeth. If it is necessary to stop the machine during a finishing cut, the power feed should never be thrown out, nor should the workpiece be fed back under the cutter unless the cutter is stopped or the workpiece lowered. Never change feeds while the cutter is rotating.

4.11 PLAIN MILLING

General

Plain milling, also called surface milling or slab milling, is milling flat surfaces with the milling cutter axis parallel to the surface being milled. Generally, plain milling is done with the workpiece surface mounted parallel to the surface of the milling machine table and the milling cutter mounted on a standard milling machine arbor. The arbor is well supported in a horizontal plane between the milling machine spindle and one or more arbor supports.

Mounting the Workpiece

The workpiece is generally clamped directly to the table or supported in a vise for plain milling. The milling machine table should be checked for alignment before starting to cut. If the workpiece surface to be milled is at an angle to the base plane of the piece, the workpiece should be mounted in a universal vise or on an adjustable angle plate. The holding device should be adjusted so that the workpiece surface is parallel to the table of the milling machine.

Selection of Cutter

A careful study of the drawing must be made to determine what cutter is best suited for the job. Flat surfaces may be milled with a plain milling cutter mounted on an arbor. Deeper cuts may generally be taken when using narrow cutters than with wide cutters. The choice of milling cutters should be based on the size and shape of the workpiece. If a wide area is to be milled, fewer traverses will be required using a wide cutter. If large quantities of metal are to be removed, a coarse tooth cutter should be used for roughing and a finer tooth cutter should be used for finishing. A relatively slow cutting speed and fast table feed should be used for roughing, and a relatively fast cutting speed and slow table feed used for finishing. The surface should be checked for accuracy after each completed cut.

Setup

A typical setup for plain milling is illustrated in Fig. 4.25. Note that the milling cutter is positioned on the arbor with sleeves so that it is as close as practical to the milling machine spindle while maintaining sufficient clearance between the vise and the milling machine column. This practice reduces torque in the arbor and permits more rigid support for the cutter.

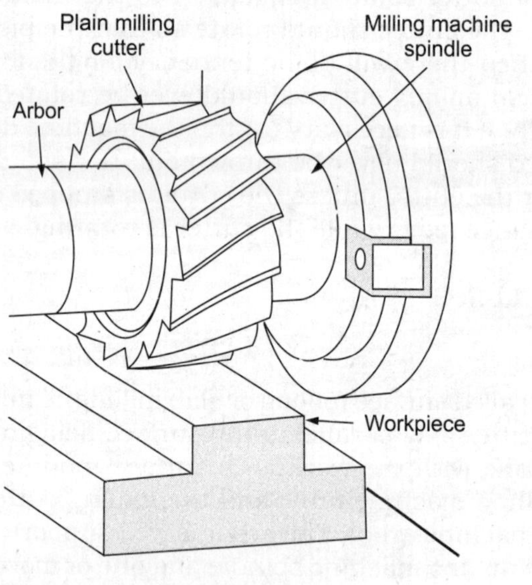

Fig. 4.25 Plain milling

4.12 ANGULAR MILLING

General

Angular milling, or angle milling, is milling flat surfaces which are neither parallel nor perpendicular to the axis of the milling cutter. A single angle milling cutter is used for angular surfaces, such as chamfers, serration's, and grooves. Milling dovetails (Fig. 4.26) is a typical example of angular milling.

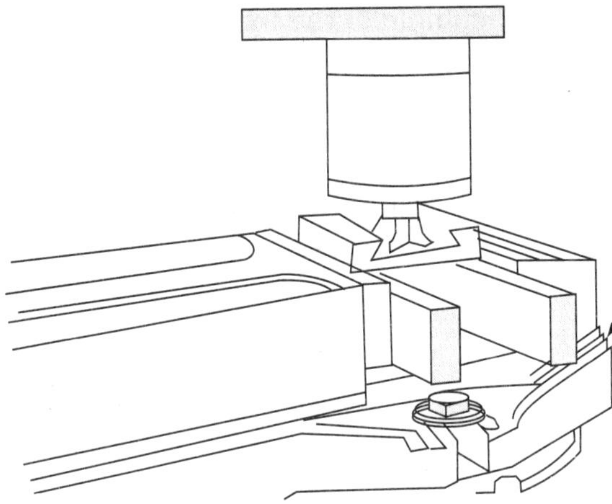

Fig. 4.26 Angular milling

Milling Dovetails

When milling dovetails, the usual angle of the cutter is 45°, 50°, 55°, or 60° based on common dovetail designs.

When cutting dovetails on the milling machine, the workpiece may be held in a vise, clamped to the table, or clamped to an angle plate. The tongue or groove is first roughed out using a side milling cutter, after which the angular sides and base are finished with an angle milling cutter.

In general practice, the dovetail is laid out on the workpiece surface before the milling operation is started. To do this, the required outline should be inscribed and the line prick-punched. These lines and punch marks may then be used as a guide during the cutting operation.

4.13 STRADDLE MILLING

When two or more parallel vertical surfaces are machined at a single cut, the operation is called straddle milling. Straddle milling is accomplished by mounting two side milling cutters on the same arbor, set apart at an exact spacing. Two sides of the workpiece are machined simultaneously and final width dimensions are exactly controlled.

4.13.1 Milling A Hexagon

Straddle milling has many useful applications in production machining. Parallel slots of equal depth can be milled by using straddle mills of equal diameters. Figure 4.27 illustrates a typical example of straddle milling. In this case a hexagon is being cut, but the same operation may be applied to cutting squares or splines on the end of a cylindrical workpiece. The workpiece is usually mounted between centers in the indexing fixture or mounted vertically in a swivel vise. The two side milling cutters are separated by spacers, washers, and shims so that the distance between the cutting teeth of each cutter is exactly equal to the width of the workpiece area required. When cutting a square by this method, two opposite sides of the square are cut, and then the spindle of the indexing fixture or the swivel vise is rotated 90°, and the other two sides of the workpiece are straddle milled.

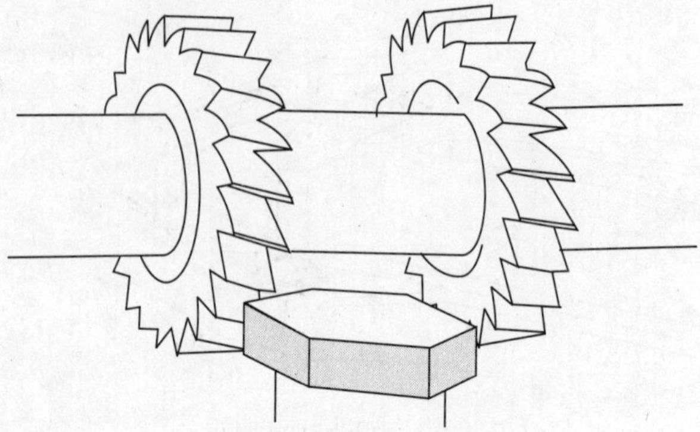

Fig. 4.27 Straddle milling

4.14 FACE MILLING

4.14.1 General

Face milling is the milling of surfaces that are perpendicular to the cutter axis, as shown in Fig. 4.28. Face milling produces flat surfaces and machines work to the required length. In face milling, the feed can be either horizontal or vertical.

In face milling, the teeth on the periphery of the cutter do practically all of the cutting. However, when the cutter is properly ground, the face teeth actually remove a small amount of stock which is left as a result of the springing of the workpiece or cutter, thereby producing a finer finish.

It is important in face milling to have the cutter securely mounted and to see that all end play or sloppiness in the machine spindle is eliminated.

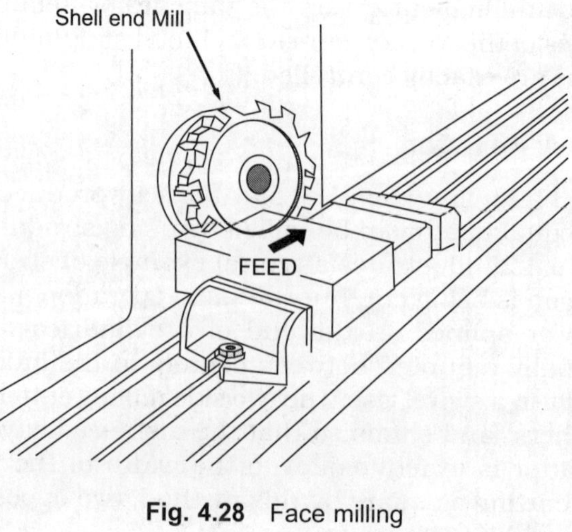

Fig. 4.28 Face milling

4.14.2 Mounting the Workpiece

When face milling, the workpiece may be clamped to the table or angle plate or supported in a vise, fixture, or jig.

Large surfaces are generally face milled on a vertical milling machine with the workpiece clamped directly to the milling machine table to simplify handling and clamping operations.

Angular surfaces can also be face milled on a swivel cutter head milling machine (Fig. 4.29). In this case, the workpiece is mounted parallel to the table and the cutter head is swiveled to bring the end milling cutter perpendicular to the surface to be produced.

During face milling operations, the workpiece should be fed against the milling cutter so that the pressure of the cut is downward, thereby holding the piece against the table. Whenever possible, the edge of the workpiece should be in line with the center of the cutter. This position of the workpiece in relation to the cutter will help eliminate slippage.

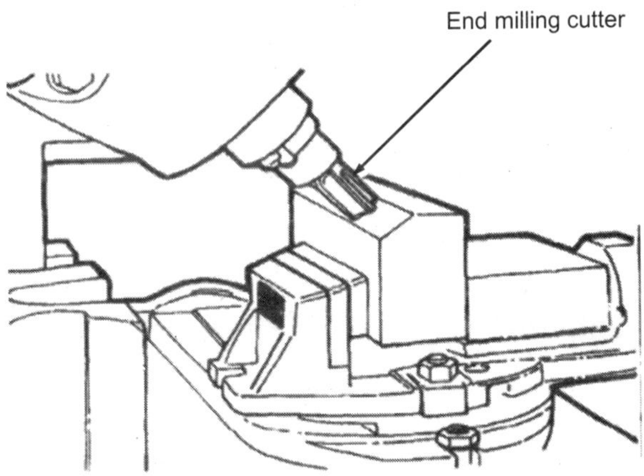

Fig. 4.29 Angular face milling

4.14.3 Depth of Cut

When setting the depth of cut, the workpiece should be brought up to just touch the revolving cutter. After a cut has been made from this setting, measurement of the workpiece is taken. At this point, the graduated dial on the traverse feed is locked and used as a guide in determining the depth of cut.

When starting the cut, the workpiece should be moved so that the cutter is nearly in contact with its edge, after which the automatic feed may be engaged.

When a cut is started by hand, care must be taken to avoid pushing the corner of the workpiece between the teeth of the cutter too quickly, as this may result in cutter tooth breakage. In order to avoid wasting time during the operation, the feed trips should be adjusted to stop the table travel just as the cutter clears the workpiece.

4.15 GANG MILLING

Gang milling is the term applied to an operation in which two or more milling cutters are mounted on the same arbor and used when cutting horizontal surfaces. All cutters may perform the same type of operation or each cutter may perform a different type of operation. For example, several workpieces need a slot, a flat surface, and an angular groove. The best method to cut these would be gang milling as shown in Fig. 4.30. All the completed workpieces would be the same. Remember to check the cutters carefully for proper size.

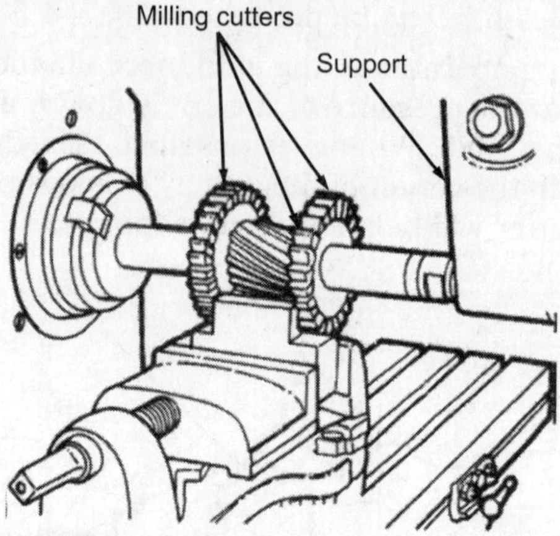

Fig. 4.30 Gang milling.

4.16 FORM MILLING

Form milling is the process of machining special contours composed of curves and straight lines, or entirely of curves, at a single cut. This is done with formed milling cutters, shaped to the contour to be cut. The more common form milling operations involve milling half-round recesses and beads and quarter-round radii on workpieces (Fig. 4.31). This operation is accomplished

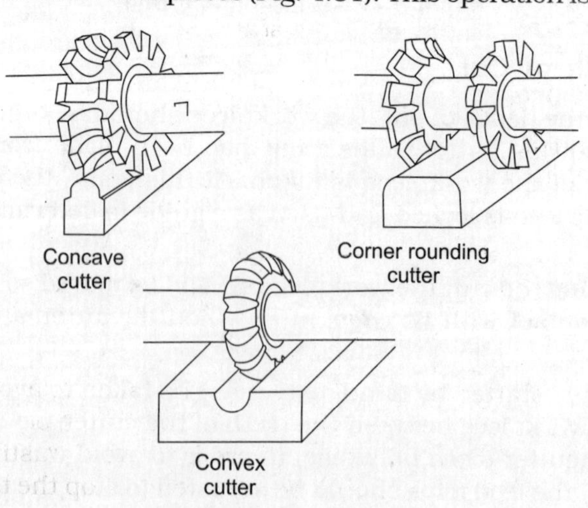

Fig. 4.31 Form milling

by using convex, concave, and corner rounding milling cutters ground to the desired circle diameter. Other jobs for formed milling cutters include milling intricate patterns on workpieces and milling several complex surfaces in a single cut such as are produced by gang milling.

4.17 FLY CUTTING

General

Fly cutting, which is also called single point milling, is one of the most versatile milling operations. It is done with a single-point cutting tool shaped like a lathe tool bit. It is held and rotated by a fly cutter arbor. You can grind this cutter to almost any form that you need, as shown in Fig. 4.32. Formed cutters are expensive. There are times when you need a special form cutter for a very limited number of parts. It is more economical to grind the desired form on a lathe-type tool bit than to buy a preground form cutter, which is very expensive and usually suitable only for one particular job.

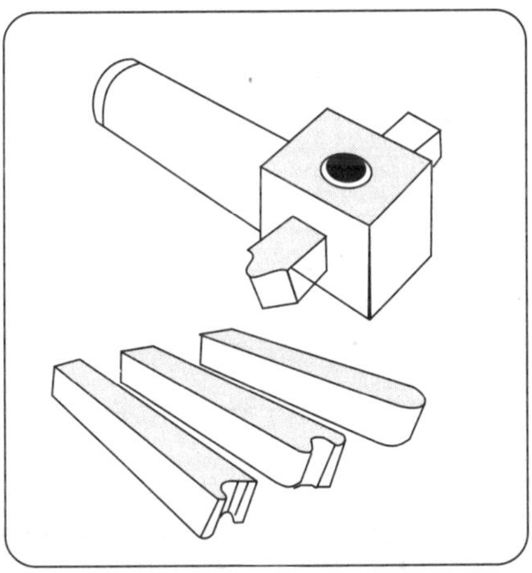

Fig. 4.32 Fly cutter arbor and special-formed cutters.

Gear Cutting

The single-point or fly cutter can be used to great advantage in gear cutting. All that is needed is enough of the broken gear to grind the cutting tool to the proper shape. It can also be used in the cutting of splines and standard and special forms.

Flat Surfaces

Another type of fly cutter, which differs mainly in the design of the arbor, can be used to mill flat surfaces as in plain or face milling (Fig. 4.33). The arbor can easily be manufactured in the shop using common lathe tool bits. This type of fly cutter is especially useful for milling flat surfaces on aluminium and other soft non-ferrous metals, since a high quality finish can be easily obtained. Boring holes with this type of fly cutter is not recommended. The arbor is so short that only very shallow holes can be bored.

4.18 KEY WAY MILLING

Key ways are grooves of different shapes cut along the axis of the cylindrical surface of shafts, into which keys are fitted to provide a positive method of locating and driving members on the shafts. A key way is also machined in the mounted member to receive the key.

The type of key and corresponding key way to be used depends upon the class of work for which it is intended. The most commonly used types of keys are the Woodruff key, the square-ends machine key, and the round-end machine key (Fig. 4.33).

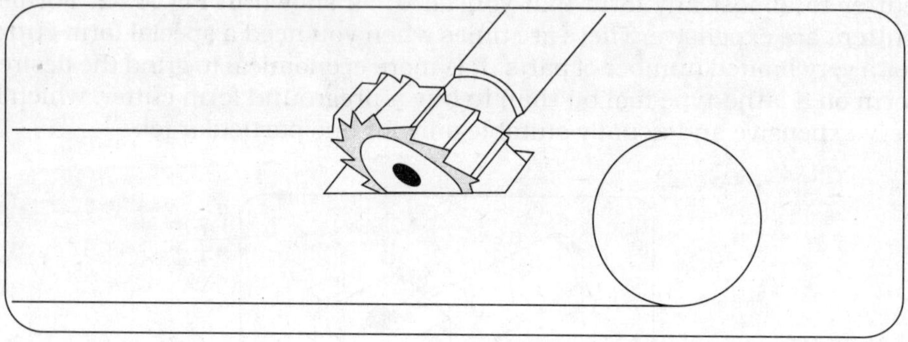

Fig. 4.33 Woodruff key slot.

4.18.1 Woodruff Key

The Woodruff keys are semi cylindrical in shape and are manufactured in various diameters and widths. The circular side of the key is seated into a key way which is milled in the shaft. The upper portion fits into a slot in a mating part, such as a pulley or gear. The Woodruff key slot milling cutter (Fig. 4.34) must have the same diameter as that of the key.

Woodruff key sizes are designated by a code number in eighths of an inch, and the digits preceding the last two digits give the width of the key in thirty-seconds of an inch. Thus, a number 204 Woodruff key would be 4/8 or 1/2 inch in diameter and 2/32 or 1/16 inch wide, while a number 1012 Woodruff key would be 12/8 or 1/2 inches in diameter and 10/32 or 5/16 inch wide. Woodruff keys commonly used and pertinent information applicable to their machining.

For proper assembly of the keyed members to be made, a clearance is required between the top surface of the key and the key way of the bore. This clearance may be from a which the last two digits indicate the diameter of the key in minimum of 0.002 inch to a maximum of 0.005 inch. Positive fitting of the key in the shaft key way is provided by making the key 0.0005 to 0.001 inch wider than the key way.

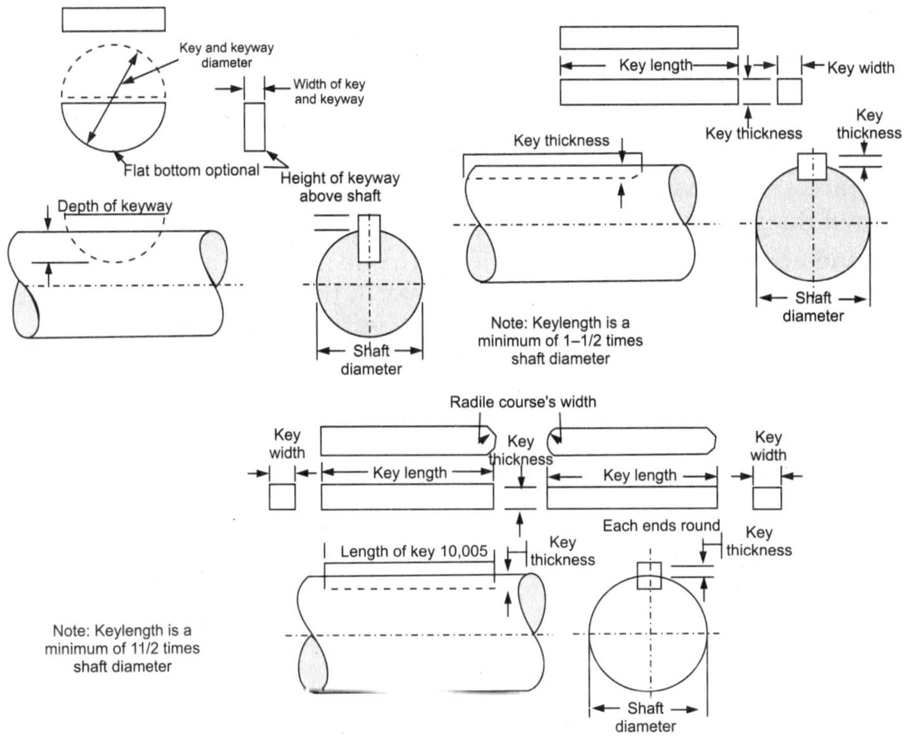

Fig. 4.34 Key way milling.

4.18.2 Square-End Machine Key

Square-ends machine keys are square or rectangular in section and several times as long as they are wide. For the purpose of interchangeability and standardization, these keys are usually proportioned with relation to the shaft diameter in the following method:

- Key width equals approximately one-quarter of the shaft diameter.
- Key thickness for rectangular section keys (flat keys) equals approximately 1/6 of the shaft diameter.
- Minimum length of the key equals 1½ times the shaft diameter.
- Depth of the key way for square section keys is 1/2 the width of the key.
- Depth of the key way for rectangular section keys (flat keys) is 1/2 the thickness of the key.

The length of each key is not included because the key may be of any length as long as it equals at least 1½ times the shaft diameter.

Round-end machine keys (Fig. 4.33). The round-ends machine keys are square in section with either one or both ends rounded off. These keys are the same as square-ends machine keys in measurements.

4.18.2.1 Milling Cutters Used for Milling Key ways

Shaft key ways for Woodruff keys are milled with Woodruff key slot milling cutters (Fig. 4.33). The Woodruff key slot milling cutters are numbered by the same system employed for identifying Woodruff keys. Thus, a number 204 Woodruff key slot cutter has the proper diameter and width for milling a key way to fit a number 204 Woodruff key.

Square-end key ways can be cut with a plain milling cutter or side milling cutter of the proper width for the key

Round-end key ways must be milled with end milling cutters (Fig. 4.35) so that the rounded end or ends of the key may fit the ends of the key way. The cutter should be equal in diameter to the width of the key.

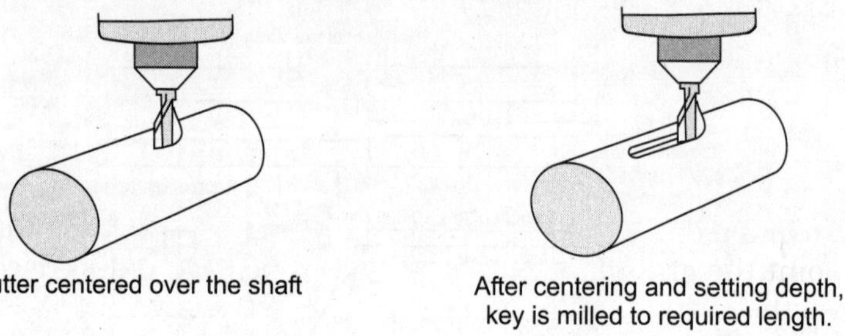

Fig. 4.35 Round-end key way.

Alignment of Milling Cutters

When milling key ways, the shaft may be supported in the vise or chuck, mounted between centers, or clamped to the milling machine table. The cutter must be set centrally with the axis of the workpiece. This alignment is accomplished by using one of the following methods:

When using a Woodruff key slot milling cutter, the shaft should be positioned so that the side of the cutter is tangential to the circumference of the shaft. This is done by moving the shaft transversely to a point that permits the workpiece to touch the cutter side teeth. At this point the graduated dial on the cross feed is locked and the milling machine table is lowered. Then, using the cross feed graduated dial as a guide, the shaft is moved transversely a distance equal to the radius of the shaft plus 1/2 the width of the cutter.

End mills may be aligned centrally by first causing the workpiece to contact the periphery of the cutter, then proceeding as in the paragraph above.

Milling Woodruff Key Slot

The milling of a Woodruff key slot is relatively simple since the proper sized cutter has the same diameter and thickness as the key. With the milling cutter located over the position in which the key way is to be cut, the workpiece should be moved up into the cutter until you obtain the desired keyseat depth. The work may be held in a vise, chuck, between centers, or clamped to the milling machine table. Depending on its size, the cutter is held in an arbor or in a spring collet or drill chuck that has been mounted in the spindle of the milling machine.

Milling Key slot for Square-End Machine Key

The workpiece should be properly mounted, the cutter centrally located, and the workpiece raised until the milling cutter teeth come in contact with the workpiece. At this point, the graduated dial on the vertical feed is locked and the workpiece moved longitudinally to allow the cutter to clear the workpiece. The vertical hand feed screw is then used to raise the workpiece until the cutter obtains the total depth of cut. After this adjustment, the vertical adjustment control should be locked and the cut made by feeding the table longitudinally.

Milling Key way for Round-End Machine Key

Rounded key ways are milled with an end milling cutter of the proper diameter. As in the case of square-ends machine key key ways, the workpiece should be properly mounted and the cutter centrally located with respect to the shaft. The shaft or cutter is then positioned to permit the end of the cutter to tear a piece of thin paper held between the cutter and the workpiece. At this point the graduated feed dial should be locked and used as a guide for setting the cutter depth. The ends of the key way should be well marked and the workpiece moved back and forth making several passes to eliminate error due to spring of the cutter.

4.19 T-SLOT MILLING

Cutting T-slots in a workpiece holding device is a typical milling operation. The size of the T-slots depends upon the size of the T-slot bolts which will be used. Dimensions of T-slots and T-slot bolts are standardized for specific bolt diameters.

Selection of Milling Cutters

Two milling cutters are required for milling T-slots, a T-slot milling cutter and either a side milling cutter or an end milling cutter. The side milling cutter (preferably of the staggered tooth type) or the end milling cutter is used to cut a slot in the workpiece equal in width to the throat width of the T-slot and equal in depth to slightly less than the head space depth plus the throat depth). The T-slot milling cutter is then used to cut the head space to the prescribed dimensions.

Milling the T-Slot

The position of the T-slot is laid out on the workpiece. The throat depth is determined by considering the thickness of the workpiece and the maximum and minimum dimensions allowable.

A side milling cutter or an end milling cutter is then selected. The cutter should be of proper size to mill a slot equal in width to the throat width prescribed for the T-slot size desired. Cut a plain groove equal to about 1/16″ less than the combined throat depth and head space depth.

Select a T-slot milling cutter for the size T-slot to be cut. T-slot milling cutters are identified by the T-Slot bolt diameter and remanufactured with the proper diameter and width to cut. Position of T-slot milling cutter over the edge of the workpiece and align it with the previously cut groove. Feed the table longitudinally to make the cut. Flood the cutter and workpiece with cutting oil during this operation. Fig. 4.36 shows a T-slot milling cutter and dimension locations for T-slots.

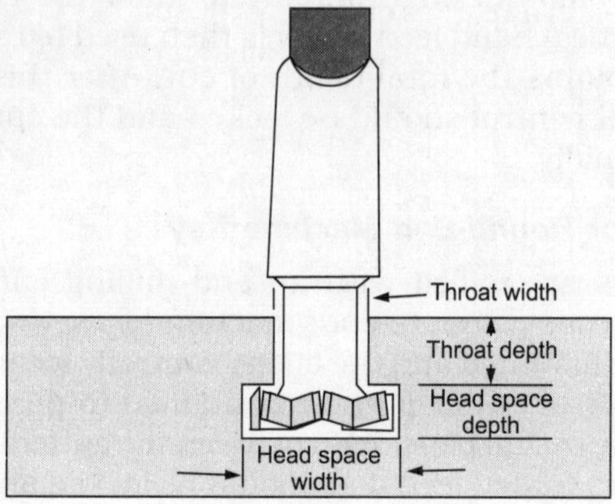

Fig. 4.36 T-slot milling

4.20 SAWING AND PARTING

Metal slitting saw milling cutters are used to part stock on a milling machine. Fig. 4.37 illustrates parting solid stock. The workpiece is being fed against the rotation of the cutter. For greater rigidity while parting thin material such as sheet metal, the workpiece may be clamped directly to the table

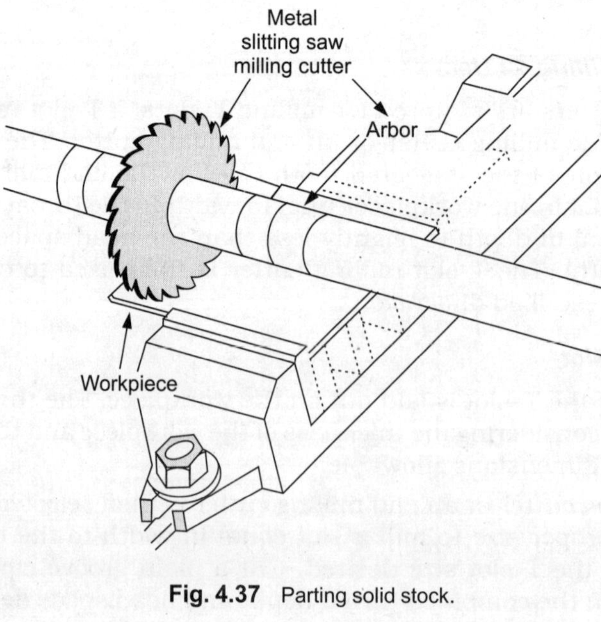

Fig. 4.37 Parting solid stock.

with the line of cut over one of the Table T-slots. In this case, the workpiece should be fed with the rotation of the milling cutter (climb milling) to prevent it from being raised off the table. Every precaution should be taken to eliminate backlash and spring in order to prevent climbing or gouging the workpiece.

4.21 HELICAL MILLING

A helix may be defined as a regular curved path, such as is formed by winding a cord around the surface of a cylinder. Helical parts most commonly cut on the milling machine include helical gears, spiral flute milling cutters, twist drills, and helical cam grooves. When milling a helix, a universal index head is used to rotate the workpiece at the proper rate of speed while the piece is fed against the cutter. A train of gears between the table feed screw and the index head serves to rotate the workpiece the required amount for a given longitudinal movement of the table. Milling helical parts requires the use of special formed milling cutters and double-angle milling cutters. The calculations and formulas necessary to compute proper worktable angles, gear adjustments, and cutter angles and positions for helical milling are beyond the scope of this manual.

Gear Cutting

Gear teeth are cut on the milling machine using formed milling cutters called involute gear cutters. These cutters are manufactured in many pitch sizes and shapes for different numbers of teeth per gear (Table 4.7, Appendix A).

If involute gear cutters are not available and teeth must be restored on gears that cannot be replaced, a lathe cutter bit ground to the shape of the gear tooth spaces may be mounted in a fly cutter for the operation. The gear is milled in the following manner:

NOTE:

This method of gear cutting is not as accurate as using an involute gear cutter and should be used only for emergency cutting of teeth which have been built up by welding.

Fasten the indexing fixture to the milling machine table. Use a mandrel to mount the gear between the index head and footstock centers. Adjust the indexing fixture on the milling machine table or adjust the position of the cutter to make the gear axis perpendicular to the milling machine spindle axis. Fasten the cutter bit that has been ground to the shape of the gear tooth spaces in the fly cutter arbor. Adjust the cutter centrally with the axis of the gear. Rotate the milling machine spindle to position the cutter bit in the fly cutter so that its cutting edge is downward.

Align the tooth space to be cut with the fly cutter arbor and cutter bit by turning the index crank on the index head.

Proceed to mill the tooth in the same manner as milling a key way.

Conventional vs Climb Milling

The direction of feed influenced machining performance.

4.22 CONVENTIONAL MILLING– "UP" MILLING

(Feed movement opposite to tool rotation)

Conventional milling (up milling) is preferred when milling castings or forgings with very rough surfaces.

Characteristics of a up milling:

- Width of chip starts from zero and increases.
- Tooth meets the workpiece at the bottom of cut.
- Upward force tends to lift up workpiece.
- More power required-rubbing provoked by chip beginning at minimum width.
- Surface finish marred (spoiled) due to the chips being carried upward by tooth.
- Chips fall in front of cutter–chip disposal difficult.
- Faster wear on tool than climb milling.
- Application (use) of coolant is difficult.
- More vibrations produced during process.
- Process is suitable for thick jobs machining.

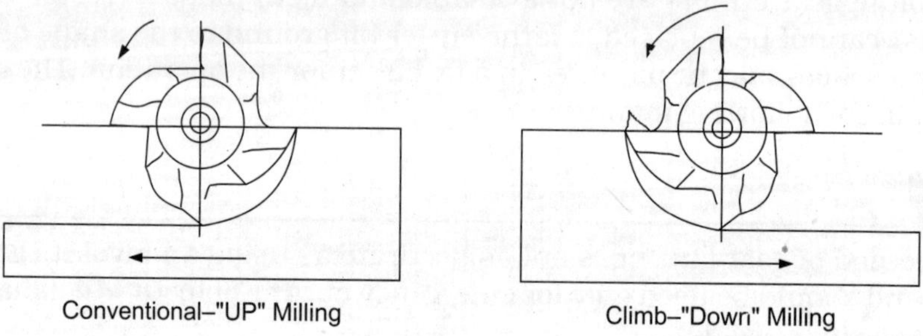

Deflection of end mills

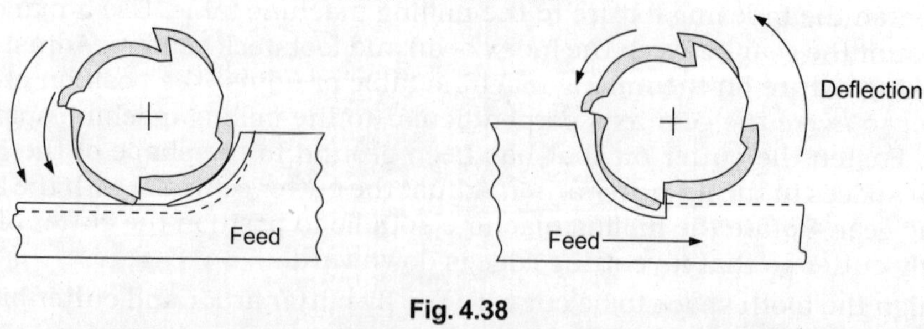

Fig. 4.38

4.23 CLIMB MILLING—"DOWN" MILLING

(Feed movement and tool rotation same direction.)

Climb milling (down milling) is preferred when milling heat treated alloys, stainless steel-reduces work hardening. Climb milling way cause chipping in milling hot rolled materials due to hardened layer on the surface.

Characteristic of a down milling:

- Width of chip starts at maximum and decreases.
- Tooth meets workpiece at top of cut.
- Easier chip disposal – chips removed behind cutter.
- Less wear – increases tool life up to 50%.
- Improved surface finish – chips less like to be carried by the tooth.
- Less power required – cutter with high rake angle can be used.
- Climb milling exerts a downward force on workpiece – fixtures simple and less costly.
- Application (use) of coolant is easy.
- Less vibrations produced during process.
- Process is suitable for thin jobs machining.

4.24 CHIP THICKNESS IN MILLING OPERATION

Chip thickness can be determined for specially two milling process:

(i) Slab milling
  - Up milling
  - Down milling

(ii) Face milling
  - Symmetrical milling
  - Unsymmetrical milling

1. Slab milling:

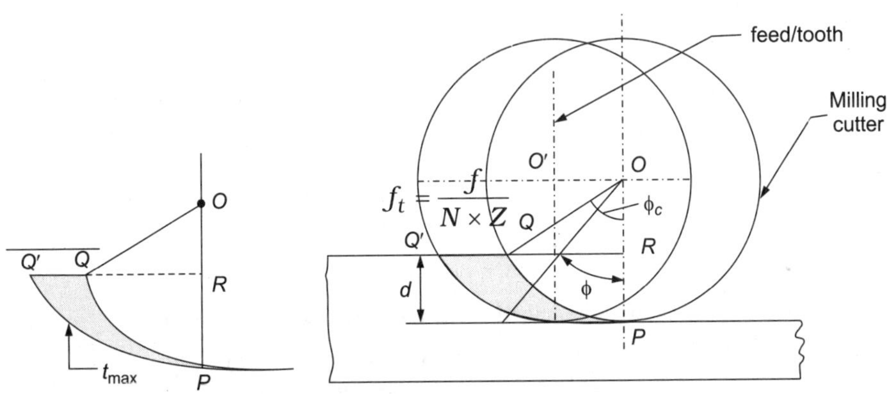

Fig. 4.39 Cross-section of chip generated by milling

f = table feed rate, N = speed (rpm)
d = depth of cut, D = cutter diameter
z = root teeth on cutter.

geometry of chips generated by milling basically depends upon the path of cutting edge of the cutter during the process as cutter is rotating with the uniform speed and workplace feed linearly, a trochoid generated by the tip of cutting tool.

$$PQ = \text{Length of chip}$$

for up milling $$PQ = \frac{D \times \phi}{2} + \frac{f\sqrt{d(D-d)}}{\pi D}$$

for down milling $$Q'P = \frac{D\phi}{2} - \frac{f\sqrt{d(D-d)}}{\pi D}$$

Now assume only one tooth is in contact at a time. Then

$$OR = \left(\frac{D}{2} - d\right), \quad OQ = \frac{D}{2}, \quad QR = \sqrt{OQ^2 - OR^2}$$

$$QR = \sqrt{\left(\frac{D}{2}\right)^2 - \left(\frac{D}{2} - d\right)^2} = \sqrt{d(D-d)}$$

tooth contact QC angle is given by

$$\cos \phi_c = \frac{D - 2d}{D}$$

If t_x is the chip thickness then $t_x = f_t \sin \phi$

where f_t = feed/tooth

table feed rate is given by $f_t = \dfrac{f}{N \times Z}$

where, Z = no. of teeth in cuts,

N = r.p.m. of cutter

or $$t = \frac{f \sin \phi}{N \times Z}$$

for maximum chip thickness $\phi = \phi_c$.

$$\sin \phi_c = \frac{2}{D}\sqrt{d(D-d)} \quad \text{From } \Delta OQR$$

$$t_{max} = \frac{2 \times f \times \sqrt{d(D-d)}}{N \times Z \times D}$$

Maximum cross-section area of chip thickness

$$A_{max} = W \times t_{max} = \frac{2 \times W \times f \times \sqrt{d(D-d)}}{N \times Z \times D}$$

Mean cross-section area

$$A_{mean} = \frac{A_{max}}{2} = \frac{W \times f\sqrt{d(D-d)}}{N \times Z \times D}$$

OBJECTIVE TYPE QUESTIONS

1. Knee is a part of
   (a) milling m/c (b) shaper m/c
   (c) plainer (d) lathe m/c
2. Arbor of a milling machine is used to hold which one of following
   (a) spindle (b) cutting tool
   (c) mandrel (d) over arm ***[UPSC-2009]***
3. Identify the process of removing metal by a milling cutter which is rotated against the direction of travel of the workpiece
   (a) saw milling (b) up milling
   (c) down milling (d) end milling
4. Gang milling is a
   (a) milling operation combineds with turning
   (b) milling process for generating hexagonal surface
   (c) process of gear cutting
   (d) process in which two or more cutters are used simultaneously ***[UPSC-2006]***
5. In milling operation two side milling cutters are mounted with a desired distance between them so that both side workpiece can be milled up simultaneously. This setup is called
   (a) side milling (b) straddle milling
   (c) gang milling (d) string milling
6. In plain milling machine table can be moved
   (a) vertically (b) longitudingly
   (c) crosswise (d) all of above
7. Which one of the following machining process is carried out by employing gang milling
   (a) grooves (b) threads
   (c) steps on prismatic (d) bores ***[UPSC-2009]***
8. In which milling machine the table can be tilted in a vertical plane by providing a swivel arrangement at the knee
   (a) Hand milling machine (b) univesal milling machine
   (c) vertical milling machine (d) plain milling machine
   (e) universal milling machine
9. Internal or external threads of different pitches can be produced by
   (a) plano-miller (b) pantograph milling machine
   (c) profile machine (d) planetary milling machine
10. Down milling is also known as
    (a) end milling (b) conventional milling
    (c) climb milling (d) face milling
11. Up milling is also known as
    (a) face milling (b) conventional milling
    (c) end milling (d) climb milling
12. The operation of generating groves around the pherisphery of a cylinderical or conical workpiece is known as
    (a) saw milling (b) profile milling
    (c) hellical milling (d) gang milling

13. A milling cutter having 8 teeth is rotating at 150 rpm, if the feed/tooth is 0.1 mm, what whould be the table speed in mm/min

(a) 75 (b) 100
(c) 120 (d) 190

14. Helical milling is the process of

(a) machining several surfaces at a single stroke
(b) producing groves around the pheriphery of a conical/cylindrical workpiece
(c) producing narrow groves or slots on workpiece
(d) reproducing an outline of template on a workpiece

15. Part of a milling machine upon which are mounted the milling cutter is known as

(a) bracket (b) tool post
(c) arbor (d) clipper box

16. Process of metal removal by rotating the cutter in the direction of travel of workpiece is known as

(a) saw milling (b) down milling
(c) up milling (d) face milling
(e) straddle milling

17. A plain milling of mild steel plate produces

(a) continous chips with bult up edge
(b) continuous chips without bult up edge
(c) irregular shaped discotinus chips
(d) regular shaped discontinuous chips.

18. Narrow groove or slots can be produced on a wrokpiece by a milling operation is knonw as

(a) end milling (b) saw milling
(c) straddle milling (d) face milling
(e) helical milling

19. A CNC vertical milling machine has to cut a straight slot of 10 mm width and 2 mm depth by a cutter of 10 mm diameter between points (0, 0) and (100, 100) on the *XY* plane (dimensions are in mm) the feed rate used for milling is 50 mm/min. Milling time for the slot (in seconds) is

(a) 120 (b) 170
(c) 180 (d) 240 ***[GATE-2012]***

ANSWERS

1.	(a)	**2.**	(b)	**3.**	(b)	**4.**	(d)	**5.**	(d)	**6.**	(d)	**7.**	(c)
8.	(e)	**9.**	(d)	**10.**	(c)	**11.**	(b)	**12.**	(c)	**13.**	(c)	**14.**	(b)
15.	(c)	**16.**	(b)	**17.**	(b)	**18.**	(b)						

QUESTIONS

1. Define milling operation in detail. What are the various work holding devices used in milling? Explain their relative application and disadvantage? ***[UPTU–2004-05]***

2. What is the function of indexing head as a milling machine attachment giving example. Explain its working? ***[UPTU–2003-04]***

3. Describe the set up that can be used for milling cams in a milling machine with the help of a neat sketch. ***[UPTU–2008-09]***

4. Differentiate the following:
   (i) Compound and differential indexing,
   (ii) Horizontal and vertical milling machine. ***[UPTU–2006-07]***
5. What is a universal milling machine? Draw at least one part which cannot be machined on a horizontal milling machine. List various type of milling cutter available and draw any two. ***[UPTU–2005-06]***
6. What are different type of milling cutters? Draw the sketch of a milling cutter & show different angles provides on its teeth. ***[UPTU–2007-08]***
7. What are the differentiate between the up milling and down milling?
8. Explain the characteristics that distinguish a milling process from other machining process. ***[UPTU–2008-09]***
9. Classify the different types of milling machine. Explain what are the following operations: boring, reaming, tapping, counter boring, sinking and countersinking. ***[UPTU–2010-11]***
10. Drive the expression for the area cross-section of a chip in slab milling ignoring the different cut.

$$A_{max} = \frac{2W \times f\sqrt{d(D-d)}}{N \times Z \times D}$$

where f = feed rate of table, N = speed in r.p.m., d = depth of cut, D = cutter diameter, Z = root teeth on cutter.

11. Describe the milling machine specification.
12. What is the function of indexing head as a milling machine attachment? Giving examples. Explain its working? ***[UPTU–2002-03]***
13. Show that the maximum chip thickness t_{max} in slab milling operation is given by

$$A_{max} = \frac{2W \times f\sqrt{d(D-d)}}{N \times Z \times D}$$ ***[UPTU-2002-03]***

where f = table feed rate, N = speed in r.p.m, D = Diameter of cutter. Z = root teeth on cutter.

14. Explain the construction feature of a diving head used in milling also explain the method for obtaining compound indexing. ***[UPTU–2002-03]***
15. A 100 mm diameter cutter having 8 teeth cut steel at 30 m/min depth of cut is 4 mm and total feed is 150 mm/min. Find the length of chip in up and down milling operation. ***[UPTU–2004-05]***

[**Hint:** Chip length $l_s = \frac{D\phi}{2} + \frac{f \times \sqrt{d(D-d)}}{\pi D}$ for up milling.

$$l_s = \frac{D\phi}{2} - f \times \frac{\sqrt{d(D-d)}}{\pi D}$$ for down milling.

❑❑❑

Chapter

5

Drilling and Boring

5.1 DRILL MACHINE

Drill machine is a machine tool, operated by power to perform a hole in a object with the twist drill. This twist drill is associated with spindle and rotating at very high speed. In drill machine pressure (force) is applied by hand and for producing hole. A multipoint (twist drill) cutting tool is used.

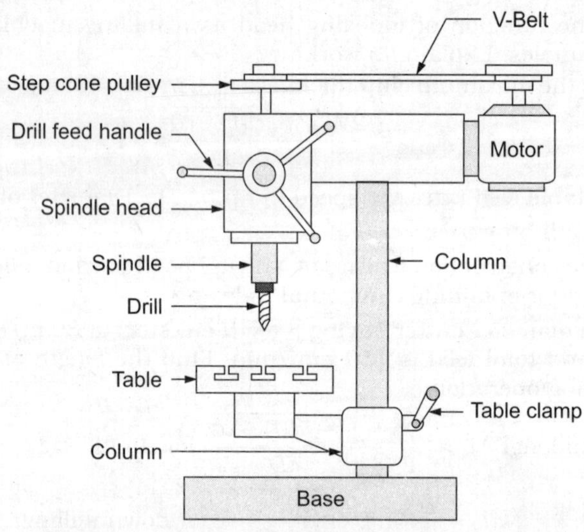

Fig. 5.1 Block diagram of drill machine

Main Parts of a Drill Machine:

(i) Spindle,

(ii) Table,

(iii) Column,

(iv) Base,

(v) Drill chuck,

(vi) Head.

5.1.1 Types of Drill Machine

1. Hand drill machine.
2. Up right drill machine.
3. Bench or sensitive drill machine.
4. Portable drill machine.
5. Radial drill machine.
6. Gang drill machine.
7. Multi spindle drill machine.
8. Automatic drill machine.

5.1.1.1 Hand Drill Machine

Hand drill machine is shown in the fig. They are very small in the size and entire machine is gripped in hand and this hand drill machine usually used to hole in a wall etc. different size and shapes of hand drill machine are available with a set of drills. Suitable to generate hole in one or two numbers, not suitable for mass production and holed by this machine is not so accurate. Cost of hand drill machine is also less.

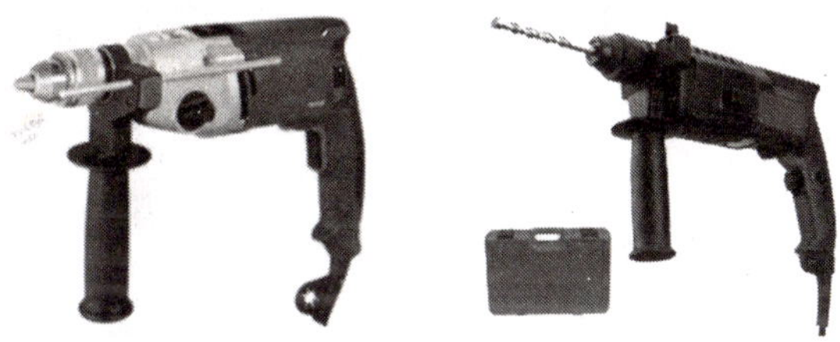

(a) Hand drill machine (b) Hand drill machino

Fig. 5.2 Type of hand drill machine

5.1.1.2 Upright Drill Machine

A upright drill machine is shown in the fig. 5.3 this type of drill machine consist vertical column, base, motor, table etc. Motor and spindle box are fixed on column and table also at column which can move up and down position. This type of machine is used to produce hole in a bigger size as compared to hand drill machine. Hole accuracy is also high and machine is

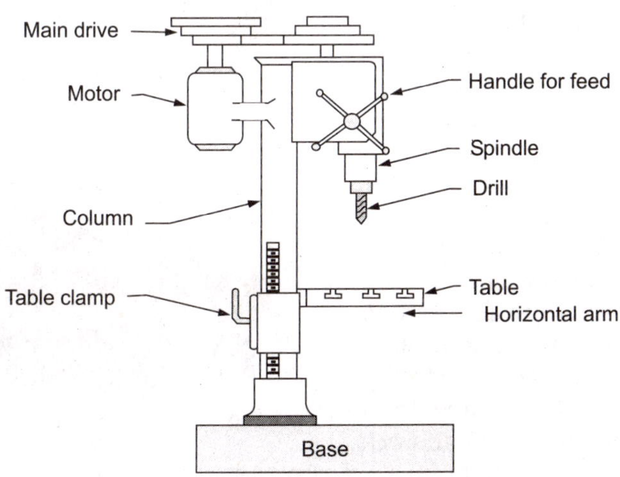

Fig. 5.3 Upright drill machine.

suitable for mass production. Usually this type of drill machine can produced a hole from ϕ1.5 mm to ϕ15 mm and these drill machine operate at **high speed.**

5.1.1.3 Piller Type Drill Machine

A piller type drill machine is shown in Fig. 5.4. This piller type drill machine is a large version of a bench drill or sensitive drill machine. This drill machine consists a piller, table, base, geared stock. Table used in this machine can rotate at any direction. Spindle is rotated by a power motor with several speed (cone. pull arrangement). This type of drill machine frequently used in industries for light duty object and it is very suitable for mass production with accuracy.

Fig. 5.4 Piller type drill machine

5.1.1.4 Portable Dril Machine

A portable drill machine is shown in Fig. 5.5. It is quite different with hand drill machine as this drill machine can transport from one place to another place hence known as portable drill machine and associated with all facility same as other drill machine. Unique feature of this drill machine that it can transport from one place to another place and suitable to mass production and accuracy of hole surface is also very good. This portable drill machine is used to produced small size of hole in a workpiece. Machine are operated by a motor some time operated by a universal motor (AC or DC motor). Speed of portable drill machines are very fast.

Fig. 5.5 Portable dril machine

5.1.1.5 *Radial Drill Machine*

Figure 5.6 shows a radial drill machine. A radial drill machine is used to produce hole in a heavy and bigger size jobs. One of the unique feature of radial drill machine that it can rotate in all three axis, i.e., X-axis, Y-axis and Z-axis.

Radial drill machine basically consist a very heavy base with heavy column. Column is mounted on the base and supported the arm which can be raised or lowered as well as swing around the column. Drill head of the machine is positioned using motor power and same travel the swinging arm. In radial drill machine workpiece remain stationary while cutting tool (twist drill) rotates.

Main advantage of radial drill machine

- it is very suitable to hole in a very heavy and bulky workpiece in any position.
- machine can rotate in all three directions.

Fig. 5.6 Actual Radial drill machine

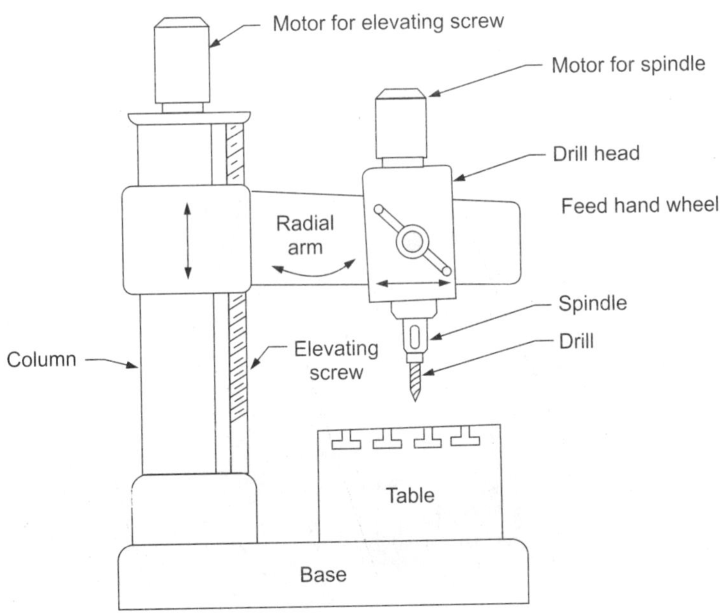

Fig. 5.7 Block diagram of a radial drill machine

5.1.1.6 *Gang Drill Machine*

Gang drill machine block diagram is shown in the Fig 5.8 as the name implies this type of drill machine consist a tool holder which carrying a numbers of drills. One of the unique feature of gang drill that in a single stroke a numbers of holes can be performed in a workpiece. This type of drill is suitable for making the job which consist a numbers of holes.

Fig. 5.8 Gang drill

5.1.1.7 *Multistation Drill Machine*

Multistation drill machine is shown in the fig. It is quite different with gang drill machine. In multistation drill machine all drill machines are arranged in a line as shown in fig. In multistation drill machine only one powerfull motor is used to run the all drill. One of the unique feature of multistation drill machine that at a time large numbers of workpiece can be drill. In fig. at a time four workpiece can be drill. Process is very suitable for mass production.

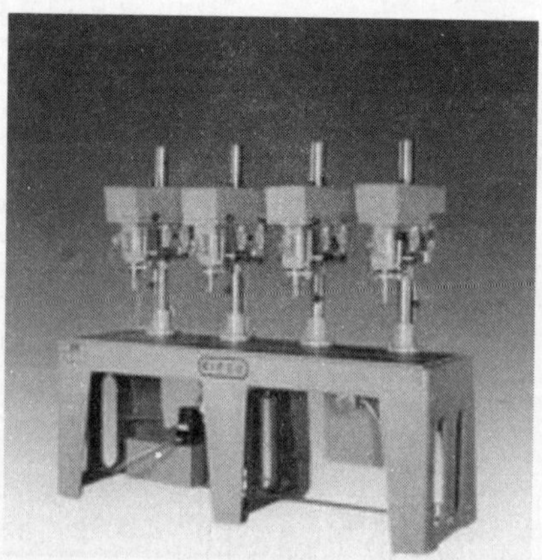

Fig. 5.9 Multistation drill machine

5.2 TWIST DRILLS

5.2.1 Definition

Drill: Drill can be defined as a rotary end cutting tool having one or more cutting lips, and having one or more helical or straight flutes for the passage of chips and the admission of a cutting fluid.

Twist drills are the most common cutting tools used with drilling machines. Twist drills are designed to make round holes quickly and accurately in all materials. They are called twist drills mainly because of the helical flutes or grooves that wind around the body from the point to the neck of the drill and appear to be twisted (Fig. 5.10). Twist drills are simply constructed but designed very tough to withstand the high torque of turning, the downward pressure on the drill, and the high heat generated by friction.

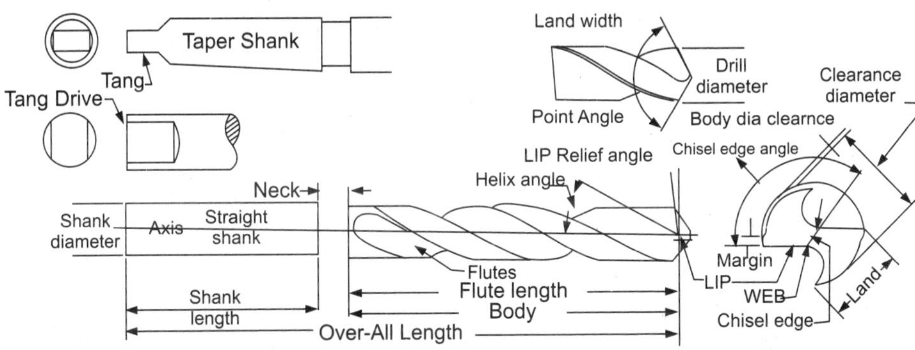

Fig. 5.10 Twist drill

Common twist drill sizes range from 0.0135 (wire gage size No. 80) to 3.500 inches in diameter. Larger holes are cut by special drills that are not considered as twist drills. The standard sizes used in the United States are the wire gage numbered drills, letter drills, fractional drills, and metric drills. Twist drills can also be classified by the diameter and length of the shank and by the length of the fluted portion of the twist drill.

5.2.2 General Classifications

A. Classification Based on Construction

1. **Solid Drills:** Those made of one piece of material such as high speed steel.
2. **Tipped Solid Drills:** Those having a body of one material with cutting lips made of another material brazed or otherwise bonded in place.
3. **Composite Drills:** Those having cutting portions mechanically held in place.

B. Classification Based on Methods of Holding or Driving

1. **Straight Shank Drills:** Those having cylindrical shanks which may be the same or different diameter than the body of the drill; the shanks may be made with or without driving flats, tang, grooves or threads.

2. **Taper Shank Drills:** Those having conical shanks suitable for direct fitting into tapered holes in machine spindles, driving sleeves or sockets; tapered shanks generally have a driving tang.
3. **Taper Shank Square Drills:** Those having tapered shanks with four flat sides for fitting a rachet or brace.
4. **Shell Core Drills:** Core drills mountable on arbors specifically designed for the purpose; commonly used with shell reamer arbors.
5. **Threaded Shank Drills:** Those made with threaded shanks generally used in close center multiple spindle applications or portable angle drilling tools.
6. **Beaded Shank Bits:** Drills with flat shanks having raised beads parallel to the axis.
7. **Flat Drill:** A drill whose flutes are produced by two parallel or tapered flats
8. **Flat (Spade) Drill:** A removable cutting drill tip usually attached to a special holder designed for this purpose; generally used for drilling or enlarging cored holes.
9. **Gun Drill:** Special purpose straight flute drills with one or more flutes used for deep hole drilling; they are usually provided with coolant passages through the body; they may be either solid or tipped.
10. **Half-Round Drill:** A drill with a transverse cross-section of approximately half a circle and having one cutting lip.
11. **Multiple-Margin Drill:** A drill whose body diameter clearance is produced to leave more than one margin in each land.
12. **Step Drill:** A multiple diameter drill with one set of drill lands which are ground to different diameters.
13. **Subland Drill:** A type of multiple diameter drill which has independent sets of lands in the same body section for each diameter.
14. **Taper Drill:** A drill with part or all of its cutting flute length ground with a specific taper to produce tapered holes; they are used for drilling the original hole or enlarging an existing hole.

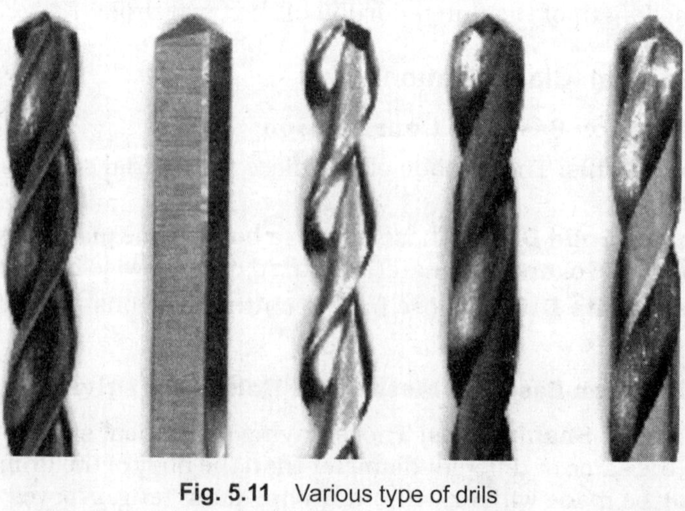

Fig. 5.11 Various type of drils

C. Classification Based on Number of Flutes

1. **Single-Flute Drills:** Those having only one flute commonly used for originating holes.
2. **Two-Flute Drills:** The conventional type of twist drill used for originating holes.
3. **Three-Flute Drills (Core Drills):** Drills commonly used for enlarging and finishing, drilled, cast, or punched holes; they will not produce original holes.
4. **Four-Flute Drills (Core Drills):** Used interchangeably with three-flute drills; they are of similar construction except for the number of flutes.

D. Classification Based on Hand of Cut

1. **Right-Hand Cut:** When viewed from the cutting point the counterclockwise rotation of a drill in order to cut; the great majority of drills are made "right hand".
2. **Left-Hand Cut:** When viewed from the cutting point the clockwise rotation of a drill in order to cut.

5.2.3 Nomenclature of Twist Drills and Other Terms Relating to Drilling

Axis: The imaginary straight line which forms the longitudinal center line of the drill.

Back Taper: A slight decrease in diameter from front to back in the body of the drill.

Body: The portion of the drill extending from the shank or neck to the outer corners of the cutting lips.

Body Diameter Clearance: That portion of the land that has been cut away so it will not rub against the walls of the hole.

Built-Up Edge: An adhering deposit of nascent material on the cutting lip or the point of the drill.

Cam Relief: The relief from the cutting edge to the back of the land, produced by a cam actuated cutting tool or grinding wheel on a relieving machine.

Chip Breaker: Nicks or Grooves designed to reduce the size of chips; they may be steps or grooves in the cutting lip or in the leading face of the land at or adjacent to the cutting lips.

Chipping: The breakdown of a cutting lip or margin by loss of fragments broken away during the cutting action.

Chisel Edge: The edge at the end of the web that connects the cutting lips.

Chisel Edge Angle: The angle included between the chisel edge and the cutting lip, as viewed from the end of the drill.

Clearance: The space provided to eliminate undesirable contact between the drill and the workpiece.

Clearance Diameter: The diameter over the cut away portion of the drill lands.

Crankshaft or Deep Hole Drills: Drills designed for drilling oil holes in crankshafts, connecting rods and similar deep holes; they are generally made with heavy webs and higher helix angles than normal.

Cutter Sweep: The section formed by the tool used to generate the flute in leaving the flute.

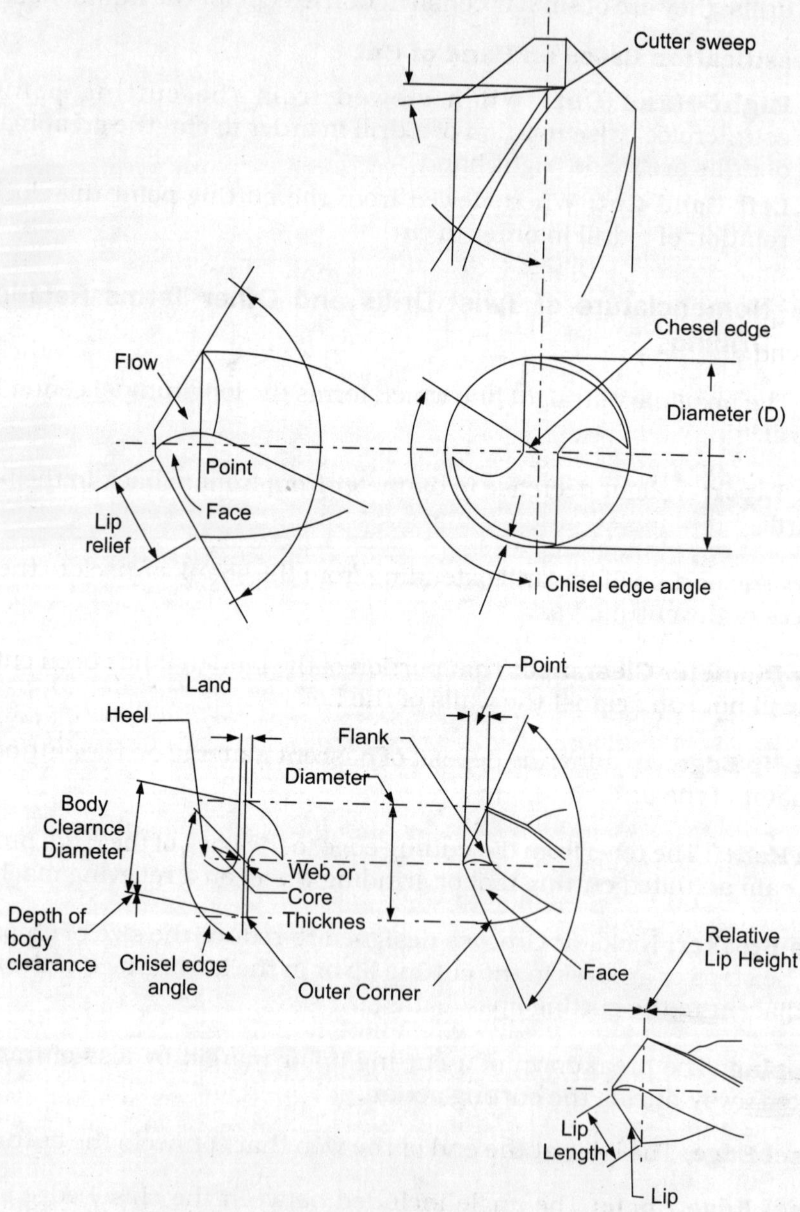

Fig. 5.12 Drill angles (Twist drill nomeclature)

Double Margin Drill: A drill whose body diameter clearance is produced to leave more than one margin on each land and is normally made with margins on the leading edge and on the heel of the land.

Drift: A flat tapered bar for forcing a taper shank out of its socket.

Drift Slot: A slot through a socket at the small end of the tapered hole to recieve a drift for forcing a taper shank out of its socket.

Drill Diameter: The diameter over the margins of the drill measured at the point.

Flutes: Helical or straight grooves cut or formed in the body of the drill to provide cutting lips, to permit removal of chips, and to allow cutting fluid to reach the cutting lips.

Flute Length: The length from the outer corners of the cutting lips to the extreme back end of the flutes; it includes the sweep of the tool used to generate the flutes and, therefore, does not indicate the usable length of the flutes.

Gage Line: The axial position on a taper where the diameter is equal to the basic large end diameter of the specified taper.

Guide: A cylindrical portion, following the cutting portion of the flutes, acting as a guide to keep the drill in proper alignment; the guide portion may be fluted, grooved, or solid.

Heel: The trainling edge of the land.

Helical Flutes: Flutes which are formed in a helical path around the axis.

Helix Angle: The angle made by the leading edge of the land with a plane containing the axis of the drill its value lies from 24° to 30°.

Land: The peripheral portion of the body between adjacent flutes.

Land Width: The distance between the leading edge and the heel of the land measured at a right angle to the leading edge.

Lead: The axial advance of a leading edge of the land in one turn around the circumference.

Lips: The cutting edges of a two flute drill extending from the chisel edge to the periphery.

Lip Relief: The axial relief on the drill point.

Lip Relief Angle: The axial relief angle at the outer corner of the lip; it is measured by projection into a plane tangent to the periphery at the outer corner of the lip it is kept 12° to 45°.

Margin: The cylindrical portion of the land which is not cut away to provide clearance.

Neck: The section of reduced diameter between the body and the shank of a drill Oil Grooves: Longitudinal straight or helical grooves in the shank, or grooves in the lands of a drill to carry cutting fluid to the cutting lips Oil Holes or Tubes: Holes through the lands or web of a drill for passage of cutting fluid to the cutting lips.

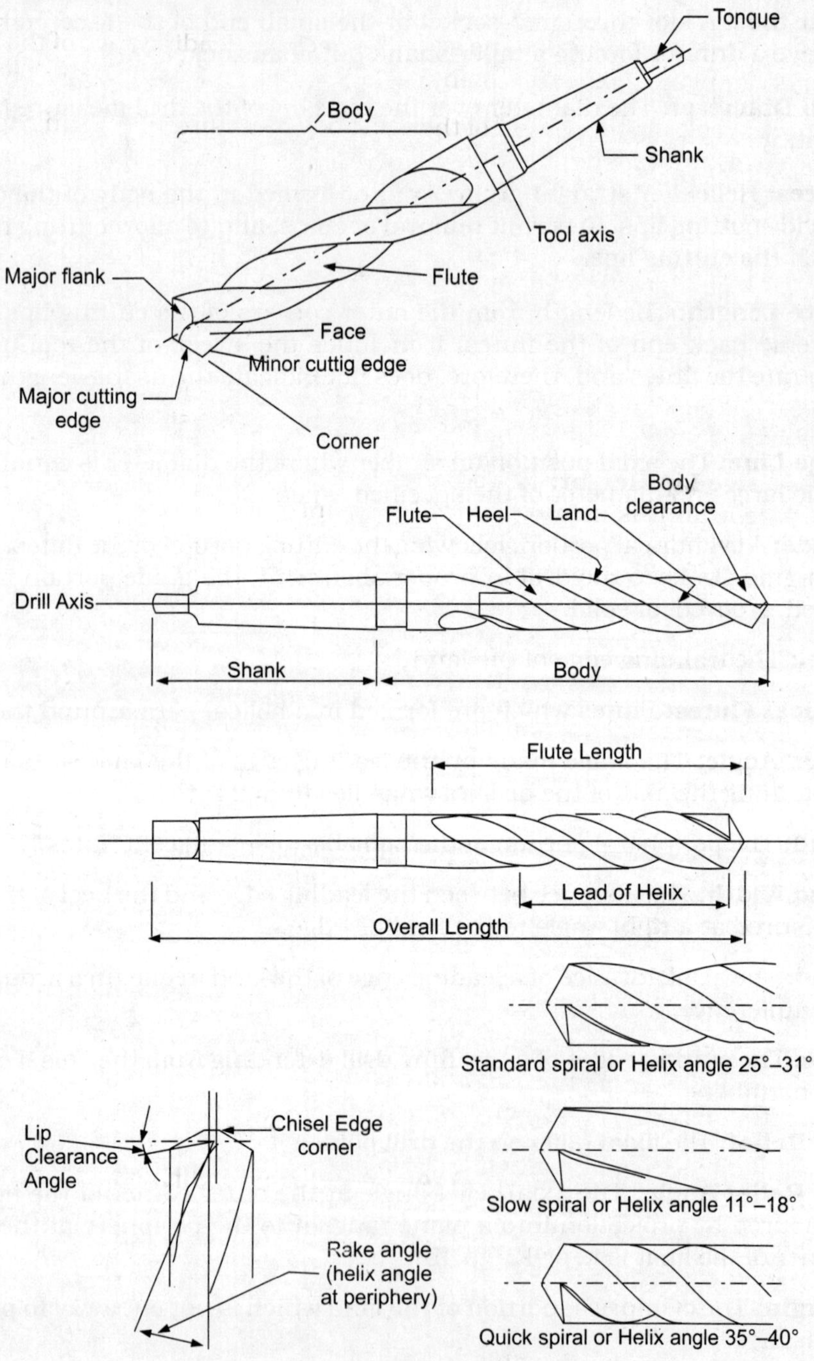

Fig. 5.13 Twist Drill [3]

Overall Length: The length from the extreme end of the shank to the outer corners of the cutting lips; it does not include the conical shank end often used on straight shank drills, nor does it include the conical cutting point used on both straight and taper shank drills.

Periphery: The outside circumference of a drill.

Peripheral Rake Angle: The angle between the leading edge of the land and an axial plane at the drill point.

Pilot: A cylindrical portion of the drill body preceding the cutting lips; it may be solid, grooved, or fluted.

Point: The cutting end of a drill, made up of the ends of the lands and the web; in form it is resembles a cone, but departs from a true cone to furnish clearance behind the cutting lips.

Point Angle: The angle included between the cutting lips projected upon a plane parallel to the drill axis and parallel to the two cutting lips. For medium hard steel its value lies from 116° to 118° and 125° for C.I. and 60° for wood. This angle referred as side cutting edge angle of a single point tool.

Relative Lip Height: The difference in indicator reading on the cutting lip of the drill; it is measured at a right angle to the cutting lip at a specific distance from the axis of the tool.

Relief: The result of the removal of tool material behind or adjacent to the cutting lip and leading edge of the land to provide clearance and prevent rubbing (heel drag).

Shank: The part of the drill by which it is held and driven.

Sleeve: A tapered shell designed to fit into a specified socket and to receive a taper shank smaller than the socket.

Socket: The tapered hole in a spindle, adaptor, or sleeve, designed to receive, hold, and drive a tapered shank.

Straight Flutes: Flutes which form lands lying in an axial plane.

Tang: The flattened end of a taper shank, intended to fit into a driving slot in a socket.

Tang Drive: Two opposite parallel driving flats on the extreme end of a straight shank.

5.3 COMBINED DRILLS AND COUNTERSINKS

5.3.1 Definition

Combined Drill and Countersink: Single or double-end cutting tool, having helical or straight flutes, and having a drill portion and an adjacent integral countersink portion, primarily used to produce center holes in work that will be held between machine centers.

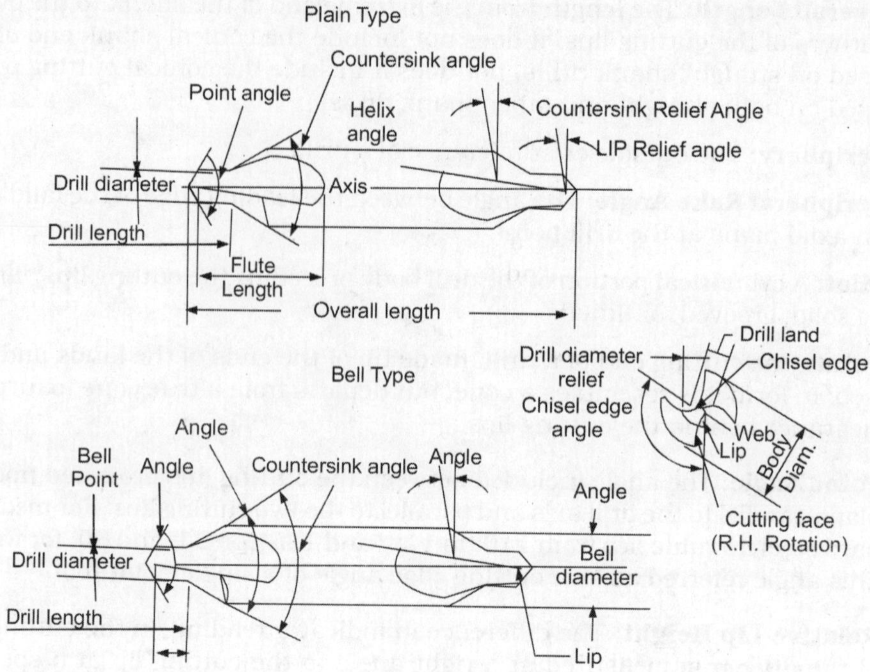

Fig. 5.14 Combined and cobintersinks drill

5.3.2 General Classifications

A. Classification Based on Construction

1. **Solid Combined Drills and Counter sinks:** Those made of one piece of material such as high speed steel.
2. **Tipped Solid Combined Drills and Counter sinks:** Those having a body or drill portion of one material with cutting edges or lips, or both, made of another material brazed or otherwise bonded in place.

B. Classification Based on Type

1. **Plain Type Combined Drills and Countersinks:** Those having a drill portion and a single adjacent integral countersink portion
2. **Bell Type Combined Drills and Countersinks:** Those having a drill portion and an adjacent integral countersink portion, plus an additional secondary conical section to provide clearance for the bearing surface.

C. Classification Based on Hand of Cut

Same as twist drills.

5.3.3 Nomenclature of Combined Drills and Countersinks

Axis: The imaginary straight line which forms the longitudinal center line of the combined drill and countersink.

Back Taper: A slight decrease in diameter from the front to back in the drill length.

Bell Angle: The included angle of the secondary conical section providing clearance or protection for the countersink angle conical surface (normally 120°).

Bell Diameter: The diameter at the intersection of the countersink portion and the bell portion at the leading edge of the land

Body: The central portion of the tool by which it is held or driven

Countersink Angle: The included angle of the countersink portion (normally 60°).

Countersink Relief Angle: The angle between a plane at right angles to the axis of the tool, and a line tangent to the surface of the countersink portion at the intersection of the countersink portion and the body and at the leading edge of the land; for the bell type, the relief angle is measured at the intersection of the bell portion and the body.

5.3.4 Process Parameters

Depth of cut: The depth of the hole generated by the drilling process.

Feed: The rate that the drill advances into the material, generally measured in distance per flute.

Speed: The cutting speed is usually measured at the periphery of the drill in surface feet or meters per minute.

Thrust: The axial force required to drill.

Torque: The twisting moment required to drill.

Surface Finish: The roughness of the walls of the drilled hole; a measure of the hole quality.

Example 5.1

Calculate time required for drilling of 20 mm diameter hole in a 15 mm thick M.S. Plate. A feed rate of 0.15 mm per revolution and a lip angle of 118° for the Twist drill may be assumed. Cutting speed for M.S. = 25 m/min. *[UPTU 2011-12]*

Solution: Given $P = 15$ mm, $f = 0.2$ mm/rev, $V = 25$ m/min, $D = 20$ mm

$$V = \frac{\pi DN}{1000} \quad N = \frac{V \times 1000}{\pi \times D} = \frac{25 \times 1000}{3.14 \times 20} = 398.1 \text{ R.P.M}$$

We know that time taken for drilling is given by

$$t = \frac{60 \times p}{N \times f} = \frac{60 \times 15}{398.1 \times 0.2} = 11.3 \text{ min} \quad \text{or} \quad 678.2 \text{ sec} \quad \textbf{Ans.}$$

OBJECTIVE TYPE QUESTIONS

1. Metal removal take place during drilling operation by
(a) extresion (b) shearing
(c) compression (d) shearing and compression
(e) shearing and extresion

2. A twist drill is
(a) front cutting tool (b) end cutting tool
(c) side cutting tool and end cutting tool
(d) front and side cutting tool

3. A twist drill is a
(a) single point cutting tool (b) two point cutting tool
(c) three point cutting tool (d) multi point cutting tool

4. A twist drill is specified by its shank, material and
(a) diameter (b) flute size
(c) lip angle (d) length of body

5. Drilling is an example of
(a) orthogonal cutting (b) oblique cutting
(c) simple cutting (d) uniform cutting

6. A standard ground drill has a point angle of
(a) 90° (b) 110°
(c) 118° (d) 126°
(e) 130°

7. Optimum value of helix angle of a drill is
(a) 10° (b) 20°
(c) 30° (d) 60°

8. A standard twist drill is made out of
(a) Mild steel (b) carbon steel
(c) high speed steel (d) cast iron

9. Rake angle in a drill
(a) remains constant
(b) irrelevant to the drilling operation
(c) increases from centre to periphery
(d) decreases from centre to periphery

10. Drill spindles are provided with a standard taper is known as
(a) chapman taper (b) brown and sharper taper
(c) morse taper (d) sellers taper

11. Helical grooves which extend to full length of drill body are known
(a) shank (b) lips
(c) flutes (d) cutting edge

12. Match list I (type of drill) and list II application and select the correct answer using the edges given below the list

	List I		**List II**
A	Straight shank	1.	soft material
B	taper shank	2.	deep hole
C	single flute	3.	general prupose
D	high helix	4.	small hole diameter

Code:

	A	B	C	D
(a)	4	3	1	2
(b)	3	4	1	2
(c)	3	4	2	1
(d)	4	3	2	1

13. A portable drill machine is specified by
(a) maximum thickness of job it can drill
(b) maximum diameter of drill it can hold
(c) maximum travel of spindle
(d) spindle speed and feed

14. A 15 mm drilling machine means that it can drill a hole
(a) having cross-sectional area of 15 mm^2
(b) maximum diameter of 15 mm
(c) in 15 mm thick plate
(d) non of above

15. For drilling the soft material, brass or copper a used
(a) tapered shank twist drill
(b) side cutting tool
(c) front cutting tool
(d) end cutting tool
(e) straight fluted drill

16. Point angle of drill for soft material is
(a) less than 118°
(b) equal to 118°
(c) greater than 118°
(d) equal to 127°

17. A tool used to withdraw a drill from the sleeve is called
(a) drill drawn
(b) drift
(c) drill puller
(d) drill remover

18. Through which mechanism spindle of drill machine gets up and down motion
(a) cross belt system
(b) compound gear train
(c) rack and pinion arrangement
(d) cross-slide system

19. What drilling time is required for producing a hole in a mild steel sheet of 25 mm thickness using a high speed drill of 25 mm diameter the cutting speed and feed for drill are 20 m/min and 0.25 mm/rev respectively
(a) 0.314 min
(b) 0.236 min
(c) 0.438 min
(d) 0.443 min
[UPSC-2009]

20. Twist fluted drills are perferred as
(a) flutes made the tool light weight
(b) holes are cut accurately and efficiently
(c) there is swift movement of drill in the workpiece
(d) chips move out automatically

21. Cutting angle of a flat drill varies from
(a) 10° – 20°
(b) 30° – 50°
(c) 60° – 90°
(d) 90° – 120°

22. Clerance/relief cutting edge angle of a flat drill varies
(a) 3° – 8°
(b) 45° – 60°
(c) 70° – 90°
(d) 90° – 118°

23. Which drill is used for brass material
(a) zero helix angle
(b) high helix angle
(c) low helix angle
(d) any helix angle

24. Which drill is used for aluminimum

(a) high helix angle (b) zero helix angle

(c) low helix angle (d) any helix angle

25. Drill machine is adopted for mass production is

(a) deep hole drill m/c (b) multi spindle drill m/c

(c) radial drill m/c (d) gang drill m/c

26. A drill machine which can drill a number of holes simultaneously on a truck engine housing is

(a) Radial drill m/c (b) universal drill m/c

(c) multi spindle drill m/c (d) gang drill m/c

27. In a single pass drilling operation, a through hole of 15 mm diameter is to be drilled in a steel plate of 50 mm thickness. Drill spindle is 500 rpm feed is 0.2 mm/rev and drill point angle is 118°. Assuming 2 mm clearance at approach and exit, the total drill time (in seconds) is

(a) 35.1 (b) 32.4

(c) 31.2 (d) 30.1 **[GATE-2012]**

ANSWERS

1. (e)	**2.** (d)	**3.** (d)	**4.** (a)	**5.** (b)	**6.** (c)	**7.** (c)
8. (c)	**9.** (b)	**10.** (c)	**11.** (c)	**12.** (c)	**13.** (b)	**14.** (b)
15. (e)	**16.** (a)	**17.** (b)	**18.** (c)	**19.** (a)	**20.** (d)	**21.** (d)
22. (a)	**23.** (b)	**24.** (a)	**25.** (d)	**26.** (c)		

Hint 19. $V = \dfrac{\pi DN}{1000}$, $N = \dfrac{1000 \times 20}{3.14 \times 25} = 254.78$

$$\text{Drill time } t = \frac{60 \times P}{N \times f} = \frac{60 \times 25}{254.78 \times 0.25} = 23.55 \text{ min}$$

QUESTIONS

1. With the help of a suitable sketch, describe the geometry of twist drill and also explain, how drill sizes are designated. ***[UPTU–2002-03]***
2. How drill size designated? Briefly explain. ***[UPTU–2003-04]***
3. Explain the different types of holes commonly used in engineering components and the process used to produce them. ***[UPTU–2004-05]***
4. Define the drilling operation in detail.
5. Classify the drills in detail.
6. Describe the various angle in drill with their function.
7. How many types of drill machine you know?
8. Differentiate between drilling and boring operation?
9. Draw a neat block diagram of a drill machine and explain the same.
10. Describe the boring operation in detail.
11. Classify the boring machine.
12. Describe the boring tools, boring bar & boring head with the help of a neat sketch.

❑❑❑

UNIT-III

Grinding and Super Finishing

Chapter 6

Grinding

6.1 GRINDING

Grinding is a metal removal process carried out by the rotating wheel (at very high speed) known as a grinding wheel. This grinding wheel consist the abrasive particles which helps in removal of metal. In grinding process metal removal take place in the form of fine particles and it is one of its unique feature.

Feature of Grinding Process

- Metal removal take place in the form of fine particles.
- High degree of surface roughness, $R_a = 0.15$ to $1.3\ \mu m$, can be achieved.
- Closer dimensional tolerance up to 0.3 to 0.5 mm can be achieved.
- Process can be applied on all types of material.
- Discontinues chips are produced.

6.2 GRINDING WHEELS

Grinding wheel is widely used in grinding machines. These wheels are expendable wheels composing of an abrasive compound. These grinding wheels are formed out of an aluminium disc or solid steel by way of attaching the particles to the exterior surface. In general, this grinding wheel is normally prepared using the mold of hard-pressed coarse particles matrix that are allied together to make a firm circular shape. Based on the planned use of the wheel, various profiles and cross-sections are also obtainable.

Fig. 6.1 Grinding wheel

In common other materials are made use of with vitrified bonding agent like **diamond** and **silicon carbide**. A variety of materials are used in production grinding and primary things available in today's market are wheels of different grade, various structure, varied abrasives, a range of grain sizes, and mixed link. Abrasive is in fact a cutting material like all other manufactured diamonds, **cubic boron nitride (CBN), ceramic, aluminium oxide (Al_2O_3)** and **zirconia aluminium oxide**. It is advisable to choose an abrasive considering the rigidity of the material that is about to slice.

Fig. 6.2 Various shapes and types of grinding wheels.

The wheel structure defines the wheel density that is the abrasive and bond versus airspace. Comparatively it is found a wheel which is of less density cuts freely and this plays a major effect on the finishing of the surface. The above nature is due to better chip clearance of the lesser density wheel. With a wheel that is of less density, we can obtain a wider or deeper cut, by means of less coolant.

Grade of a wheel can be determine by studying how closely the abrasive sticks to the bond. The grade has effect on about all aspects of grinding like flow of coolant, speed of wheel, depth of grinding and feed rate ranges. It is the grain which decides the size of a physical abrasive in any wheel. Though a grain which is of greater size cut swiftly without any strain, it would we be still able to get only a poor finish of surface. To get a fine and precision surface finish make use of ultra-fine sized grain. Wheel bonding agent is what decides how a wheel holds an abrasive and also has its impact on wheel speed, its finish and coolant flow, etc.

Wheel manufacturing is a process that is precise and strictly controlled not only because of the security risk of a spinning disc but also depends on the consistency of composition that is required to stop that particular disc from explosion due to the high stress released at the time of revolving.

These grinding wheels have small level of sharpening by itself. To get the best possible use, clad it with the help of grinding dressers. Dressing here means taking out of existing abrasive coating, this exposes a new and razor-sharp surface to the surface of work. Turning method is followed to make the grinding surface flat. It is done for the purpose of getting a parallel grinding surface to the respective table or to the orientation of the plane which could produce perfect surface.

The cup or plain wheel rest profusely on the arbors, perfectly sized metal discs at the side of the wheel and apply the essential hold tightly force to transmit the revolving motion. Paper blotter dissolves the force uniformly within the surface of the wheel.

6.3 TYPES OF GRINDING WHEELS

6.3.1 Straight Grinding Wheels

Fig. 6.3 Straight grinding wheel

Straight wheel are the most common mode of wheel that is found on pedestal or bench grinders. This is widely used for centreless & cylindrical surface grinding operations. As it is used only on the periphery, it forms a little concave surface on the piece. This is used to gain on several tools like chisels. The size of these wheels differs to a great extent, width and diameter of its face obviously depends on the category of its work, machines grinding power.

6.3.2 Cylinder or Wheel Ring

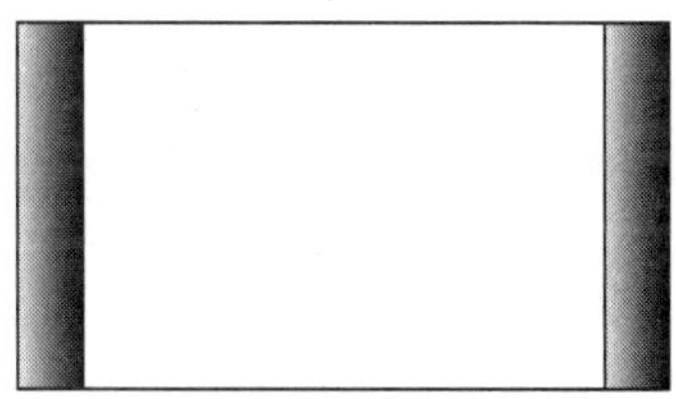

Fig. 6.4 Cylinder

A cylinder wheel has no center mounting support but has a long and wide surface. Their width is up to 12" and is used purely in horizontal or vertical spindle grinders. This is used to produce flat surface, here we do grinding with the ending face of the wheel.

6.3.3 Tapered Grinding Wheels

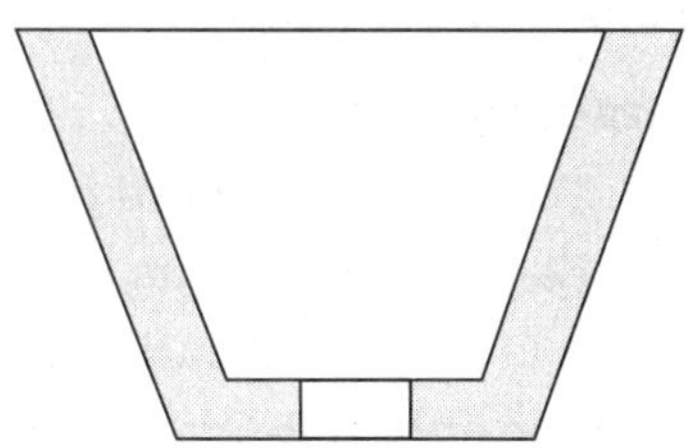

Fig. 6.5 (*a*) Tapered grinding wheel.

Tapered grinding wheel is a straight wheel that tapers externally towards the mid point of the wheel. As this is stronger than straight wheels, it accepts advanced lateral loads. Straight wheel with tapered face is chiefly used for gear teeth, grinding thread, etc.

6.3.4 Straight Cup

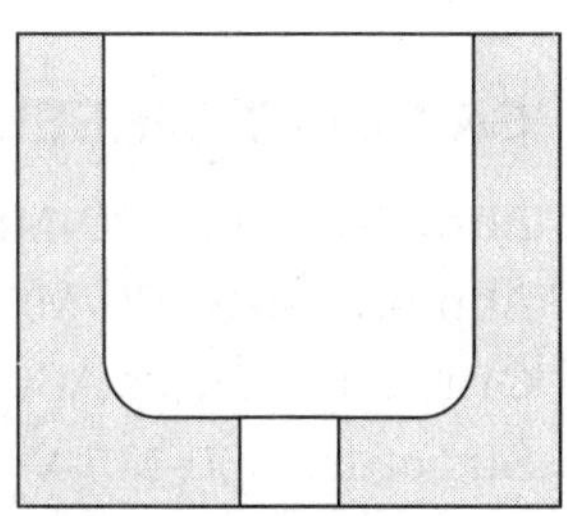

Fig. 6.5 (*b*) Straight cup

This straight cup wheels forms an option for cup wheels in cutter and tool grinders, having an extra radial surface of grinding is favorable.

6.3.5 Dish Cup

In fact this is used primarily in jig grinding and cutter grinding. It is a very thin cup-style grinding wheel which permits grinding in crevices and slot.

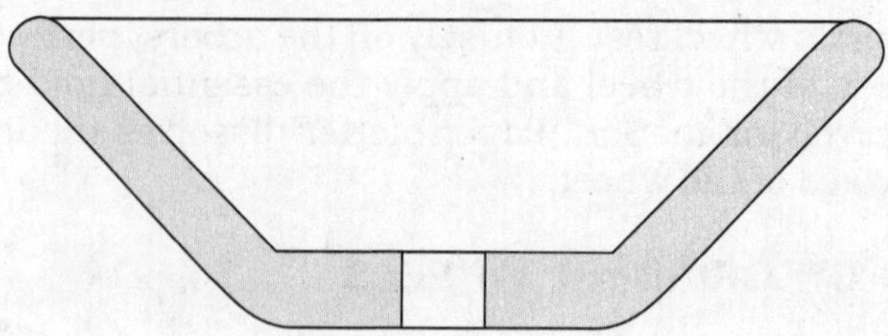

Fig. 6.6 Dish cup

6.3.6 Saucer Grinding Wheels

Saucer grinding wheel is an exceptional grinding profile used for grinding twist drills and milling cutters. This finds wide usage in non-machining areas, as this saw filers are used by saucer wheels to maintain saw blades.

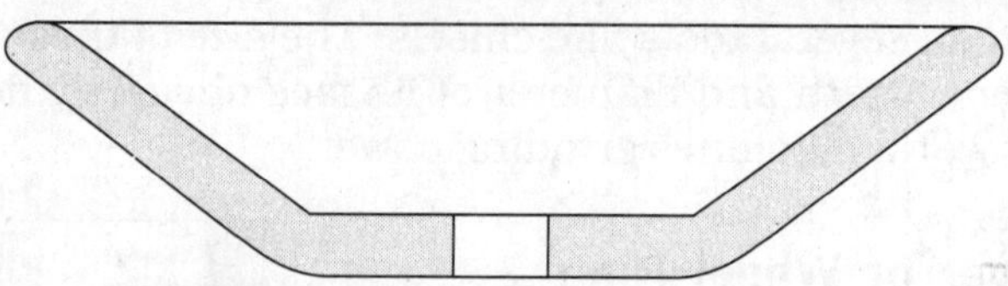

Fig. 6.7 Dish cup

6.3.7 Diamond Grinding Wheels

In diamond wheels industrial diamonds remain bonded to the edge. This is used to grind hard materials like concrete, gemstones and carbide tips. A slitting saw is designed for slicing gemstones like hard materials.

Characteristics:

- To remove material from a workpiece in an abrasive action
- Each grain acts as a cutting tool
- It is a chip formation process
- The only self-sharpening cutting tool
- Consider as abrasive machining

6.4 GRINDING WHEEL COMPONENTS

- Abrasive Type A60-M5-V
- Abrasive size A60-M5-V
- Grade/Hardness A60-M5-V
- Structure A60-M5-V
- Bond A60-M5-V

Specification: Indian standard specification for grinding wheel has been prepared with a view to establishing a system of marking of going gringing wheels.

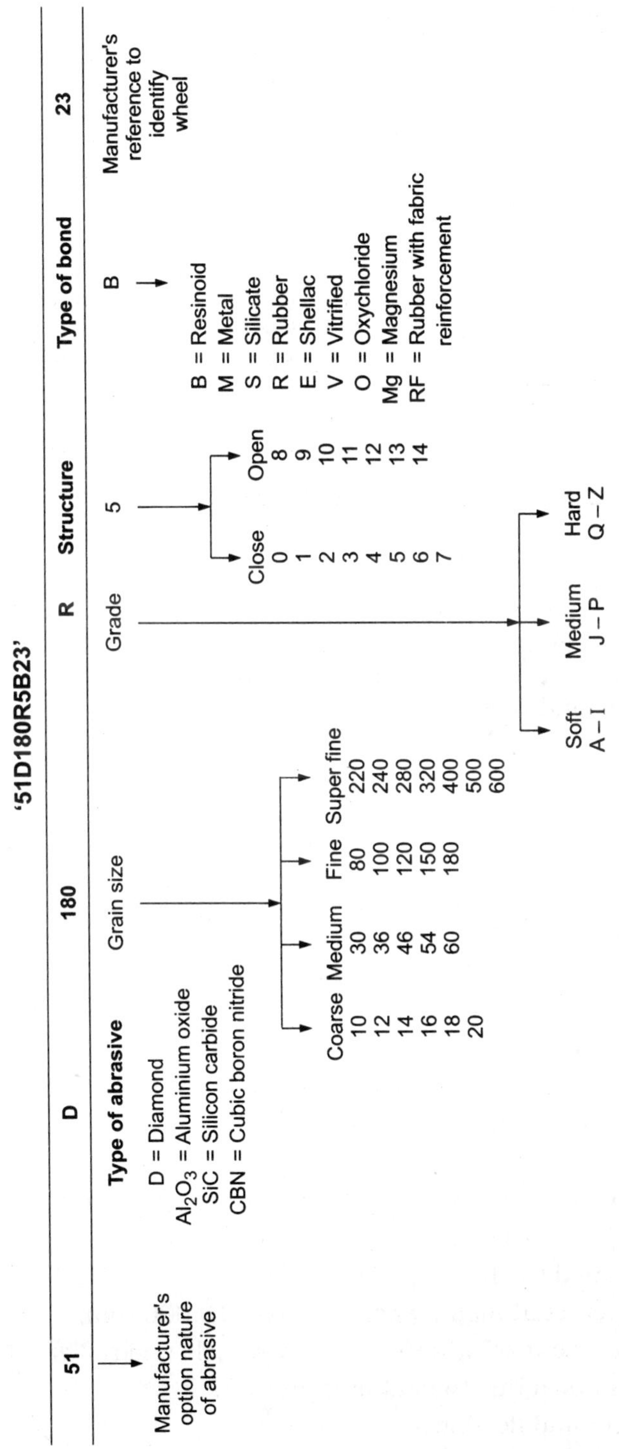

Fig. 6.8 Specification of a grinding wheel.

6.4.1 Grade of Grinding Wheel

In grinding wheel grade is referred to the hardness of the bond and its usually designated by the alphabet letter and grade of wheel usually depends upon the types of bond, abrasive gain and wheel structure.

From letter *A* to *I* – soft grade

J to *P* – Medium grade

Q to *Z* – hard grade.

In metal cutting process hard wheels are used for soft materials and soft wheel are used to hard materials.

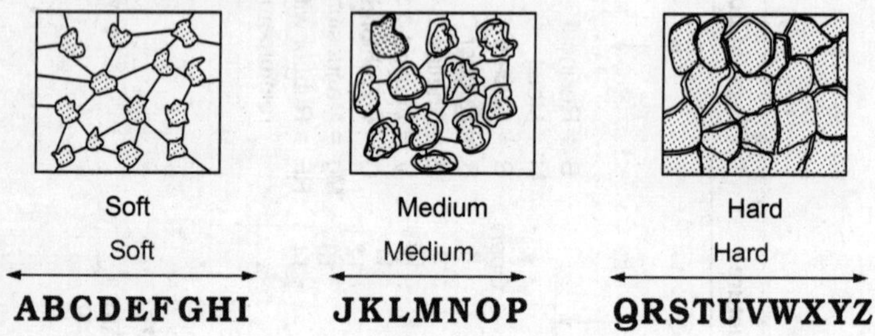

6.4.2 Bonds

The primary object of bond to adhere abrasive grain together under the action of grinding force. Material used as the "GLUE" to hold the abrasive grains together bond has no cutting action.

Commonly bonds are:

V-Vitrifieds
B-Resin
R-Rubber
E-Shellac
P-Epoxy

6.4.2.1 Types of Bonds

A. Vitrified:

- Designated by the letter "V"
- Glass or porcelain
- Normal operating speed is 6,500 rpm
- Breaks down due to pressure
- Very rigid, stiff
- uniform structure due to wet mixing of various components.

B. Resin:

- Designated by the letter "B"
- Can operate at higher speeds (9,500-16,000 rpm)
- Suited for cut-off wheels, foundry wheels and roll grinding
- Breaks down due to heat and pressure
- Stronger and flexible
- fabricated by synthetic or organic resin.

C. Shellac:

- Designated by the letter "E"
- Very cool cutting
- Used for roll grinding
- Resilient, very forgiving
- For very high finishing.

D. Silicate band:

- Designated by the letter "S".
- Bond are light in grey in colour.
- No used to produced very hand band.
- Silicate soda in the base material.

Structure

Structure of a grinding wheel shows the spacing between the grains and structure may be open or dense and it is represented by numbers.

Al_2O_3-Aluminium oxide
C-Silicon carbide
Z-Zirconia

Close	Open
0	8
1	9
3	10
4	11
5	12
6	13
7	14

Manufactures Abrasive Code

Coarse	*Medium*	*Fine*	*Very Fine*
8	30	80	220
10	36	100	240
12	46	120	280
14	60	150	320
16	180	20	400
24			500
			600

6.5 ABRASIVE GRAIN TYPES

- Aluminium Oxide-Al_2O_3
- Silicon Carbide-Sic
- Cubic Boron Nitride (CBN)
- Synthetic Diamond

6.5.1 Abrasive Grain Properties

- Friability—Ability to fracture under pressure so that as the cutting edges become dull, the grain breaks off and exposes sharp new cutting edges.
- **Hardness** — Ability to scratch or penetrate the workpiece.
- **Size** — Sizing is accomplished by the size of the screen through which the grain will pass.

6.5.2 Abrasive Manufacturing

Manufacture Process:

- Bauxite + Coke + Iron Borings
- Cylindrical furnace 7' dia × 6' high
- Heated by electrodes to 3,700° for 16 to 36 hours
- Poured into molds
- Cooled, broken, crushed to size.

6.5.3 Abrasive Grain Types

Aluminium Oxide (Al_2O_3) — Types

- Regular
- Brown in colour
- Strongest and toughest
- Used in severe operations and on soft steels
- Semifraible
- Light brown or blue/gray in colour
- Cooler cutting than regular brown aluminium oxide
- Used on harder material
- Friable
- White (pink,ruby and gray) in colour
- Fractures very easily
- Used on hard steels and alloy steels.

Silicon Carbide — Types (SiC)

- Regular
- **Black in colour:** Used on iron base materials, marble, ceramics, aluminium, brass and copper
- **Green:** More friable than black
- Suited for carbide and other very hard materials
- Review Pacific Specifics Handbook
- Abrasive and Bond designations.

Zirconia and Ceramic — Types

- Friable
- Faster removal action
- Free cutting action
- Less pressure needed
- Tough
- Longer in life
- Form Holding

Ceramic Abrasive

- Longer Life
- Aggressive removal rates
- Cooler cutting
- Good form holding
- Best on hard materials over 50 RC

6.6 GRADE/HARDNESS

Definition The strength with which the bond holds the abrasive grain together n Indicated by a letter. A (Softest) -------------------------- Z (Hardest)

Soft	Medium	Hard
ABCDEFGHI	**JKLMNOP**	**QRSTUVWXYZ**

Grade/Hardness

Softer grade wheel	**Harder grade wheel**
• Faster cutting	• Slower cutting
• Less chatter	• More chatter
• Lower amps/power	• Higher amps/power
• Lower wheel life	• Longer wheel life
• Corse finish	• Better finish
• Less form holding	• Greater form holding

6.7 STRUCTURE

Definition: Indicates the density of the wheel (or) the spacing of grain and bond Indicated by a number:

- 4-6 (Dense)
- 7-9 (Open)
- 10 + (Induced Porosity)

6.8 MECHANICS OF GRINDING

Grinding is a metal removal process in the form of power by the multipoint cutting tool in grinding process metal removal basically occurs due to the abrasive particles which are firmly joint together here the mechanism of grinding specially surface grinding and cylindrical grinding are discuss:

6.8.1 Surface Grinding

Process of metal removal by grinding is same as milling machine. Figure 6.9 shows chip formation by a single abrasive grit.

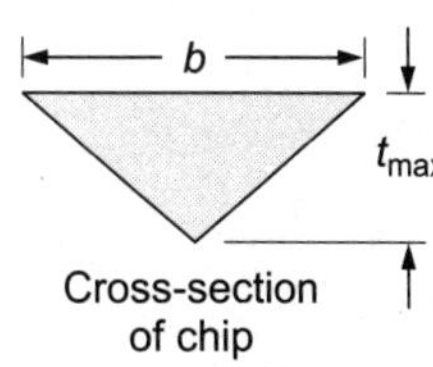

Fig. 6.9

Let, $\quad v$ = work speed,

V_W = Grinding wheel speed,

d = in feed.

Average length of chip ($L_S = \frac{D}{2} \times Q$) (from figure)

D = Diameter of grinding wheel

Q = angle

$$\cos\theta = \frac{OT}{OP} = \frac{OR - TR}{OP} = \frac{\frac{D}{2} - d}{\frac{D}{2}}$$

or
$$d = \frac{D}{2}(1 - \cos\theta)$$

Similarly,
$$PT = \frac{D}{2}\sin\theta = \frac{D}{2} \times \sqrt{(1 - \cos^2\theta)}$$

$$= \frac{D}{2} \times \sqrt{1 - \left(1 - \frac{d}{\frac{D}{2}}\right)^2} = \sqrt{D \times d - d^2}$$

L_S = are length of PR = chord length PR

$$L_S = PR = \sqrt{PT^2 + d^2} = \sqrt{D \times d}$$

or
$$L_S = \left(1 + \frac{V}{V_W}\right)\sqrt{D \times d}$$

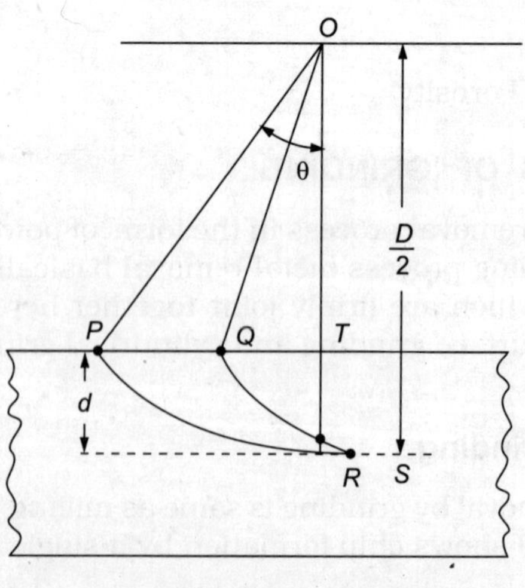

Fig. 6.10 Surface grinding operation

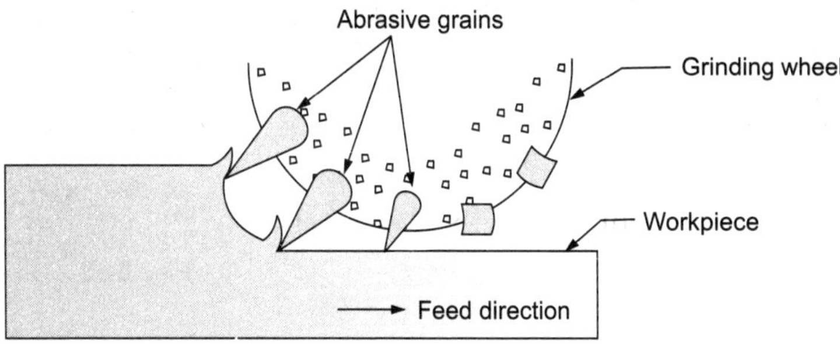

Fig. 6.11 Chipformation in surface grinding

6.8.2 Maximum Chip Thickness (t_m) in Surface Grinding

Suppose, V_t = table speed,

V_N = grinding wheel surface speed,

K = number of abrasive grains/unit area of grinding wheel surface

b = width of cut for surface grinding

N = grinding wheel r.p.m.

d = depth of cut for surface grinding

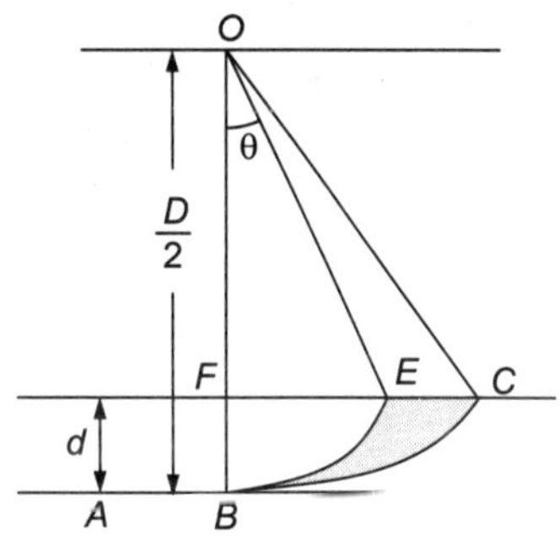

Fig. 6.12

from figure: $$\cos\theta = \frac{OF}{\frac{D}{2}} = \frac{1-d}{\frac{D}{2}}$$

$$\cos\theta = \frac{D-2d}{D} = 1 - \frac{2d}{D}$$

but $$CF^2 = \left(\frac{D}{2}\right)^2\left[1-\left\{1-\frac{d}{D/2}\right\}^2\right]$$

or $$CF = \sqrt{D\times d - d^2}$$

l = arc length of BC = chord length BC

$$l = BC = \sqrt{CF^2 + d^2} = \sqrt{Dd}$$

but $$\sin\theta = \sqrt{1-\cos^2\theta} \quad \text{or} \quad \sin\theta = 2\sqrt{\frac{d}{D}+\frac{d^2}{D^2}}$$

feed is given by $f = \dfrac{V_t}{V_W}$ and maximum chip thickness is given by

$$t_{\max} = \frac{f\sin\theta}{bc}$$

Now, Fig. 6.12 $$t = CE\sin\theta = 2\,CE\sqrt{\frac{d}{D}-\frac{d^2}{D^2}}$$

Neglecting $\frac{d^2}{D^2}$ as $\frac{d}{D}$ is very small

$$CE = \frac{V_t}{K.N}$$

$$t = \frac{2V_t}{K.N} \times \sqrt{\frac{d}{D}} \qquad K = \pi DbC$$

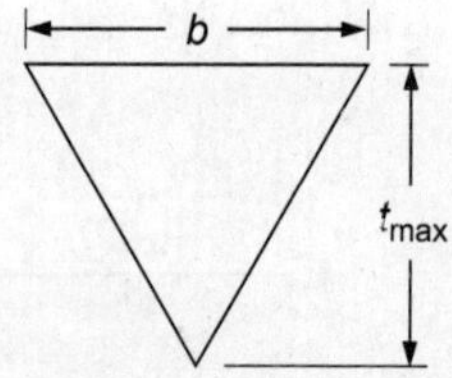

Fig. 6.13 Cross-section of chip

$$t = \frac{2}{b.c} \times \frac{V_t}{V_W} \times \frac{\sqrt{d}}{D}$$

Proved

6.8.3 Mechanism of Grinding Wheel Wear

Figure 6.14 shows the mechanism of grinding wheel wear. It is some what like the creep curve. This curve consist the three zones, i.e.,

(i) initial rapid wear zone,
(ii) constant rate wear zone or maximum life period of wheel,
(iii) rapid wheel wear zone.

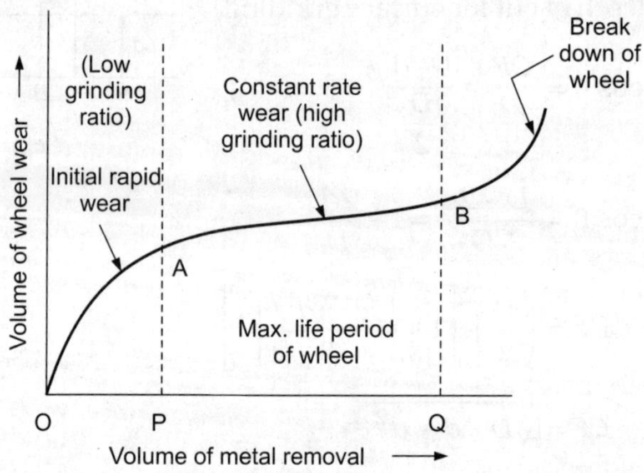

Fig. 6.14 Wheel wear curve

In 1st stage initial rapid wear of wheel take place but once the break particles are out, grinding wheel gets stabilized and wear take place at uniform rate.

2nd stage of shows the maximum life period of wheel at this stage constant rake of wear take place. During this stage the continuous load is applied on wheel (continuous metal removal take place).

3rd stage of curve shows that life of grinding wheel is over and it needs redressing. Life of grinding wheel is referred as grinding time between the two consecutive re-dressing grinding ratio is an important factor when the performance of grinding wheel is considered and it is denoted by G_r.

$$G_r = \frac{\text{Amount of metal removed}}{\text{Amount of wheel war}} = \frac{\Delta_w}{\Delta_t}$$

or in other words it is expressed as ratio of the volume of metal removed to the volume worn away. Ideally grinding wheel should be dressed at the end of the middle region of uniform wear.

For maximum life of a grinding wheel, grinding ratio should be high, during the leading thrust force also effect the grinding ratio. It is very clear from Fig. 6.15 that wheel removed rate becomes large and hence the ratio becomes small when the thrust force is excessive.

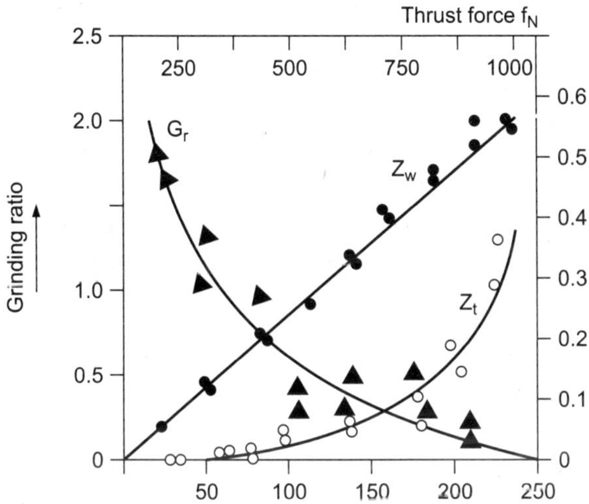

Fig. 6.15 Effect of thrust force f_t on the metal and wheel removed rates Z_w and Z_t. [3]

Specification of Grinding Wheel

According to Indian Standard of Institution designation of grinding wheel is very essential and grinding wheels are produced by the B.I.S. specification. Every grinding wheel consist the same special letter (Alphabet, NUM).

6.9 SELECTION OF A GRINDING WHEEL

Proper selection of a grinding wheel is very essential task to achieve the optimum grinding efficiency.

Following factors are considers during the selection of a grinding wheel:

- Material to be ground-type and hardness
- Stock to be removed
- Workpiece geometry and surface finish required
- Grinding machine-type, horsepower and condition
- Wheel speeds and feeds
- Grinding contact area
- Grinding fluid-wet or dry
- Severity of operation
- Dressing method

6.9.1 Specification of Selection Factors

A. Type of material

- Steels (aluminium oxide)
- Low tensile strength materials and non-metallic materials (Silicon carbide).

B. Hardness of material

Hard materials	**Soft material**
• Finer grit sizes	• Coarse grit size
• Friable abrasives	• General purpose abrasive
• Softer grade wheels	• Harder grade wheel
• Sharper abrasives	

C. Stock to be removed

Light stock removal	**Heavy stock removal**
• Friable grains	• Tough grains
• Finer grit sizes	• Coarse grit size
• Softer grade wheels	• Harder grade wheel

D. Workpiece geometry

• Flat/uninterrupted surface (no form)	• Form/interrupted surface
• Finer grit sizes	• Coarser grit sizes
• Softer grade wheels	• Hard grade wheels
• Open structures	• Tighter structures

E. Surface finish

• Better finish	• Coarse finish
• Finer grit sizes	• Coarse grit sizes
• Harder grade wheels	• Softer grade wheels
• Less friable abrasives	• More friable abrasives

F. Machine horsepower

• Higher horsepower	• Low horse power
• Harder grade wheels	• Softer grade wheels
• Tougher abrasives	• Friable abrasive
• Wheel speed	
• Faster wheel surface feet per minute	• Slower wheel surface feet per minute
• Softer grade wheels	• Harder grade wheels

G. Grinding contact area

• Large contact area	• Small contact area
• Softer grade wheels	• Harder grade wheels
• Coarser grit size	• Finer grit size
• Friable abrasives	• Tougher abrasives

H. Wet or Dry Applications

- Wet
- Harder grade wheels
- Tougher abrasives
- Open structures

- Dry
- Softer grade wheels
- Friable abrasives
- Tighter structures

6.9.2 Grinding Operations

- Surface grinding
- Bench grinding
- Cutter grinding thread grinding
- Crank and cam grinding
- Internal grinding
- Foam grinding
- Snagging
- Saw grinding
- Centerless grinding

6.10 STANDARD WHEEL SHAPES

Standard wheel shapes special wheel faces wheel care and safety:

- Grinding wheel care
- Inspection
- Visually inspect for chips, cracks or damage to wheel
- Vitrified only
- Ring test with wooden or plastic handle
- Wheel handling rules
- Handle wheels carefully to prevent dropping or bumping
- Do not roll wheel on its edge
- Use trucks or suitable conveyors which support heavy wheels
- Stack wheels carefully on trucks to prevent damage in transit
- Do not pile any materials or tools on top of a grinding wheel

6.11 WHEEL CARE AND SAFETY

- Wheels should be stored in proper racks, bins or drawers
- Check MOS against machine speed
- Never force a grinding wheel onto a machine spindle
- Do not use flanges that are not flat and clean
- Wear safety glasses
- Always use a wheel guard that cover at least 1/2 of wheel
- Always stand to one side when starting the wheel
- Allow a newly mounted wheel to run one minute before using
- Never grind on the side of the wheel

- Always shut off the coolant before stopping the grinder
- Never force the wheel by suddenly jamming work into it.
- Grind only materials for which the wheel is designed

6.12 TYPES OF GRINDING MACHINE

Various kinds of grinding machine may be classified as follows:

(a) surface grinding
(b) centre less grinding
(c) plain cylindrical grinding
(d) face grinder
(e) Belt grinder
(f) drill grinder
(g) bench precision grinder
(h) portable grinder
(i) universal grinder
(j) universal tool grinder
(k) tool post grinder, etc.

6.13 SURFACE GRINDING

Surface grinding is a manufacturing process which moves or grinding wheel relative a surface in a plane while a grinding wheel contacts the surface and removes a minute amount of material, such that a flat surface is created. The term surface grinding designates any process which accurately processes or grinds a surface.

Surface grinding methods include horizontal-spindle, vertical-spindle, vertical-spindle rotary grinding, horizontal spindle single disk, and vertical swivel head grinding.

Surface grinding is the most common of the grinding operations. A rotating wheel is used in the grinding of flat surfaces. However, the grinding wheel is not limited to just a cylindrical shape, but can have a myriad of options that are useful in transferring different designs to the object being worked on.

When surface grinding an object, one must keep in mind that the shape of the wheel will be transferred to the material of the object like a mirror image. It is common practice to what is called "dress" the stone. So as to make sure the shape of the wheel is what you want. This is done by using a diamond to remove the abrasive material not wanted and to give the desired geometry. Surface grinding is a finishing process that smooths the surface of metallic or non-metallic materials and gives them a more refined look.

The surface grinder is comprised of an abrasive wheel, a work-holding device known as a chuck, and a reciprocating table. Depending on the material being worked on and the desired surface finish, the wheel's abrasive material can vary from aluminium oxide, silicon carbide, diamond, and cubic boron nitride. The chuck is a device that is used to hold the material in place while it is being worked on. It can do this one of two ways; metallic pieces are held in place by a magnetic chuck, while non-metallic pieces are vacuumed in place. During the grinding process, the workpiece must be flooded with a coolant to keep it from getting too warm, to act as a lubricant, and also to keep the excess shavings from building up.

There are different types of coolants that are used depending on the material that is being worked on; the most common are water-soluble chemical fluids, water-soluble oils, synthetic oils, and petroleum-based oils. In applying the coolant, care should be taken to apply the liquid directly to the location of the grind. This ensures the coolant reaches the material and is not deflected by the grind wheels high speeds. Typical workpiece materials include cast iron and minor steel.

Parts may require surface grinding for several reasons.

The following are a few of the more common reasons:

- Produce a flat surface.
- Very accurate thickness tolerance specified.
- A very smooth surface roughness R_a is specified/required.
- Cutting tool sharpening.

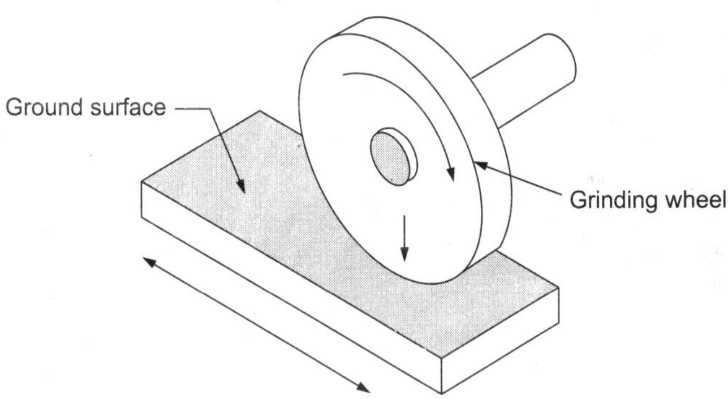

Fig. 6.16 Horizontal-spindle reciprocating table surface grinding

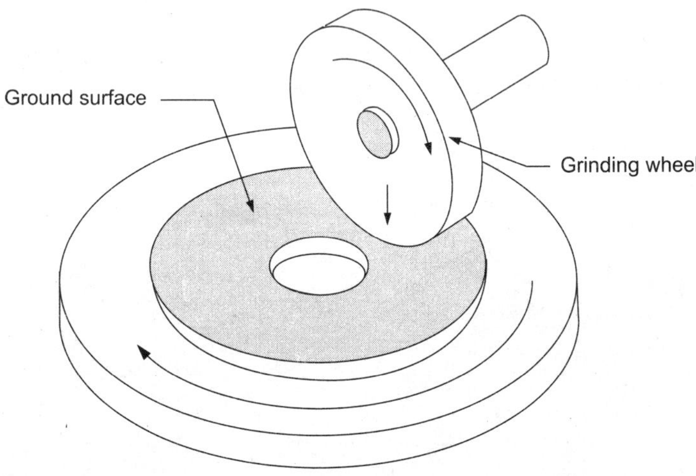

Fig. 6.17 Horizontal-spindle rotary table surface grinding

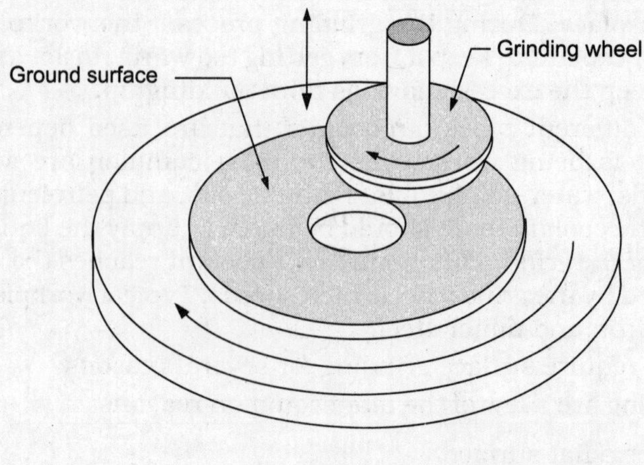

Fig. 6.18 Vertical-spindle rotary table surface grinding

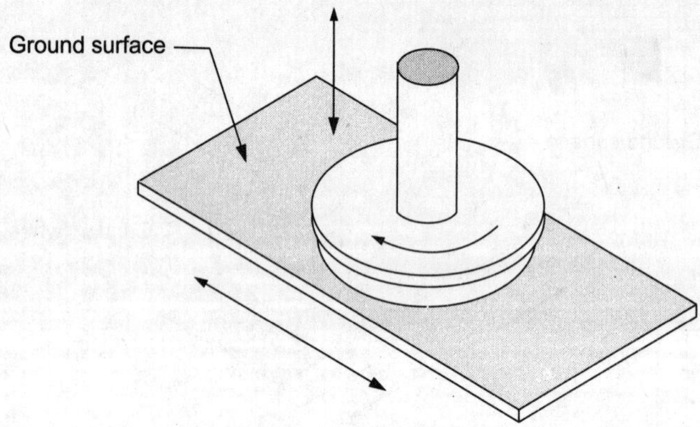

Fig. 6.19 Vertical-spindle reciprocating table surface grinding

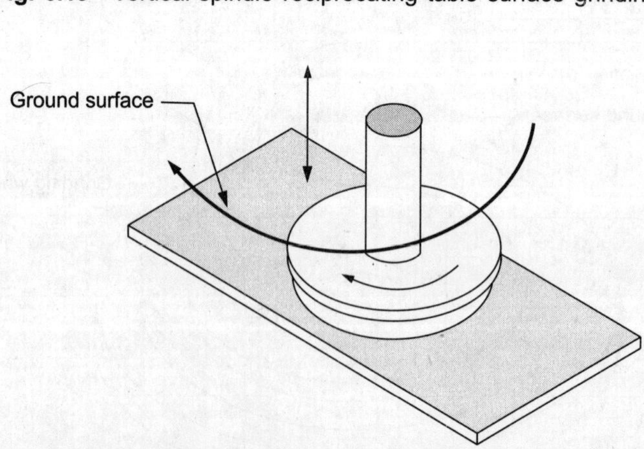

Fig. 6.20 Vertical-spindle swivel head surface grinding

6.14 CENTERLESS GRINDING

Centerless grinder is a type of outer diameter (OD) grinder that hold the piece against the wheel with a smaller wheel, known as a regulating wheel. Material is removed through grinding, similar to centered grinding except for the absence of the spindle. A large number of parts can be manufactured in a short time.

The workpiece is set up between the regulating wheel (or back up wheel) and the grinding wheel, and is supported by the work blade (or work rest). Both wheels are rotated in the same direction. The work rest is located between the wheels. The work is placed upon the work rest, and the latter,

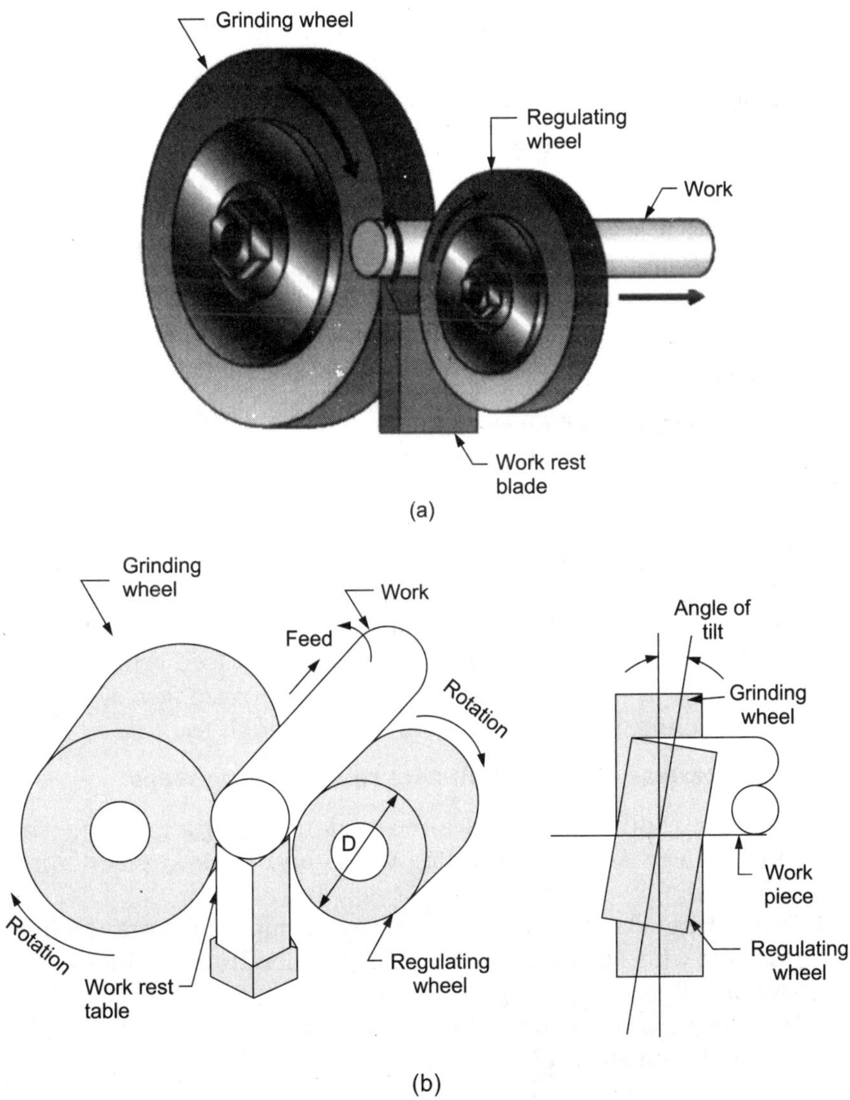

Fig. 6.21 Principle of centerless grinding

together with the regulating wheel is fed forward, forcing the works against the grinding wheel. The axial movement of the work past the grinding wheel is obtained by tilting the regulating wheel at a slight angle from horizontal.

Some of the benefits of centerless grinding include the ability to grind parts with geometries that do not allow them to be OD ground, the ability to remove three, five and other odd numbered lobing on the shaft of a part, and to maintain size beyond what is typically capable of an OD grinder due to the low overall pressures spaced out along the workpiece.

The amount of travel of the work rest the grinding wheel in a direction to its axis may be calculated by the following formula:

$$f = \pi \times D \times B \sin \theta$$

where f = amount of travel of the work rest the grinding wheel or feed in inch/min. or m/min.

D = regulating wheel diameter in mm,

N = speed of regulating wheel in r.p.m,

θ = angle of tilt in degree

angle of tilt is usually varies from 0 – 10° and speed of regulating wheel varies from 40 – 200 feet/min. In centre less grinding the stock is fed by either the through feed or in feed method.

Through Feed Method

Usually applied where workpiece being ground has no interfering shoulders. It is fed continouly between the gap of wheels and drops off the work rest on the apposite side of grinder. This process is only suitable for simple cylindrical shapes like simple shaft, piston pins, races of ball/roller bearing. Material removal in a stock varies from 0.0127 to 0508 mm. Centre less grinder is used to grinding shoulder work like gas engine inlet and exhaust value stems. In this case the work is fed in between the gap of wheel to stop and, when ground is ejected, this method is known as in feed grinding. In feed grinding is differ with through feed grinding in the way of flow of production through the machine is not continuous as in through feeding.

In feed centerless grinding consists the following steps:

(a) With gap width set to the desired work diameter regulating wheel is moved away from the grinding wheel and work is place upon the work rest.

(b) Regulating wheel is brought forward to face the work against the grinding wheel until it reaches a set position thus producing desired size of a object.

(c) Regulating wheel is moved back to its starting position and work is ejected by passing a lever.

Advantage of Centerless Grinding

(i) No required to center drilled on the ends of work hence time saving

(ii) As workpiece is not mounted on centers hence saving in loading and unloading time grinding operation is almost continuous.

(iii) Life of grinding wheel is more as metal removal stock is less in centre-less grinding process.

(iv) Production rate is very high.

(v) As no move moving parts in the centreless girding hence cost of machine is not very high.

(vi) No need of skill operator to operate the machine.

Example 6.1

Calculate the maximum uncut chip thickness during the surface grinding of 30 mm wide low carbon steel plate with a depth of cut 0.05 mm, diameter of wheel is 300 mm and rotating speed of wheel is 300 r.p.m. number of grid/mm² is 6.

Solution: Assume r_g is 20 then

$$t_{1\,max} = \sqrt{\frac{6 \times f}{\pi \times N \times D \times r_g \times C} \times \sqrt{\frac{d}{D}}}$$

given $f = 100$, $c = 6$ grid/mm^2, $D = 300$ N = 3000 r.p.m

$$= \sqrt{\frac{6 \times 100}{\pi \times 3000 \times 300 \times 20 \times 6} \times \sqrt{\frac{0.05}{300}}}$$

$$= \sqrt{0.00000177 \times 0.0129} = 0.00015 \quad \textbf{Ans.}$$

OBJECTIVE TYPE QUESTIONS

1. Grinding operation is permored to

(a) finishing (b) forming

(c) shaping (d) dressing

2. Grinding wheel is a

(a) single point cutting tool (b) multi point cutting tool

(c) both of above (d) none of these

3. By which process maximum amount of material removes

(a) super finishing (b) honing

(c) grinding (d) lapping

4. Turning of grinding wheel done by

(a) balancing the wheel (b) loading the wheel

(c) glazing the wheel (d) dressing the wheel

5. Grade of a grinding wheel i.e. strength of bond represented by A to Z where

(a) A is very hard (b) Z is very soft

(c) A is very soft (d) Z is very hard

(e) both c and d

6. Surface speed of a grinding wheel in centerless grinder is (in m/min)

(a) 50 – 60 (b) 1500 – 1000

(c) 1000 – 1500 (d) 1500 – 1800

7. In centerless grinding workpiece held by

(a) collot chuck (b) magnetic chuck

(c) universal chuck (d) none of above

8. In a grinding wheel specificate 40 A 30 H 65 B the letter H indicate here

(a) abvasive (b) grade of bond

(c) type of bond (d) structure

9. Which of the following grinding processes has highest possible speed

(a) form grinding

(b) centerless grinding

(c) rubber shellac grinding off wheel

(d) surface grinding

10. In question no (8) the letter 'B' indicate

(a) grade of bond (b) abrasive

(c) structure (d) type of bond

11. For hard material grinding type of abrasive used

(a) Fine grain (b) coarse grain

(c) medium grain (d) both (a) and (c)

12. For flat surface grinding uses

(a) surface grinding (b) form grinding

(c) thread grinding (d) cylindrical grinding

13. For low tensile material such as Aluminium, bronze, copper, brass etc. type of abrasive used

(a) corundum (b) silicon carbide

(c) Aluminium oxide (d) . garnet

(e) Boron nitride

14. Which one of the following is not used as abrasive material in grinding wheel

(a) Aluminium oxide (b) manganese oxide

(c) silicon carbide (d) cubic boron nitride *[UPCS-2009]*

15. Abrasive material used in grinding wheel selected for grinding ferrous alloys is

(a) diamond (b) boron carbide

(c) silicon carbide (d) Aluminium oxide *[GATE-2000]*

16. Hardness of a grinding wheel is determined by the

(a) hardness of the bond

(b) hardness of abrasive grains

(c) ability of the grinding wheel to penetrate the workpiece

(d) ability of the bond to retain abrasive *[GATE-2002]*

17. Soft material can not be economically ground due to

(a) frequent wheel clogging (b) low workpiece stiffness

(c) high temperature involved (d) rapid wheel wear *[UPSC-1995]*

18. Which one of the following material is used as binding material for grinding wheels?

(a) Aluminium oxide (b) Sodium silicate

(c) Silicon carbide (d) boron carbide *[UPSC-1997]*

19. Specific energy requirements in a grinding process are more than those in turning for the same metal removal rate as

(a) high heat produced during grinding

(b) high cutting velocity

(c) specific pressure between wheel and work being high

(d) larger contact area between wheel and workpiece *[UPSC-1996]*

20. Given that the peripheral speed of the grinding wheel of 100 diameter for cylindrical grinding of a steel workpiece is 30 m/sec. What is estimated rotation speed of the grinding wheel in r.p.m.

(a) 11460 (b) 95

(c) 2865 (d) 5730 *[UPSC-2009]*

21. Which amongst the following parameters influence the axial feed rate in centre less grinding 1. grinding wheel diameter 2. regulating wheel diameter 3. speed of grinding wheel 4. speed of regulating wheel 5. angle between the axes of grinding and regulating wheel

(a) 1, 4 and 5 (b) 1, 2 and 3

(c) 2, 4 and 5 (d) 3, 4 and 5 *[UPSC-1993]*

22. Structure of a grinding wheel is denoted by a number from

(a) 1 to 5 (b) 1 to 8

(c) 1 to 15 (d) 1 to 25

23. Hard grade grinding wheels are represented by latter from

(a) A – P (b) A – H

(c) H – Q (d) Q – Z

ANSWERS

1. (a)	**2.** (b)	**3.** (c)	**4.** (d)	**5.** (d)	**6.** (d)	**7.** (d)
8. (b)	**9.** (c)	**10.** (d)	**11.** (a)	**12.** (a)	**13.** (b)	**14.** (b)
15. (d)	**16.** (d)	**17.** (a)	**18.** (b)	**19.** (d)	**20.** (d)	**21.** (c)
22. (c)	**23.** (d)					

QUESTIONS

1. How are grinding wheel specified? Clearly differentiate between grade and structure of grinding wheel. *[UPTU-2002-03]*

2. Show that the chip length (L_S) is horizontal surface grinding using grinding wheel of diameter *D*, is given by *[UPTU-2002-03]*

$$L_S = \left(1 + \frac{V}{V_W}\right)\sqrt{(D \times d)}$$

where V_W = speed of grinding wheel, V = work speed and d = in feed

3. What decides the hardness of grinding wheel? Distinguish between dressing and turning of grinding wheel. *[UPTU-2002-03]*

4. What are the main difference between cylindrical and centreless grinding. *[UPTU-2002-03]*

5. What are the wear mechanism of a grinding wheel? Discuss in brief. *[UPTU-2002-03]*

6. Show that the maximum chip thickness t_m in surface grinding. Using grinding wheel of diameter *D*, is given by *[UPTU-2002-03]*

$$t_m = \frac{2V_t}{b \times C \times V_W} \times \sqrt{\frac{d}{D}}$$

7. What are the important factors which need to be considered for specifying a grinding wheel? Explain with suitable example. *[UPTU-2004-05]*

8. What are the classification method that could be used for grinding machine? Give the applications each variety of grinding machine. *[UPTU-2004-05]*

9. Discuss briefly dressing and turning of a grinding wheels. *[UPTU-2004-05]*

10. Explain three different ways in which the wear of grinding wheel can take place. What can be done to prevent them. *[UPTU-2005-06]*

11. Describe any four types of bond for bonded abrasives on a grinding wheel. *[UPTU-2005-06]*

12. How are grinding wheels specified? Take a suitable example and explain what the different terms in specification of a wheel by international standard stand for. *[UPTU-2005-06]*

13. How grinding wheels specified? Clearly differentiate between grade and structure of a grinding wheel. *[UPTU-2006-07]*

14. Explain centreless grinding. *[UPTU-2006-07]*

15. What are the various factors to be considered in selection of grinding wheel. Discuss each in detail. *[UPTU-2007-08]*

16. Explain 'Dressing' and 'Turning' of grinding wheel. *[UPTU-2008-09]*

17. What are the parameters that control the grinding process? Explain their effect on grinding process. ***[UPTU–2008-09]***

18. What are the important factors which need to be considered for specifying a grinding wheel? Explain with suitable examples. ***[UPTU–2010-11]***

19. Show that the maximum chip thickness t_m in surface grinding, using grinding wheel of diameter D, is given by t_m ***[UPTU–2010-11]***

$$t_m = \sqrt{\frac{6 \times f}{\pi \times N \times D \times r_g \times C} \times \sqrt{\frac{d}{D}}}$$

N = number of abrasive/unit area of grinding wheel surface.

b = width, d = depth of cut.

20. Define the grinding operations in detail.

21. What are the advantages, limitations and applications of a grinding wheel.

22. List of different type of bonds used in the making of grinding wheel as also mentions their applications.

❑❑❑

Chapter 7

Super Finishing Process

7.1 INTRODUCTION

Lapping and polishing are such machining process by which material is precisely removed from a workpiece (or specimen) to produce a desired dimension, surface finish, or shape. Lapping and polishing process has been applied to a wide range of materials and applications, ranging from metals, glasses, optics, semiconductors, and ceramics. Lapping and polishing techniques are beneficial due to the precision and control with which material can be removed. Surface finishes in the nanometer range can also be produced using these techniques, which makes lapping and polishing an attractive method for materials processing.

7.2 HONING PROCESS

Honing is a grinding process to finishing or rounding a hole. Honing is not a cutting/machining process. Only a finishing process by which **cylindrical hole, taper hole** and similar parts produced. Honing tool basically consist the abrasive lining with bonding material (shown in Fig. 7.2).

Honing tool slowely rotates and oscillate inside the hole. Honing operation is usually performed under an oil bath for lubrication purpose and reduce the friction between the tool and workpiece surface.

Various abrasive such as aluminium oxide **(Al_2O_3)**, **silicon carbide**, **diamond** or **cubic boron nitride** are used frequently. For cast iron and non-ferrous metal silicon carbide is used. Where as Aluminium oxide (Al_2O_3) is used for hone the steel parts. If material is very hard like ceramic or hard carbide than diamond honing tool is used and size of abrasive grain is usually from 80–600 grit. Mostly all types of material can be horned by honing process and most important application of honing operation is to hone the walls of a cylinder of an engine. Honing tool rotate at the speed of 15–40 m/min but **main draw back** of honing process that only very little quantity of material can be removed (0.5 mm for prime honing and 0.01 mm for secondary honing). Honing process may be performed by manually or by some machine.

Honing process is differ with grinding operation in manner of speed, pressure applied on tool. In honing process low speed and low pressure is applied which result the lowering the temperature where as high speed and high pressure is required in grinding operation, abrasive stones used in

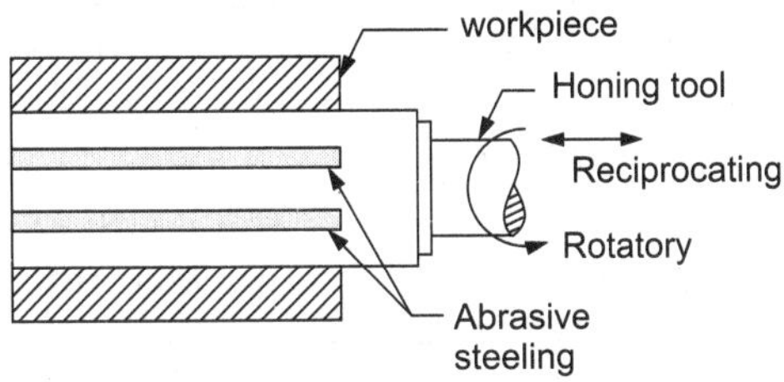

Fig. 7.1 Working principal of honing process

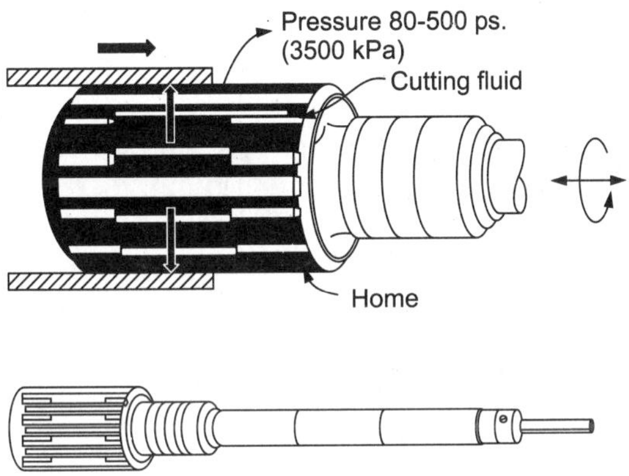

Hones

Fig. 7.2 Honing using a mandrel with honing sticks on an expanding core

honing tool cover large area of abrasive in contact with workpiece as low temperature and processes is required for honing operation hence less change of surface damage. Surface finish up to 0.1 μm can be achieved by honing process. Upto some extant honing can correct some out of roundness, taper and axial distortion of a cylindrical hole. For holding the workpiece fixture is required.

Honing Tool: Various types of honing tools are used and it basically depends upon the type of surface want to produced. Honing tool consist the four parts body, cone, conerod, and drive shaft body is used to support the honing sticks.

7.2.1 Honing Machine

Honing machines are basically two types:

(i) horizontal honing machine
(ii) vertical honing machine

7.2.1.1 Horizontal honing machine

Mostly horizontal honing machine is used to hone a very large (longer) jobs like barrel of a gun, tubes etc. This machine consist a horizontal spindle, in which the honing tool is mounted, workpiece is kept on table, hone reciprocates about its own axis and simulteously rotate too.

7.2.1.2 *Vertical honing machine*

Vertical honing machine is used to hone the small size of parts. Vertical honing machine is some what like the Drilling Machine. In this tool hold in tool holder vertically and reciprocating no motion of workpiece in vertical honing machine. Coolant is also used in vertically honing process to cool the tool as well as to carry away the chips in the form of fine particles.

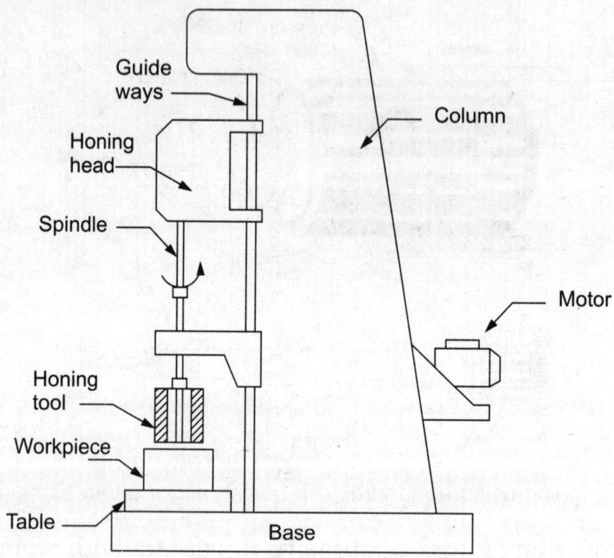

Fig. 7.3 Vertical honing machine

7.2.2 Advantages of Honing

(i) high degree of surface finish can be achived.

(ii) numbers of holes can be honed at a time by multi spindle machine.

(iii) circularity of surface upto 1.25 μm can be achieved.

(iv) as compared to any other hole finished process, high productivity at low cost.

(v) minimum heat generation and distortion of workpiece.

(vi) distribution of cutting points over a large contact area reduces heat generation during cutting.

7.2.3 Limitations of Honing

(i) Lack of straightness can not be 100% remove.

(ii) it is difficult to hone tough non-ferrous metals due to glazing or clogging of the cores of the abrasive sticks.

7.3 LAPPING

Lapping is a metal removal process to produce a **smooth, flat, unpolished** surface by the use some abrasive particles. In lapping metal removal take place in form of fine particles and process is suitable to produced dimensionally accurate parts with high degree of finish (up to 2.5 μm).

Process involves:

(i) use the loose abrasive in between the lap and workpiece
(ii) Lap and workpiece are not driven only guided in contact with each other
(iii) Relative motion between the lap and workpiece is not fixed it is changable.

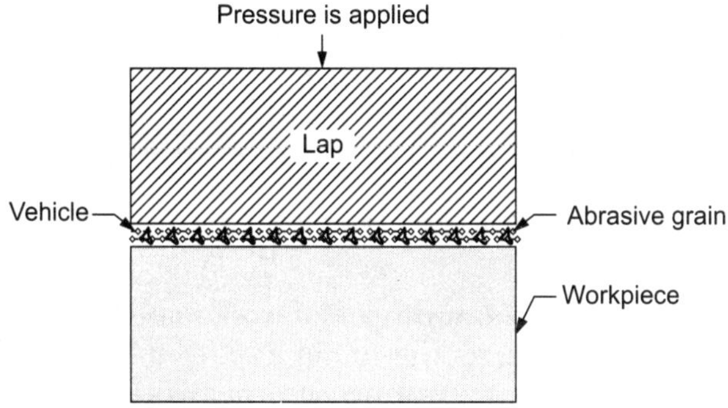

Fig. 7.4 Working principle of lapping

Lapping is performed on a **flat, spherical, cylindrical** or some special surface in contact with lap. Process is capable to removes sub surface damage caused by sawing or grinding to produce required flatness and thickness of work.

In lapping process, abrasive is applied by the means of lapping shoes or quils to the workpiece. Now lap is charge with some quantity of abrasive by spreading it over a hardend steel plate and rolling the lap into the abrasive. Sharp and fine abrasive particles becomes embeded in the lap and now it is ready to application (use). Sometimes abrasive is available in the form of bonded wheel and lapping process is same as centre less grinding and vertical spindle surface grinding.

Lapping operation can be performed in two way by **hand** or by **machine**. If lap is softer than the workpiece **silicon carbide** (SiC), **boron carbide, Aluminium oxide** (Al_2O_3), in fine size used. In lapping carbide, tools and jewels, diamond particles permanently embedded in copper laps are most successful.

7.3.1 Lapping Speed

Type of accuracy	Roughness (Rain μ)	Lapping speed (m/min)
(i) for medium accuracy	<10	200–400
(ii) for good accuracy	$10 < R_a < 12$	100–250
(iii) for very fine accuracy	$12 < R_a < 14$	10–100

7.3.2 Method of Lapping

(i) hand lapping
(ii) machine lapping

7.3.2.1 *Hand Lapping*

Hand lapping process can be performed by two ways:
(a) flat work hand lapping (b) external cylindrical hand lapping. (Ring Lapping).

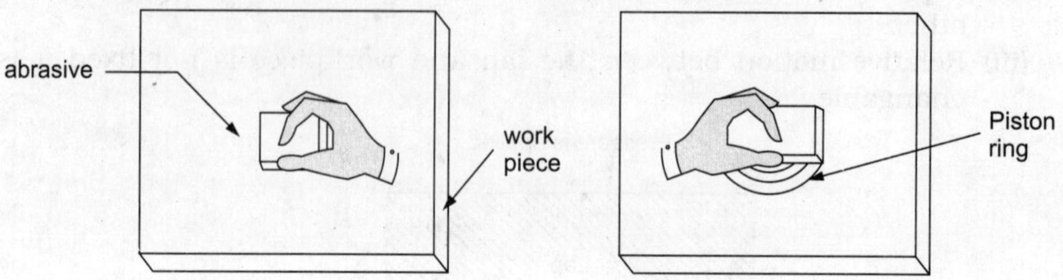

Fig. 7.5 Hand lapping

7.3.2.1.1 Flat Work Hand Lapping Flat work hand lapping is also known as flat surface hand lapping as the name implies it is used to lap the flat surfaces by hand, in hand lapping process either workpiece in motion or lap is in motion and rubbing of two surfaces in contact with each other. This process basically applied for lapping the dies, metal mould for casting etc. In some cases a lapping compound is placed in between the two surfaces and they are moved again each other.

7.3.2.1.2 External Cylindrical hand Lapping It is also known as ring lapping and used to finishing the external surface of cylindrical jobs. It is made out of cast iron of soft close grain size. Ring lap is slightly shorter than the workpiece. Ring lap is reciprocated over the external surface of workpiece abrasive and vehicle are fed through the slot.

7.3.2.2 *Machine Lapping*

Same method of lapping done by some machines. They are the following:

(i) centre less lapping machine
(ii) flat machine lapping
(iii) centre less roll lapping machine
(iv) spherical lapping machine
(v) external cylindrical lapping machine.

7.3.2.2.1 External Surface Cylindrical Lapping Machines: Upto some extant it is like a drill machine which consists a vertical spindle with carrying one upper stationary lap and lower rotating lap. Upper lap is free to float and rest on the machine which rides on the face of lower lap. Now pressure is applied by hand wheel.

7.3.3 Advantages of Lapping Process

(i) Small or thin parts can be easily lapped.

(ii) Thermal disfortion in workpiece are virtually non-existent.

(iii) Very little change of clamping distortion in the workpiece.

(iv) Brittle parts are easy to run because of gentle grinding action.

(v) Mostly all kinds of material including non-magnetic material can be handled.

7.3.4 Limitations of Lapping

(i) Metal removal rate is very low.

(ii) Process is slow as metal removal rate is low.

(iii) Process is not suitable for batch production.

7.3.5 Applications of Lapping

(i) To lapping of piston rings.

(ii) Lapping of cylindrical jobs from both external/internal sides.

(iii) Lapping of piston pins.

(iv) To lapping of valves, gears, rollers, bearing, thrust washers, and optional parts.

7.4 POLISHING

Polishing is the removal of materials to produce a **scratch-free, specular surface** using fine (<3 μm) abrasive particles. Polishing typically done at very low speeds using either polishing **cloths, abrasive films,** or **specially** designed lapping plates. Polishing with a cloth or lapping plate requires the use of free abrasive, and is a very low damage process when performed properly. Plate material and cloth material are critical when polishing a particular sample as the properties of these substrates are important in the final polish quality of the specimen. Polishing with a lapping plate is a common process used in the case of metals and hard ceramic type materials. Polishing using copper composite plates or tin/lead. Lapping plates can produce high quality surface finishes with high removal rates. In many cases the use of a polishing cloth is required, and thus the selection of a proper polishing cloth is important. Polishing cloth properties needed depend on the application. If flatness is of primary concern, short nap cloths (such as **Nylon**) are used to maintain flatness. When the final surface finish is of primary concern, longer napped cloths (such as **Rayon** and **Silk**) are used. Many cloth materials today combine the best of both worlds, allowing flatness and surface finish combined to provide maximum performance. Polyurethane pads are commonly used for final polishing processes and produce excellent flatness and surface finish. Aluminium oxides and silicon carbide of various grid size are used as abrasive.

Polishing with abrasive films also produces excellent results. The flatness of the films combined with high removal rates makes them an attractive alternative to cloth and plate polishing methods.

QUESTIONS

1. What do you understand about any two of the following:
   (a) Lapping, (b) Reaming, (c) Down milling, *[UPTU-2002-03]*
2. Distinguish between honing and lapping.
3. Explain the honing process in detail. *[UPTU-2005-06]*
4. Explain super finishing process.
5. Explain any two of the following finishing process:
   (i) Lapping, (ii) Buffing, (iii) Electro-polishing.
6. With the help of a heat sketch explain the process of honing. What are the limitation of this process?
7. Define polishing operation in detail.

❑❑❑

Limits, Fits Tolerance and Surface Roughness

Chapter **8**

8.1 INTRODUCTION

This toll includes the first group and therefore can be used to determine dimensional tolerances and deviations of machine parts.

As mentioned above, it is principally impossible to produce machine parts with absolute dimensional accuracy. In fact, it is not necessary or useful. It is quite sufficient that the actual dimension of the part is found between two limit dimensions and a permissible deviation is kept with production to ensure correct functioning of engineering products. The required level of accuracy of production of the given part is then given by the dimensional tolerance which is prescribed in the drawing. The production accuracy is prescribed with regards to the functionality of the product and to the economy of production as well.

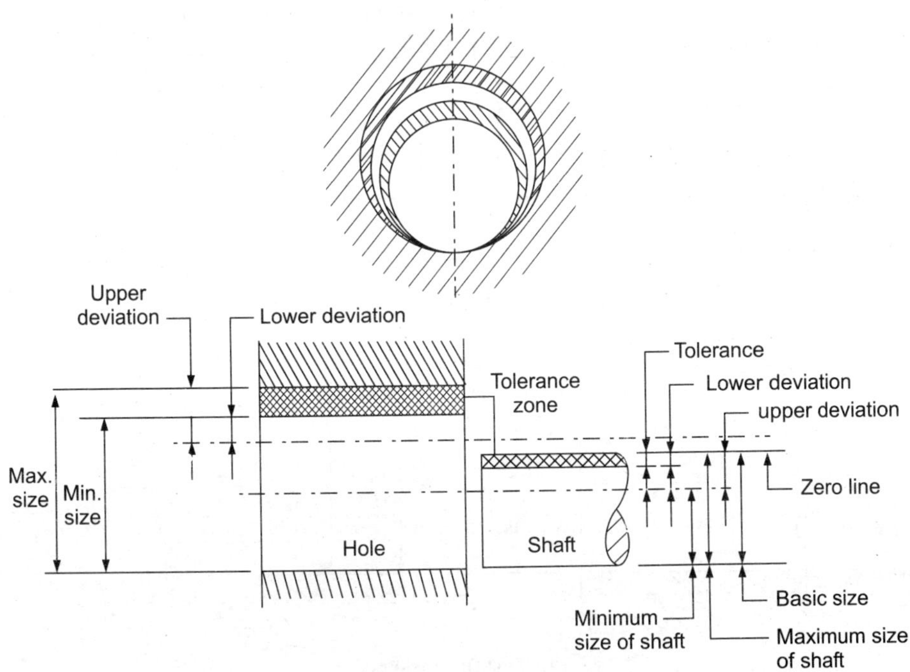

Fig. 8.1 Tolerance zone

A coupling of two parts creates a fit whose functional character is determined by differences of their dimensions before their coupling.

where: $d = D$ ... basic size

D_{max}, D_{min} ... limits of size for the hole.

d_{max}, d_{min} ... limits of size for the shaft.

ES ... hole upper deviation.

EI ... hole lower deviation.

es ... shaft upper deviation.

ei ... shaft lower deviation.

8.2 FITS

Depending on the mutual position of tolerance zones of the coupled parts.

Three types of fit can be distinguished:

A. Clearance fit B. Transition fit C. Interference fit

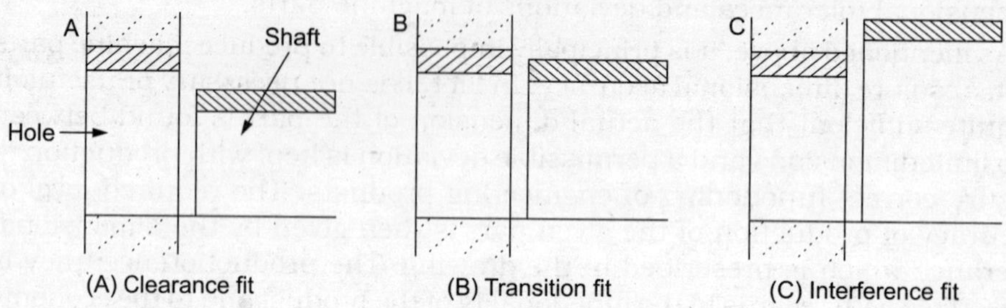

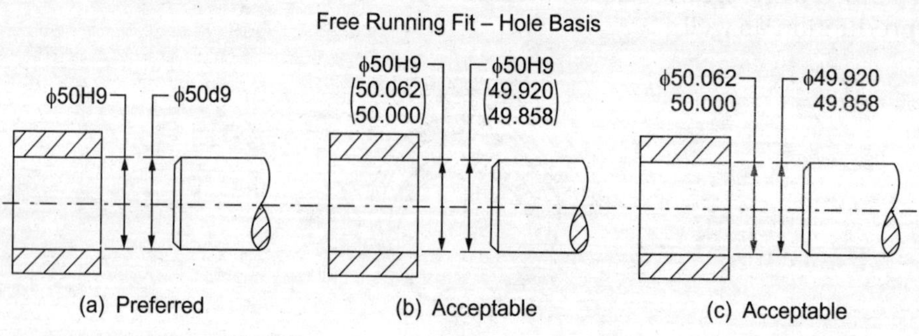

Fig. 8.2

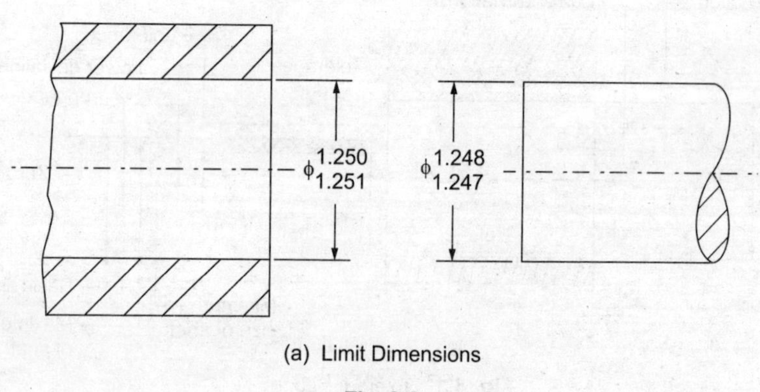

Fig. 8.3

8.2.1 Terminology for Limits and Fits

1. **Tolerance** is the total amount that a specific dimension is permitted to vary;

   It is the difference between the maximum and the minimum limits for the dimension.

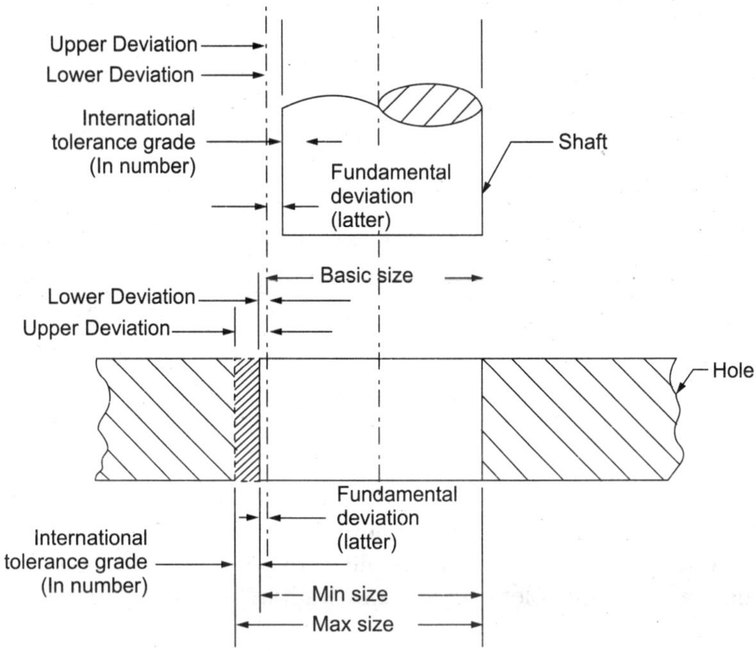

Fig. 8.3 (b)

   For example a dimension given as 1.625 ± 0.002 means that the manufactured part may be 1.627" or 1.623", or anywhere between these limit dimensions.

2. **Nominal size:** It is the designation used for general identification and is usually expressed in common fractions. For example, in the previous figure, the nominal size of both hole and shaft, which is 11/4" would be 1.25" in a decimal system of dimensioning.
3. **Basic size or basic dimension:** It is the theoretical size from which limits of size are derived by the application of allowances and tolerances.
4. **Actual size:** is the measured size of the finished part.
5. **Allowance:** is the minimum clearance space (or maximum interference) intended between the maximum material condition of mating parts.
6. **Upper deviation:** It is the algebraic difference between the maximum Limit and basic size. It is denoted by *ES* for a hole and *es* for a shaft. Upper deviation is negative when maximum Limit of size is less than basic size and positive when maximum limit of size is greater than basic size.

7. **Lower deviation:** It is the algebraic difference between the minimum limit and basic size. It is denoted by *EI* for hole and *ei* for a shaft and its is positive when minimum limit of size is more than the basic size and negative when the minimum limit of size is less than the basic size.
8. **Mean deviation:** It is the algebraic difference between the upper deviation and lower deviation.
9. **Tolerance zone:** The zone (Area) between the maximum and minimum limit size is known as tolerance zone.

8.2.2 Classification of Fits

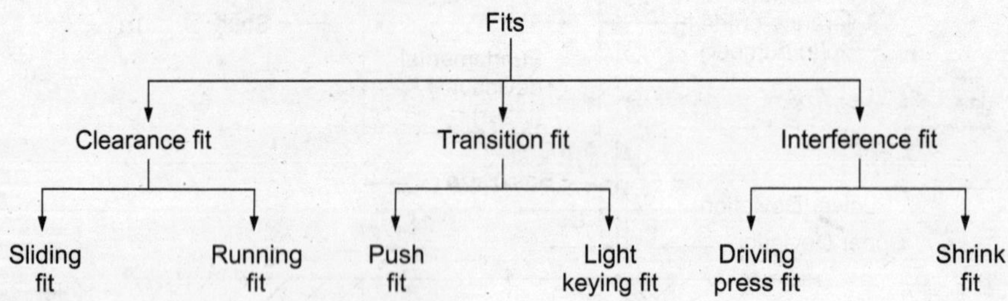

Fig. 8.4 Block diagram of types of fits

8.2.3 Fit Between Mating Parts

- Fit is the general term used to signify the range of tightness or looseness that may result from the application of a specific combination of allowances and tolerances in mating parts.

There are three Types of Fits Between Parts

1. **Clearance Fit:** An internal member fits in an external member (as a shaft in a hole) and always leaves a space or clearance between the parts.

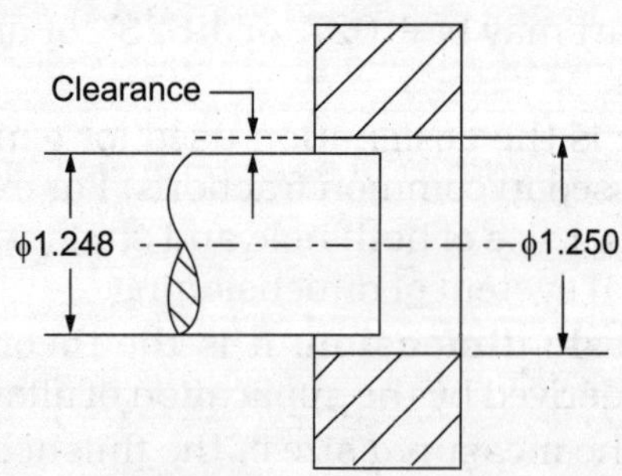

Fig. 8.5 Clearance fit

2. **Transition Fit:** may result in either a clearance or interference condition. In the Fig. 8.6, the smallest shaft 1.2503" will fit in the largest hole 1.2506", with 0.003" to spare. But the largest shaft, 1.2509" will have to be forced into the smallest hole, 1.2500" with an interference of metal of 0.009".

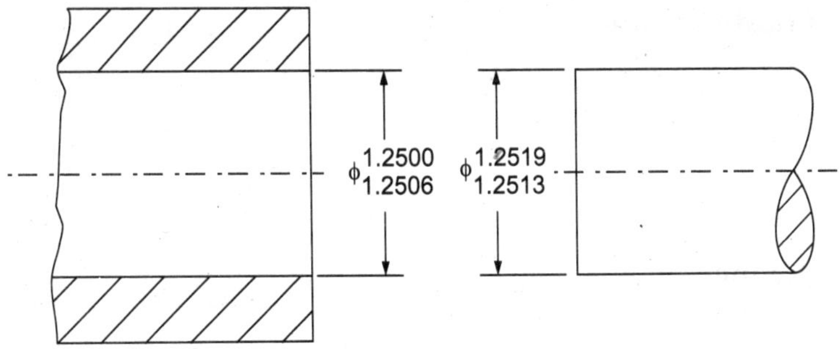

Fig. 8.6 (a) Transition fit

3. **Interference Fit:** The internal member is larger than the external member such that there is always an actual interference of material. The smallest shaft is 1.2513" and the largest hole is 1.2506", so that there is an actual interference of metal amounting to at least 0.0007". Under maximum material conditions the interference would be 0.0019". This interference is the allowance, and in an interference fit it is always negative.

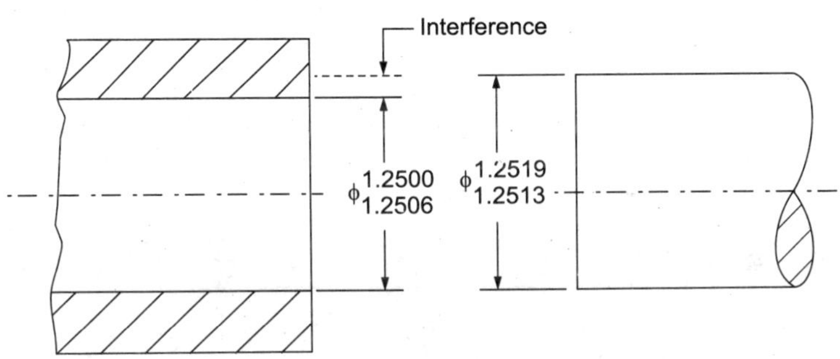

Fig. 8.6 (b) Interference fit

8.3 BASIC HOLE SYSTEM

- Minimum hole is taken as the basic size, an allowance is assigned, and tolerances are applied on both sides of and away from this allowance.

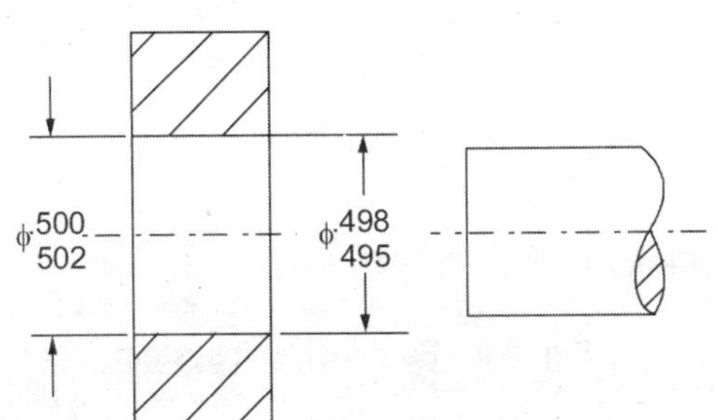

Fig. 8.7 Basis hole fit

Basis of Limits System:

(i) Hole basis system,
(ii) Shaft basis system

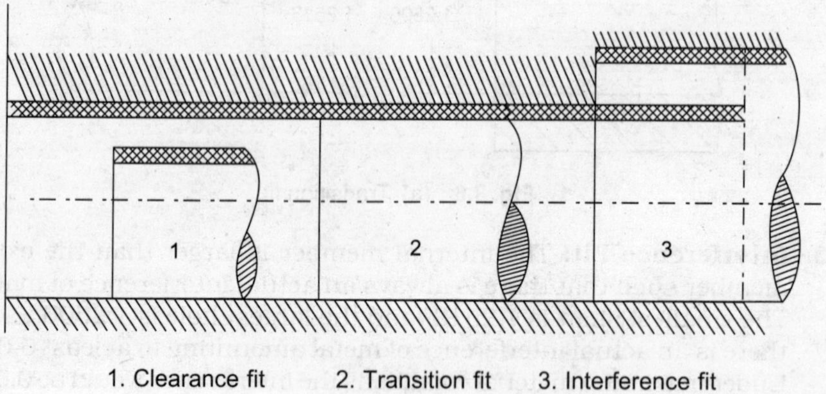

Fig. 8.8 (a) Hole basis system

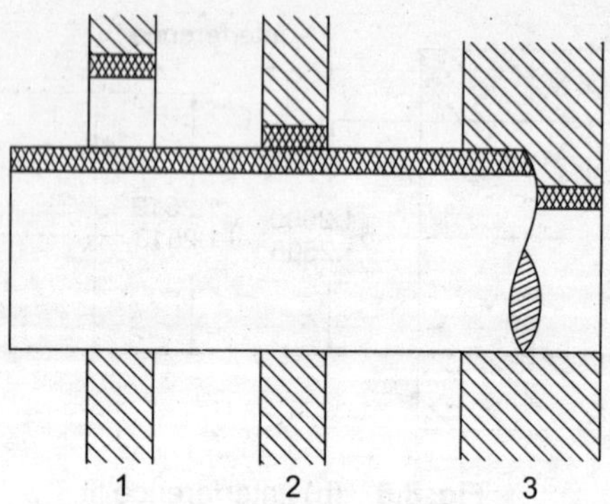

Fig. 8.8 (b) Shaft basis system

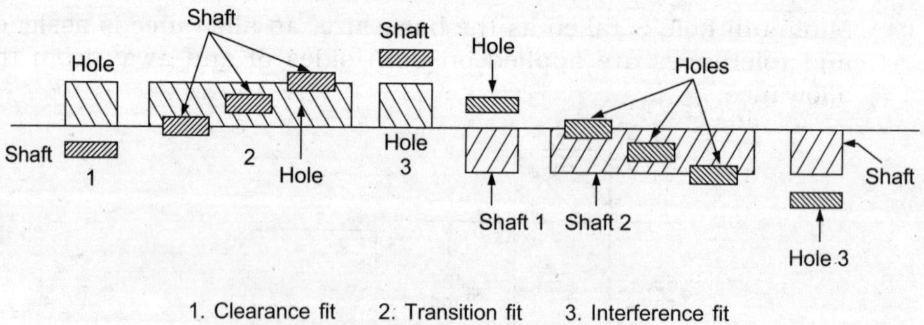

Fig. 8.9 Bases of limit system.

Indian Standard System for Limit and Fits: As per Bureau of Indian Standard (ISI) IS:919(PT-I)–1993, Limit, fits system comprises 18 grade if fundamental tolerance represented by letter symbol for both holes and shaft

(*A* to *ZC* for holes and *a* to *zc* for shaft) in diameter steps from 1 to 500 mm. A unilateral hole basics system is recommended but a unilateral or bilateral shaft basics system is also adopted if required. A 18 grade tolerance are referred as IT01, ITO and IT1 to IT16. It is known as standard is calculated by

Tolerance in microns is calculated by

$$1(\text{microns}) = 0.45 \times \sqrt[3]{D} + 0.001\,D$$

where, D = Mean diameter in mm.

value of standard tolerance corresponding to grade IT01, ITO, and IT1 is given by

$$\text{IT01},\ i = 0.3 + 0.008\,D \quad (\text{Micros})$$

$$\text{ITO},\ i = 0.5 + 0.012\,D \quad (\text{Micros})$$

$$\text{IT1},\ i = 0.8 + 0.020\,D \quad (\text{Micros})$$

The fundamental deviations for holes and shafts as shown in Fig. 8.10. It is clear that for shafts '*a*' to '*h*' be appear deviation is below the zero line and it is prove the zero line for shafts "*j*" to *zc*. Similarly, the linear deviation for holes 'A' to '*H*' is above the zero and it is below the zero line for holes '*j*' to '*ZC*'.

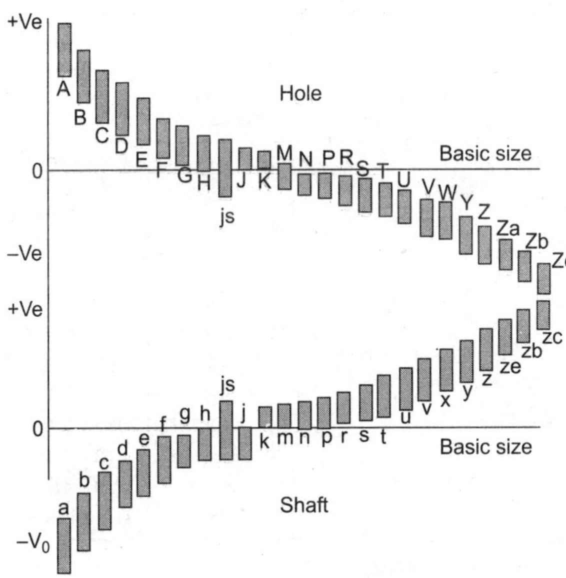

Fig. 8.10 Fundamental deviations

The shaft '*h*' for which the upper deviation is zero the "Basic Shaft". Similarly the hole "*H*" for which the lower deviation is zero is called the "Basic size". For shafts, the upper deviation is denoted by and the lower deviation by '*ei*'.

The corresponding deviations for holes are denoted by '*ES*' and '*EI*' respectively.

The fundamental deviations can be calculated with of formula given in the specifications. The deviations can be evaluated directly by using the value of the tolerance IT. Thus, for shafts,

$$ei = es - IT$$

$$es = ei + IT.$$

Similarly, for holes $$ES = EI + IT$$

The values of the deviations for the shafts are given in Table 8.1, for sizes upto 500 mm.

TABLE 8.1

Upper Deviation (es) *Lower Deviation (ei)*

Shaft Designation	**es, micronss**	**Shaft Designation**	**ei., Microns**
a	$= -(256 + 1.3D)$; $D \leq 120$	*j*5 to j8	
	$= - 3.5D$, $D \geq 120$	*k*4 to *k*7	$+0.6 \sqrt[3]{D}$
b	$= -(140 + 0.85D)$; $D \leq 160$	*k* for grades	
	$= -1.8\ D$; $D \geq 160 \leq 3$ and ≥ 8	0	
c	$= -52\ D^{0.2}$; $D \leq 40$	*m*	+(IT 7 – IT6)
	$= -(95 + 0.8D)$; $D > 40$	*n*	$+5D^{0.34}$
d	$-16D^{0.44}$	*p*	+ IT7 + 0 to 5

Upper Deviation (es) *Lower Deviation (ei)*

Shaft Designation	**es, micronss**	**Shaft Designation**	**ei., Microns**
e	$-11D^{0.41}$	*r*	geometric mean of values of *ei* for *p* and *s*.
f	$-5.5D^{0.41}$	s	+IT8 + 1 to 4; $\Delta \leq 50$
g	$-2.5D^{0.34}$		+IT7 + 0.4*D*; $D > 50$
h	0	*t*	+ IT7 + 0.63 *D*
		u	+IT7 + *D*
		v	+IT7 + 1.25*D*
		x	+IT7 + 1.6*D*
		y	+IT7 + 2*D*
		z	+IT7 + 2.5*D*
		z_a	+IT8 + 3.15*D*
		z_b	+IT9 + 4*D*
		z_c	+IT10 + 5*D*

EI = Upper deviation of shaft

ES = Lower deviation of shaft.

Example 8.1

According to basic hole system the dimensions of mating part is given below:

Hole:– **30.00** **Shaft: 29.98**

30.04 **29.95**

Calculated the tolerance for hole, for shaft and allowance.

Solution: Lower limit of hole – 30.0 mm, upper limit of hole – 30.04 mm
upper limit of shaft – 29.98, lower limit of shaft – 29.95

(a) Tolerance for hole = upper limit of hole – lower limit of hole
$= 30.04 - 30.00 =$ **0.04 mm Ans.**

(b) Tolerance for shaft = upper limit of shaft – Lower limit of shaft
$= 29.98 - 29.95 =$ **0.03 mm Ans.**

(c) Allowance = Lower limit of hole – upper limit of shaft
$= 30.00 - 29.98 =$ **0.02 mm Ans.**

Example 8.2

Calculate the tolerance, fundamental deviation and limits of size for the shaft designated as 40 H_8/f_7.

Solution: Designing the 40 H8/f7 means that 40 mm is the basic size and 8 is the tolerance grade for hole

Hole – 8 (IT8), Shaft – 7(IT7)

Step diameter 30 to 50 mm for 40 mm. Now mean diameter

$$D = \sqrt{30 \times 50} = 38.73 \text{ mm}$$

By formula standard tolerance

$$i = 0.45\sqrt[3]{D} + 0.001D$$

$$i = 0.45\sqrt[3]{38.73} + 0.01 \times 38.73 = 1.55973$$

$$i = 1.56 \text{ microns} \quad \text{or} \quad i = 1.56 \times 10^{-3} \text{ mm.}$$

from table we find the standard tolerance for hole grade 8

$= 25i = 25 \times 1.56 \times 10^{-3} =$ **0.039 mm. Ans.**

Similarly standard tolerance for shaft grade 7

$= 16i = 16 \times 1.56 \times 10^{-3} =$ **0.025 mm Ans.**

Fundamental Deviation

Fundamental deviation for hole *H*, (Lower deviation)

$$EI = 0$$

from table, we find that fundamental deviation (upper) for shaft

$$es = -5.5(D)^{.41}$$

$$= -5.5\,(38.73)^{.41} = -24.63 \quad \text{or} \quad -25\mu$$

$es = -25 \times 10^{-3} =$ **−0.025 mm Ans.**

Fundamental Deviation (Lower) for Shaft *f*

$el = es - IT = -0.025 - 0.025 =$ **−0.05 mm Ans.**

Limit Size

Lower limit for hole = Basic size = **40 mm Ans.**

Upper limit for hole = Lower limit for hole + Tolerance for hole

= 40 + 0.039 = **40.039 mm Ans.**

Upper limit for shaft = Lower limit for hole – Fundamental (upper) deviation

= 40 – 0.025 = **39.975 mm Ans.**

Lower limit for shaft = Upper limit for shaft – Tolerance for shaft

= 39.975 – 0.025 = **39.95 mm Ans.**

Example 8.3

A journal bearing with basic size 80 mm is running with close running calculate limit for shaft and bearing and maximum and minimum clearance.

Solution: Basic size of normal size = 80 mm.

From table for close running fit, represented by H8/g7, i.e., shaft g7 should be used with H8 hole.

Since 80 mm lie in 50 to 80 mm step diameter, hence mean diameter

$$D = \sqrt{50 \times 80} = \sqrt{4000} = 63 \text{ mm}$$

Standard tolerance unit $i = 0.45 \times \sqrt[3]{D} + .001\,D$

$$= 0.45 \times \sqrt[3]{63} + .001 \times 63$$

$$= 1.79 + 0.063 = 1.853\mu$$

$$= 1.853 \times 10^{-3} \text{ mm}$$

Standard tolerance for hole H of grade 8

$$= 25i = 25 \times 1.853 \times 10^{-3} = 0.046 \text{ mm}$$

Standard tolerance for shaft 'g' of grade 7

$$= 16i = 16 \times 1.853 \times 10^{-3} = 0.03 \text{ mm}$$

From table,

upper deviation for shaft g

$$es = -2.5 \times (D)^{.34} = -2.5\,(63)^{.34}$$

$$= -10\,\mu = -10 \times 10^{-3} \text{ mm}$$

Lower deviation for shaft g

$$ei = es - IT = -0.01 - 0.03 = -0.04 \text{ mm}$$

Lower limit for hole = Basic size = 80 mm

Upper limit for hole = Lower limit for hole + Tolerance for hole

= 80 + 0.046 = 80.046 mm

Upper limit for shaft = Lower limit for hole – Upper deviation for shaft

= 80 – 0.01 = 79.99 mm

Lower limit for shaft = Upper limit for shaft – Tolerance for shaft

= 79.99 – 0.03 = 79.96

Maximum clearance = Upper limit for hole – Lower limit for shaft

= 80.046 – 79.96 = 0.086

Minimum clearance = 80 – 9.99 = **0.01 mm Ans.**

8.4 SURFACE ROUGHNESS OR SURFACE TEXTURE

Surface roughness or texture is the measure if the finer surface irregularities in the surface texture and is composed of three components: roughness, waviness and form these are the result of the manufacturing process employed to create the surface.

Surface roughness average (R_a), also know as arithmetic average (AA) is rated as the arithmetic average deviation of the surface valleys and peaks expressed in micro inches or micro meters. ISO standards use the term CLA (Center Line Average). Both are interpreted identical.

where R_a is the arithmetic average value or departure from profile front and center line, the equation for four as-measured values:

$$R_a = CLA = AA = \frac{M_1 + M_2 + M_3 + M_4}{4}$$

where M_x = measure value

Root Mean Square (RMS/R_q/R_s) can be calculated by:

$$\text{RMS} = \sqrt{\frac{M_1^2 + M_2^2 + M_3^2 + M_4^2}{4}}$$

The ability of a manufacturing operation to produce a specific surface roughness depends on many factors. For example, in end mill cutting, the final surface depends on the rotational speed of the end mill cutter, the velocity of the traverse, the rate of feed, the amount and type of lubrication at the point of cutting, and the mechanical properties of the piece being machined. A small change in any of the above factors can have a significant effect on the surface produced

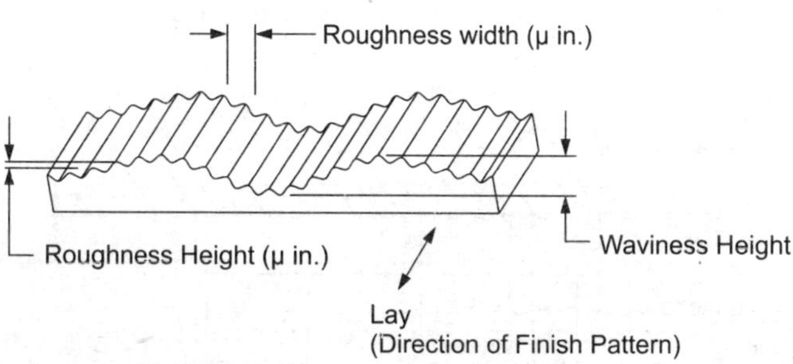

Fig. 8.11 Surface characteristic [30]

8.4.1 Specification

In the United States, surface finish is usually specified using the ASME Y14.36M standard. The other common standard is International Organization for Standardization (ISO) 1302.

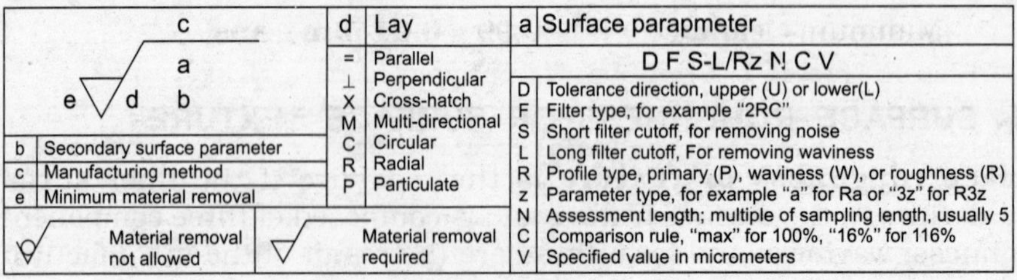

Fig. 8.12

Surface roughness, often shortened to **roughness**, is a measure of the texture of a surface. It is quantified by the vertical deviations of a real surface from its ideal form. If these deviations are large, the surface is rough; if they are small the surface is smooth. Roughness is typically considered to be the high frequency, short wavelength component of a measured surface (see surface metrology).

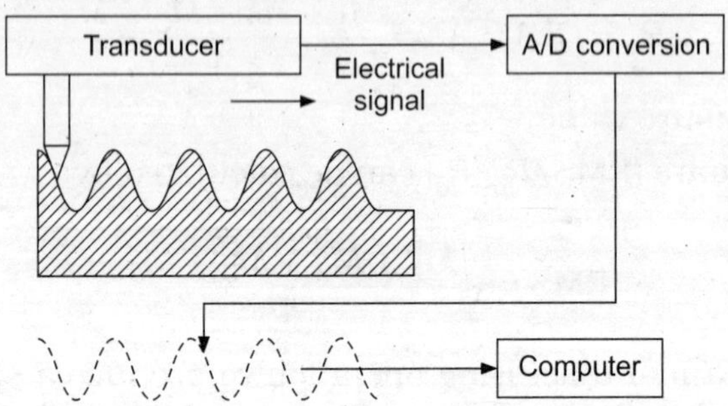

Fig. 8.13 Principle of surface roughness measurement

Roughness plays an important role in determining how a real object will interact with its environment. Rough surfaces usually wear more quickly and have higher friction coefficients than smooth surfaces (see tribology).

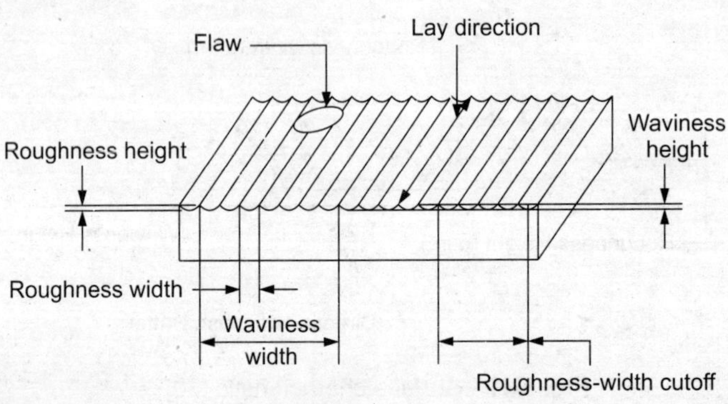

Fig. 8.14(a) Surface characteristic (Courtesy, ANSI B46.1-162) [30]

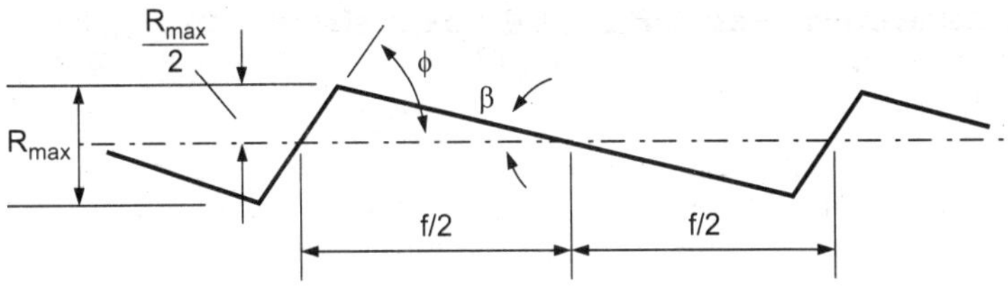

Fig. 8.14(b) Idealized model of surface roughness [30]

Roughness is often a good predictor of the performance of a mechanical component, since irregularities in the surface may form nucleation sites for cracks or corrosion.

Although roughness is usually undesirable, it is difficult and expensive to control in manufacturing. Decreasing the roughness of a surface will usually increase exponentially its manufacturing costs. This often results in a trade-off between the manufacturing cost of a component and its performance in application.

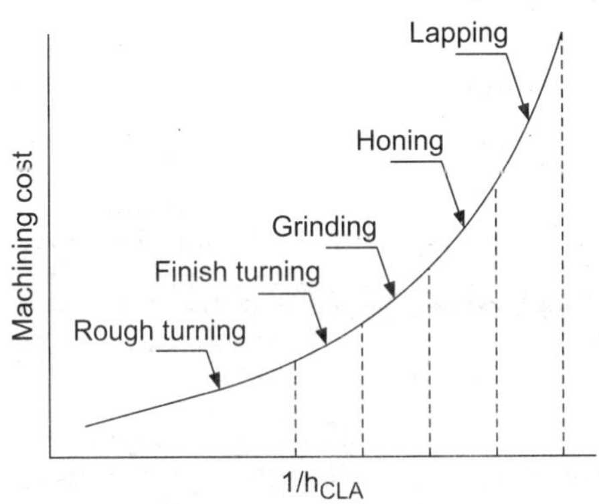

Fig. 8.15 Variation of machining cost with surface roughness [8]

OBJECTIVE TYPE QUESTIONS

1. Algebraic difference between maximum limit and basic size is known as
(*a*) Upper deviation (*b*) Fundamental deviation,
(*c*) Actual deviation, (*d*) Lower deviation.

2. In an interchangeable system of manufacturing
(*a*) components assemble together at random,
(*b*) highly skilled labour is required to perform the operation,
(*c*) Individual fitting take place,
(*d*) Components produced on traditional machine.

3. A basic shaft is one whose
(*a*) Upper deviation is zero,
(*b*) Lower deviation is zero
(*c*) Upper & lower deviation is zero
(*d*) None of above.

4. A dimension is expressed as $60^{+0.05}_{-0.02}$ the basic size is
(*a*) 60.15 mm (*b*) 60 mm
(*c*) 59.48 mm (*d*) 60.03 mm

5. Tolerance are specified on component
(*a*) to facilitate inspection (*b*) Not possible to produce exact size
(*c*) to have proper allowance (*d*) to produce desired fit.

6. A basic hole is one whose
(*a*) upper deviation is zero (*b*) Lower deviation is zero
(*c*) upper and lower deviation is zero
(*d*) none of above.

7. Surface roughness on a drawing is represented by
(*a*) dots (*b*) square
(*c*) circles (*d*) triangles.

8. As per Bureau of Indian standard 100 H6/g5 mean that
(*a*) basic size is 100 mm (*b*) Actual size is 100 mm
(*c*) difference between the basic size of actual size is 100 mm
(*d*) none of above.

9. According to the Indian standards total number of tolerance grade are
(*a*) 12 (*b*) 16
(*c*) 18 (*d*) 20

10. According to Indian standards specification 80 H6/g5 means
(*a*) tolerance grade for shaft is 6 hole is 5
(*b*) tolerance grade for hole is 6 and shaft is 5
(*c*) tolerance grade for shaft is 4 to 8 and for hole is 3 to 7
(*d*) tolerance grade for shaft is 4 to 8 and for shaft is 3 to 7.

11. System of expressing the size $30.5^{-0.05}$ mm is refer as
(*a*) universal tolerance system (*b*) unilateral tolerance system
(*c*) bilateral tolerance system (*d*) limiting dimension system.

12. System of expression the size $20^{+0.00}_{-0.004}$ mm is known as
(*a*) limiting dimension system (*b*) bilateral dimension system
(*c*) unilateral system (*d*) universal dimension system.

13. Interchangeable assembly, shaft of size $25.000^{+0.040}_{-0.010}$ mm mate with hole of size $25.000^{+0.030}_{+0.020}$ mm. The maximum interference (in microns) in the assembly is
(*a*) 40 (*b*) 30
(*c*) 20 (*d*) 10 ***[GATE-2012]***

ANSWERS

1. (*a*), **2.** (*a*), **3.** (*a*), **4.** (*b*), **5.** (*d*), **6.** (*b*), **7.** (*d*), **8.** (*a*), **9.** (*c*), **10.** (*b*), **11.** (*c*), **12.** (*c*).

QUESTIONS

1. Define the following terms limit, fit, tolerance with reference to the dimension measurement.
2. What is fit? Explain different types of fits. ***[UPTU-2007-08]***
3. Define the following terms in limit and fits: Basic size, normal size, tolerance, upper deviation. Lower deviation standard size, Allowance.

4. Explain the concept of transition, clearance, and interference fit with examples.
5. What do you understand by the terms hole basis and shaft basies in terms of assembly fit specifications? Which is preferred and why?
6. White the short notes on concept of inter changeability and what is its importance in engineering?
7. How are holes, shaft and fits designated?
8. What is zero line?
9. Differentiate between the tolerance and allowance.
10. Define the following terms:
    (i) force fit, (ii) fundamental deviation, (iii) transition fit.
11. Draw the figure for limit system and show essential conditions for a clearance fit.
12. As per Bureau of Indian Standard (BIS) in a dimensioning system, what is the mean by
    (i) Tolerance zone (ii) grade of tolerance
13. Define flow, roughness and waviness to characterize surface. Show surface profile for a rough, lapped and finish object. *[UPTU–2008-09]*
14. Give the significance of dimensional tolerance control during various machining process. *[UPTU–2008-09]*
15. What are the various methods for expressing the surface roughness?

❑❑❑

UNIT-IV

Metal Joining (Welding)

Chapter 9

Metal Joining

9.1 WELDING

Use of the welding in Today's technology is extensive. Welding is used in every branch of science and technology. Such as *computer industries, electronic industries, mechanical industries, petrochemical industries*, etc. most material available in universe can be weld easily. Welding is completed by the melting of material. By the one welding machine all types of material can not be join. For different material, different welding machine are used. Temperature (heat) play a very important role in welding. Process is not limited up to metals even some cases non-matel can be welded too.

Defination: Welding is a process of joining of two metal (same/different) with the application of heat with pressure without pressure with filler Rod without filler Rod.

9.1.1 History of Arc Welding

Electric arc welding 1st described by Davy in England in late 1809, but beginning of arc welding could become possible only the improvement in generator/dynamometer. Welding of metals by carbon arc was 1st suggested by a French person in 1881. In 1889, Zerner proposed an idea for twin carbon arc welding. In 1907 in America Strohmenger obtained another patent on flux coated electrode and 1st good welding joint was produced, he used asbestors coating with sodium silicate binder. Now a lot of changes and development have occurred for constituents of flux coating and core wire composition.

In 1928 shielding of welding (CO_2) possible by Alexander in USA. Ist Tungsten inert gas welding **(TIG)** and Metal inert gas welding **(MIG)** develop consequently in 1946 and 1948 respectively.

Joining Process: On the basics of composition of the joint, joining process are three types:

1. **Homogeneous Process:** In this joining process filler rod (electrode) is used and two same metal is joining no pressure is applied during the welding process.

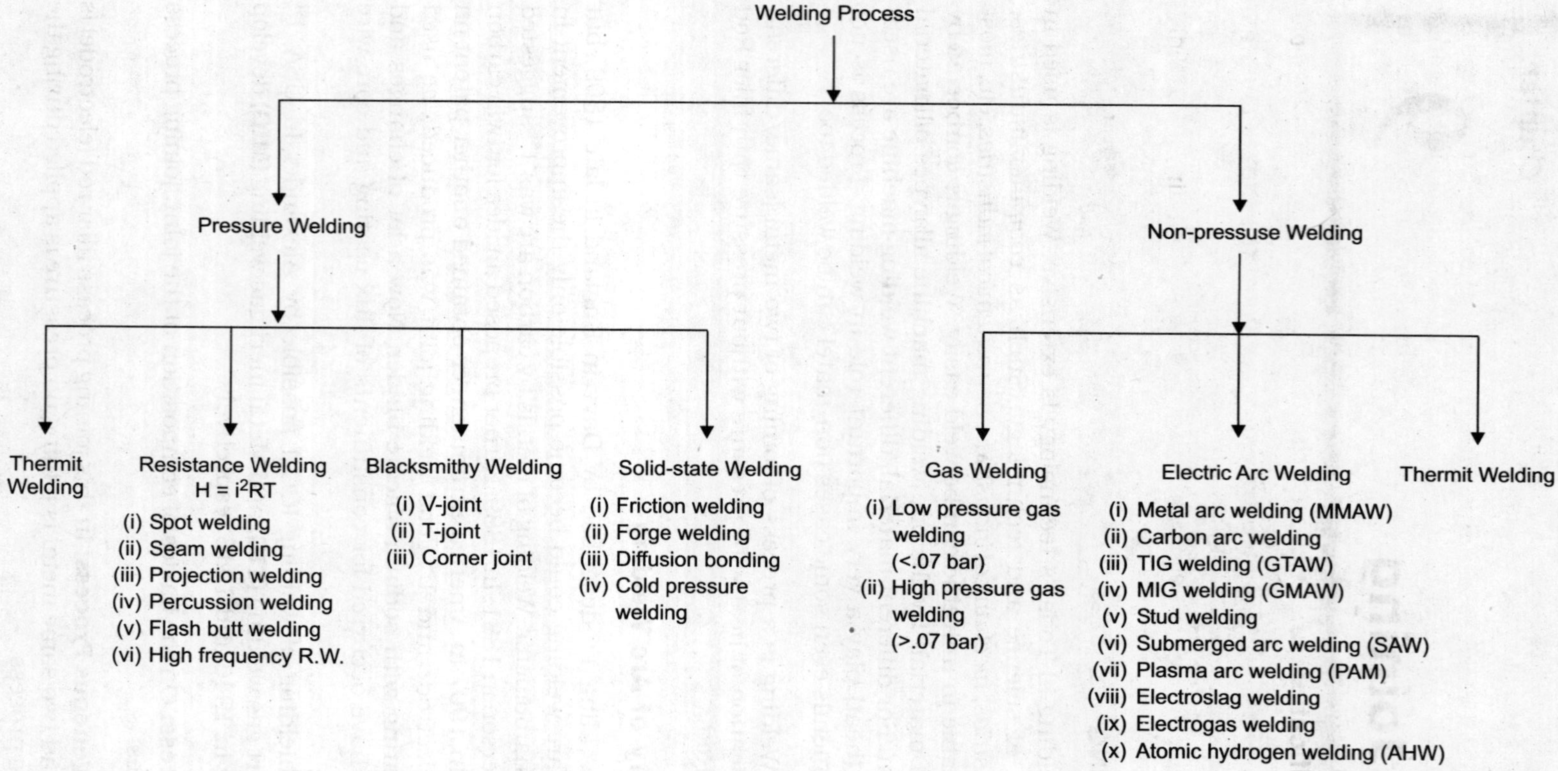

Fig. 9.1 Block diagram of welding classification [30]

2. **Hetrogeneous Process:** In this joining process filler metal is soluble in both the parent metal which themselves are soluble in each other.
3. **Autogeneous Process:** In this type of welding process no filler rod is used but pressure is extensively used. Resistance welding, forge welding, friction welding, diffusion welding are the examples of autogeneous welding.

9.1.2 Classification of Welding Process (Fig. 9.1) *[UPTU–2008-09]*

There are several ways by which we can classified the welding such as pressure bases, non-pressure bases, etc.

Other joining process are soldering, brazing, adhesive bonding, ultrasonic welding, EBW, metal spraying, etc.

9.1.3 Metals that Can be Welded

(i) Ferrous metal such as low carbon steel, high carbon steel, stainless steel, cast iron, etc.

(ii) **Non-ferrous.** Those material which are rich other can carbon such as Al, Cu, brass, bronze, nickel magnesium. zinc and its alloys.

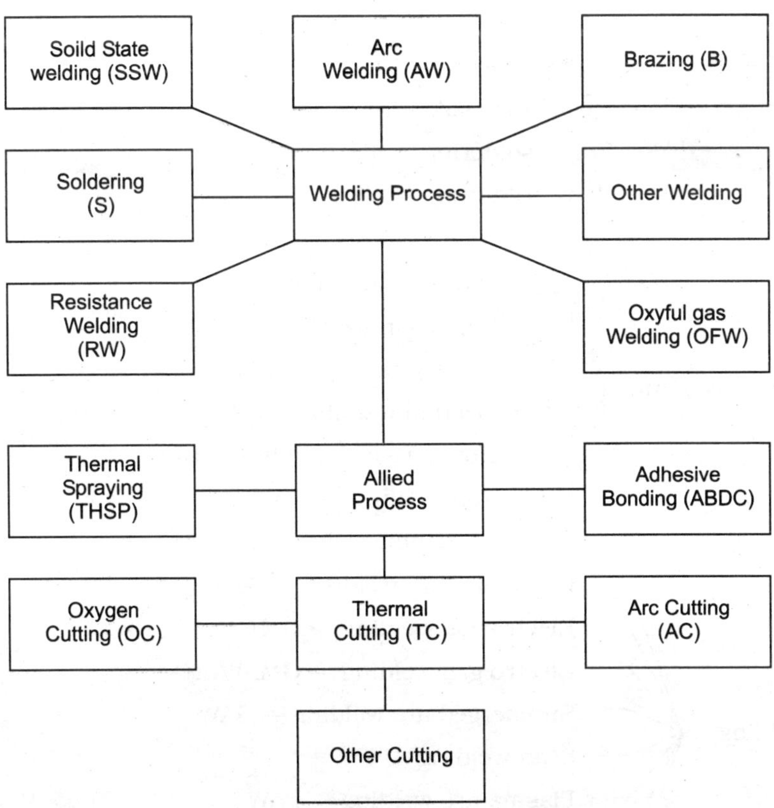

Fig. 9.2 Master chart of welding and allied process

- Oxy fuel gas welding
  - Air acetylene welding — AAW
  - Oxy-acetylene welding — OAW
  - Oxy-hydrogen welding — OHW
  - Pressure gas welding — PGW
- Brazing
  - Arc brazing — AB
  - Block brazing — BB
  - Diffusion brazing — DFB
  - Dip brazing — DB
  - Flow brazing — FLB
  - Furnace brazing — FB
  - Induction brazing — RB
  - Resistance brazing — RB
  - Torch brazing — TB
- Soldering
  - Dip soldering — DS
  - Furnace soldering — FS
  - Induction soldering — IF
  - Infrared soldering — IRS
  - Iron soldering INS
  - Resistance soldering — RS
  - Torch soldering — TS
  - Wave soldering — WS
- Solid state welding
  - Cold welding — CW
  - Diffusion welding — DFW
  - Explosion welding — EXW
  - Forge welding — FOW
  - Friction welding — FRW
  - Hot pressure welding — HPW
  - Roll welding — ROW
  - Ultrasonic welding — USW
- Arc welding
  - Gas metal arc welding — GMAW
  - Electroslag welding — GMAW
  - Electro gas welding — GMAW-EG
  - Submerged arc welding — SAW
  - Stud welding — SW
  - Plasma arc welding — PAW
  - Shielded metal arc welding — SMAW
  - Flux cored arc welding — FCAW

9.1.4 Advantages of Welding

(i) Welding joint is stronger as base metal.
(ii) Machining is possible on welding joint such as (grinding, shapering. Turning etc.).
(iii) Welding joints efficiency upto 100%.
(iv) Appearance of welding joint is very good.
(v) Alteration and amendment is possible in existing structure use by welding.
(vi) A large number of similar and dissimilar metal can be welded easily.
(vii) Welding joints are light in weight.
(viii) Welding is a permanent joint.

9.1.5 Limitations of Welding

(i) Skill person is needed for welding purpose.
(ii) Ultraviolet rays and infrared rays generate during the welding process which are harmful for operator's skin as well as eyes.
(iii) Distortion due to high temperature is possible in workpiece.
(iv) Thermal stresses developed in workpiece heat treatment process is required to relieve the thermal stresses.
(v) Weld heat produces metallurgical change in workpiece.
(vi) Edge preparation is required before welding which is time and labour consuming process.
(vii) Special jig and fixtures are required for welding purpose.

9.1.6 Comparison of Welding with Other Joining Process

(i) Welding joints are lighter in weight as compared to riveting and casting (process).
(ii) Welding is more economical and much faster process as composed to casting and riveting.
(iii) Welding joint have more tensile strength and rigid as compared to riveting and casting.
(iv) Less cost involvement in welding as compared to riveting and casting.
(v) Welding produces 100% efficient joint which is not possible by any other joining process.
(vi) Jig-fixture not so much required as compared in riveting and casting.
(vii) Welding is a permanent joint while riveting and bolting are not permanent joint (temperary joint).

9.1.6.1 Similarities and Differences Between Castings and Welding Process *[UPTU-2009]*

Similarties

(i) In welding and casting process heat is required.
(ii) Both process are melting process. Means in welding as well as casting process material melts.

(iii) Welding and casting both are joining process.
(iv) Both welding and casting process are permanent joining process.
(v) Both process welding and casting produces various defects such as pin holes, blow holes, porosity, etc.
(vi) Machining process such as facing, grinding, etc., process can be easily performed on welded and casted surface.
(vii) Welding and casting both process required skill operator (labour).

9.1.6.2 Differences

(i) In welding process three zone (HAZ, grain refine and transission) produces while in casting only one zone produce.
(ii) Two dissimilar metals cannot be casted very easily while two different metals can be easily welded.
(iii) Set up cost is very high for casting purpose which is not applicable for welding process.
(iv) Welding process is more economical and much faster as compared to casting.
(v) Portable welding equipment is available while portable casting equipment is not available.

9.1.7 Applications of Welding

(i) Automobile industries
(ii) Building construction
(iii) Railroad industries
(iv) Pressure vessels and tank (cylinders)
(v) Aircraft industries
(vi) Storage tank
(vii) Pipeline industries
(viii) Petrochemical industries
(ix) Earth moving equipment and machinery
(x) Shipbuilding industries, etc.

(i) Automobile and Transport Industries. Where in cars, truck, jeeps and many other transportation machine and Equipment are fabricated.

(ii) Building Industries. Welding greatly use in building industries for joining frames of doors and windows, reinforcement in concrete works, railings and stair cases when building is a steel frame construction comprising of steel roofing frames covered with asbestos sheets or Galvanized iron sheet; welding has still greater role to play in joining the structural components for joining building, frames and trusses.

(iii) Railroad Industries. Railroad industries is another important field where welding is involved in the production of locomotive under frames, boggies, trolleys, railway bridges, electric frame network, signal, equipment, lighting towers, plat form shades, storage tanks and bodies, etc.

(iv) Pressure Vessels and Tank. Pressure vessels and tanks are widely used in various industries for carrying/storage the fuel and other liquid material these are made by welding together the bend steel plates. Oil gas and water storage tanks are also steel fabricated.

(v) Aircraft Industries. In aircraft industries welding is extensively used for joining the aircraft component of **alloy steel**, **stainless steel** and **aluminium alloys**, no compromise for welding in aircraft industries beside the fabrication of the **aircraft body**, **frames**, **mounts**, **fuel storages tank**, **duct** and fittings. Welding is used for the production of allied equipment that assist aircraft in operations and maintenance like material handling system, transport means for man and language, shades, etc.

(vi) Petrochemical Industries. Chemical and petroleum industries make goods use of welding for the fabrication of plant and machinery, stainless steel vessels and storage tanks besides many other structure.

(vii) Earth Moving Equipment Industries. Earth moving equipment manufacturing industries largly depends on the welding for the fabrication of earth moving machine like **bulldozers**, **loaders**, **trenches** and **drilling rigs** for all oil exploration and water tube wells and other related machine.

(viii) Material Handling Equipment Industries. In this field welding is used for fabrication of **overhead cranes**, **jib cranes** and **tower cranes**, etc., along with their auxiliary equipment like **trolley**, **lifting aids**, etc.

9.2 GAS WELDING

- Gas welding is another method of joining process.
- It is a non-pressure welding process.
- As well as gas welding is a fusion welding process.

It join the metal with the help of heat, combustion of air (O_2) and fuel gas [hydrocarbon base gas] (C_2H_2, H_2, propane and butane) mixture.

9.2.1 Principle of Oxy-acetylene Welding

When Acetylene (C_2H_2) mixed with oxygen (O_2) in appropriate quantity in welding torch and burn. Flame generated at torch of tip is sufficient to heat the metal or melt the metal and thus joint can be formed during the gas welding process.

Different types of flame generated at the torch of tips are:

1. Neutral flame
2. Oxidizing flame
3. Carburizing flame

9.2.1.1 Neutral Flame

When one part of C_2H_2 (acetylene) is mixing with one part of oxygen [O_2] and resultant mixture is burning at the torch of tip then a flame will generate that and flame is define as **neutral flame**.

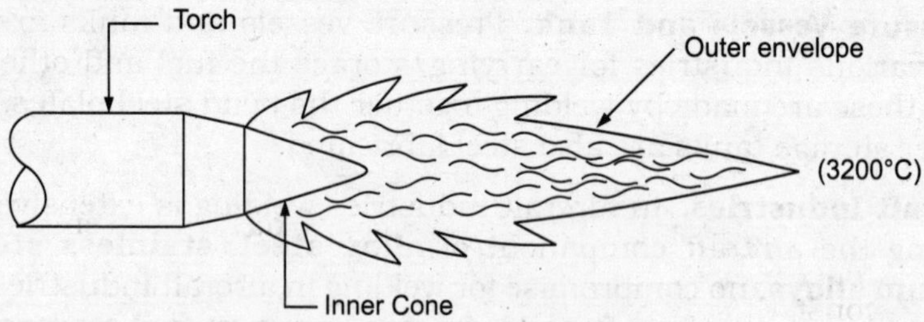

Fig. 9.3 Neutral flame at torch

This flame consist the two envelope, i.e., outer envelope and other is inner cone and the temperature of inner cone is **3200°C.**

$$C_2H_2 : O_2 = 1 : 1$$

A neutral flame is named so because it effects **no chemical change** in the molten metal and therefore no oxidation of molten metal.

This type of flame **(neutral)** is commonly applicable for joining the different materials, i.e.,

(i) Mild steel
(ii) Cast iron
(iii) Stainless steel
(iv) Copper
(v) Aluminium and alloys

9.2.1.2 Oxidizing Flame

Once the neutral flame is establish and now the oxygen quantity is increases then oxidizing flame stablish ***or*** when more than the one part of oxygen (O_2) is mixing with one part of Acetylene (C_2H_2) in the gas torch and burning a

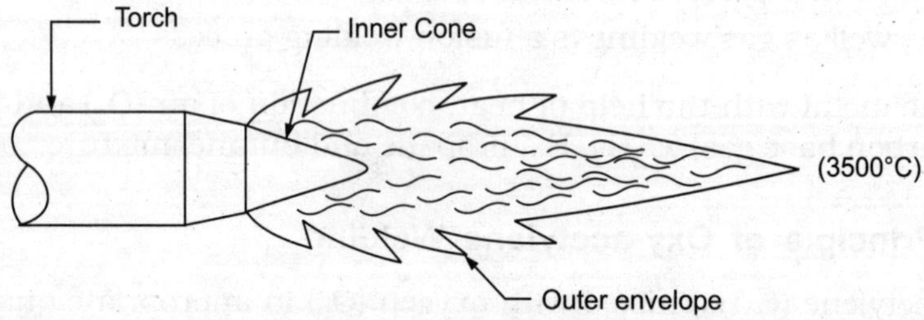

Fig. 9.4 Oxidizing flame at torch

$$O_2 : C_2H_2 = 1.5 : 1$$

flame is generating at the torch tip resultant flame is define as oxidizing flame. This flame also consist a outer envelope and inner cone but now the temperature of inner cone is 3500°C means it is hotter than neutral flame. An oxidizing flame burns with a decided loud roar. This type of flame is limited for use only for

(i) Copper base metal
(ii) Zinc base metal.

It is not applicable for steel welding.

9.2.1.3 Carburizing Flame or Reducing Flame

Once the neutral flame is establish at tip of torch and now volume (quantity) of acetylence gas (hydrocarbon based) is increased then the flame generated at the torch of tip is refer ***carburizing flame or*** when more than the one part of Acetylene (C_2H_2) is burning with one part of oxygen (O_2) then resulting flame is defined as carburizing flame or reducing flame. Temperature of carburizing flame is very low as compared to all the three flame, i.e., 3150°C this flame consist three cone, i.e., outer envelope, intermediate cone and inner cone.

A reducing flame is used for annealing, i.e., to soften then steel sheet.

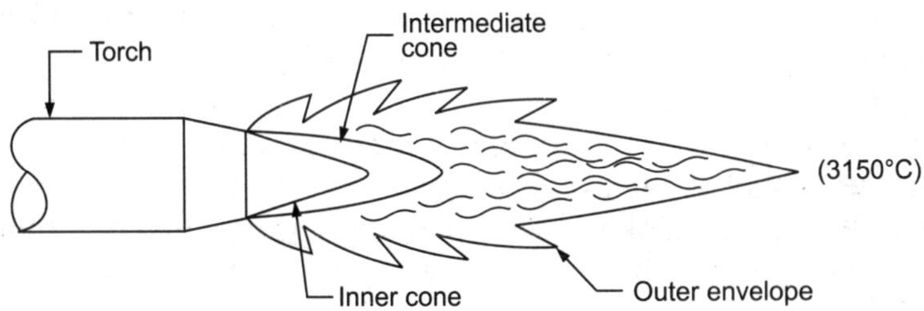

Fig. 9.5 Carburizing flame at torch

9.2.2 Availability of O_2 and C_2H_2 Gases

In oxy-acetylene welding flame is obtained by burning C_2H_2 which is a fuel gas and O_2 supports the combustion of C_2H_2. When mixed with oxygen and ignited acetylene burns explosively undergoing oxidation to CO_2 and water.

$$2C_2H_2 + 5O_2 \rightarrow 4CO_2 + 2H_2O + 620 \text{ calories.}$$

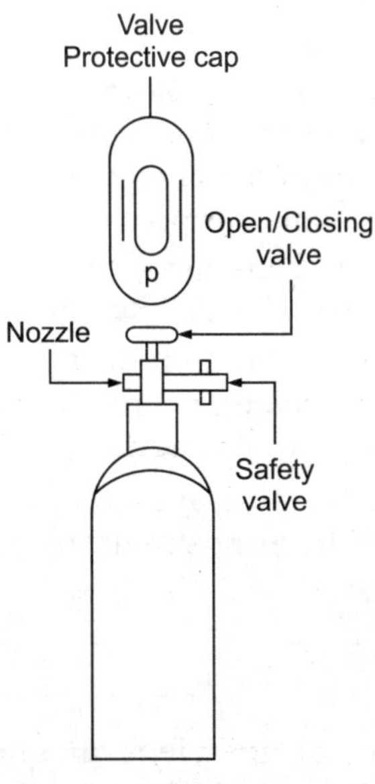

Fig. 9.6 Oxygen cylinder

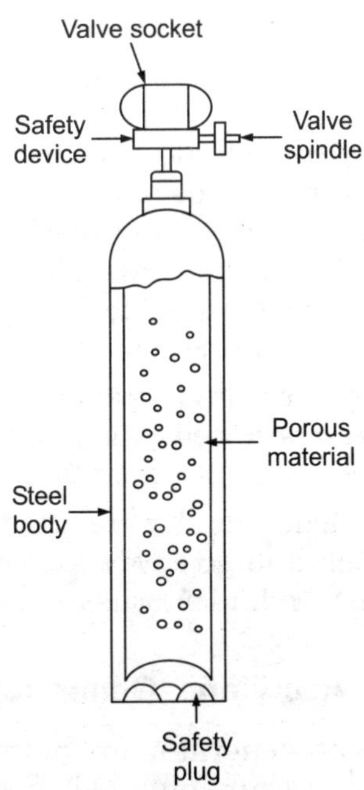

Fig. 9.7 Acetylene cylinder

9.2.3 Oxygen

It is a clear gas without **colour** and with **no taste**. Oxygen produced by liquid oil process or electrolytic process and that oxygen is used for industrial purpose. In liquid air process air Ist cleared of dust and other carbonic gases and later cooled to liquid at –180°C. N_2 is separated from this process by warming slowely. In electrolytic process electric current passed through water vapour containing electrolyte such as caustic soda (NaCl) for increasing conductivity as the current passed oxygen is produced in the form of bubbles on the surface of +ve pole and H_2 on –ve pole.

Oxygen is filled in solid extruded cylinder one piece (mild steel or alloy steel) having with different capacity such as **3400 litres**, **5200 litres**, **6800 litres** and the pressure of oxygen inside the cylinder is from **120** to **140 kg/cm²**.

Oxygen cylinder are painted **black** and the valve outlets are screwed **right handed** thread. A safety cap is provided at the top of cylinder as shown in Fig. 9.6 to allow any leakage of oxygen at the time when due to increases in temperature oxygen cylinder should not be stored near by any combustible gas cylinder.

9.2.4 Acetylene

Acetylene can be obtained for welding purpose by following way:

1. Acetylene cylinders with filled of C_2H_2 gas from some gas producing plant such acetylene is called "dissolved acetylene".
2. **Acetylene Cylinder.** Acetylene can be generated at shopfloor and installed very easily. In generator calcium carbide react with water or vice-versa.

$$CaC_2 + 2H_2O \rightarrow C_2H_2 + \underset{\text{Calcium hydroxide}}{CO(OH)_2}$$

Acetylene cylinder (Fig. 9.7) are also solid extuded one piece cylinder and filled with a porous material, i.e., **acetone** which is a hydrocarbon liquid at room temperature and pressure. Acetone can dissolve acetylene about 25 times of its actual valve. Therefore at 15 atmospheric pressure which is normal pressure of dissolved acetylene this is increased by 375 times. Acetylene compressed into cylinder at pressure of 16 to 20 kg/cm^2.

Acetylene cylinders are **maroon colour** painted and their outlet valve are **left hand screwed**. It is just for difference with oxygen cylinder. Acetylene cylinder are available in **2800 litre**, and **5600 litre capacity**.

Acetylene cylinder are fitted with several fusible plug at its bottom which may melt and give way to acetylene to escape in atmosphere at **104°C** if chance to cylinder Brust.

9.2.5 Acetylene Generator

Acetylene generator are used for production of acetylene at shopfloor (for low pressure purpose). It produces acetylene upto 10 kg/cm^2 (1 bar). It is portable and gives acetylene at the rate of 15 litre/min.

Acetylene generator are basically two type:

(i) carbide to water type
(ii) water to carbide type.

Carbide to water type generator is preferred over water to carbide type as the former uses complete reaction of water and calcium carbide (CaC_2) and is more efficient. A typical calcium carbide (CaC_2) to water type acetylene generator is shown in Fig. 9.8 small size of calcium carbide particles stored in hopper at top of the generator where calcium carbide (CaC_2) is fed down to the water by a feed mechanism and acetylene is produced as a result of action of water on calcium carbide. At the bottom of generator calcium hydroxide [$Ca(OH)_2$], i.e., sludge is collected.

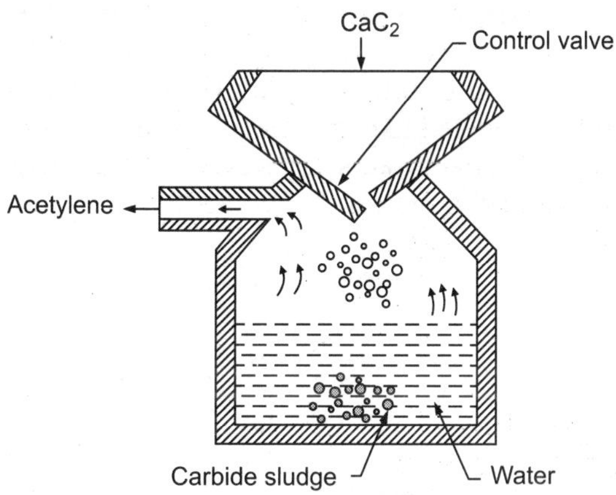

Fig. 9.8 Acetylene generator

9.2.6 Safety/Precautions of Acetylene Cylinder

Following safety points are needed for safely use of acetylene cylinder:

(i) Acetylene is highly flamable and so necked flame should not be brought close to the cylinder.
(ii) Never use the hammer for opening/closing of spindle valve.
(iii) Use the cap on valve when cylinder is not in use position.
(iv) Acetylene should not be used at a pressure more than 1 kg/cm^2 in any condition.
(v) Never do the practice to transfer gas from one cylinder to another one.
(vi) Acetylene cylinder should never used as a supporting purpose.
(vii) Keep the acetylene cylinder always in upward position.
(viii) Keep the acetylene cylinder far away from sunlight and heat.
(ix) Dragling/rolling of cylinder should avoided.

9.2.7 Backfire/Popping

Backfire is a small explosion of the flame followed by either extinguishing of flame or continued burning of gases. It is basically due to pre-ignition of gases.

Following cause of backfire:

(i) When gases come out too slowely from torch under low pressure due to small tip used. Pre-ignition or burning of gases may to be take place within torch.

(ii) If tip of torch become over heated. It may be dogged also.

(iii) Due to deposition of hot metal particles inside the torch tip in such a way torch become overheated. This result in pre-ignition of gases with in the torch.

9.2.8 Flashback

Flashback is quitely different with backfire phenomenon. A flash back occurs when the flame dissears from the tip and travels back inside the hose. **Significance** of flashback that flame makes a hissing sound. **Remedy** of this both the gases should be shut off immediately to avoid combustion within the torch, torch should be allowed to cool before restart. Flash back may also due to a clogged barral or mixer passage and excessive pressure of oxygen. Sometimes organic oxides in oxygen hose may also responsible for flashback.

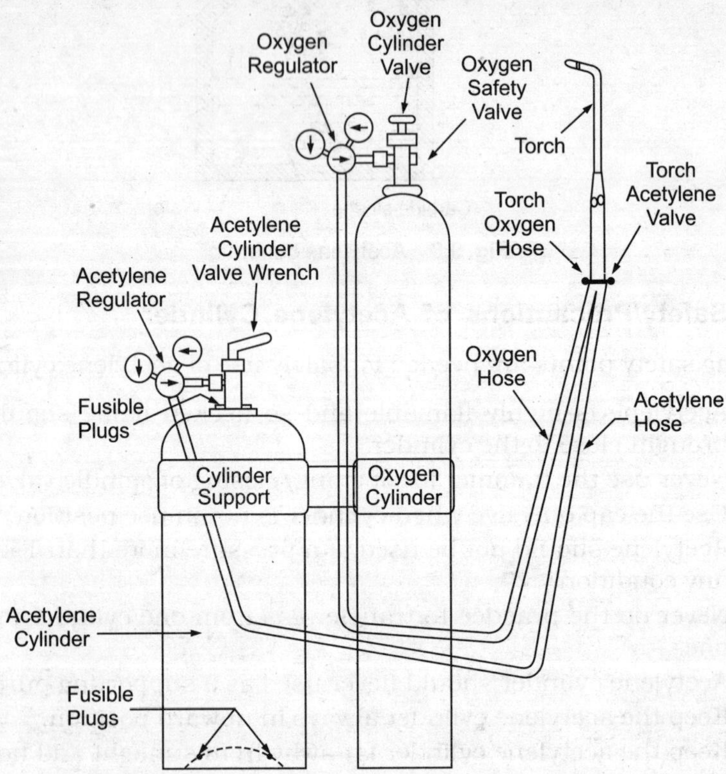

Fig. 9.9 Set up for oxy-acetylene welding process [11]

9.2.9 Arrangement of Oxy-acetylene Gas Welding

1. Oxygen and acetylene gas cylinder should be accommodated on a trolley for their easy transportation from one destination to another destination.

2. Regulator (complete) with high and low pressure gauges. Two pressure gauges are required one for oxygen and another for acetylene cylinder.
3. **Hoses.** Hose is used to supply the gases from cylinders to torch. **Red colour** hose pipe is used for transportation of acetylene and **black/ blue colour** hose pipe is used for oxygen transportation. Welding (gas). Hose has a seamless lining made from rubber compound and reinforced or wrapped with cotton piles. Hose has metal clamp or clips to attach welding hose to a nipple.
4. **Welding Torch (blow pipe).** Oxygen and acetylene have been reduced in pressure by the gas regulator are fed through their hoses to the welding torch or blow pipe. Torch function is to mixing and control the flow of gases to welding nozzle or tip of torch where gas mixture burns and produces flame.

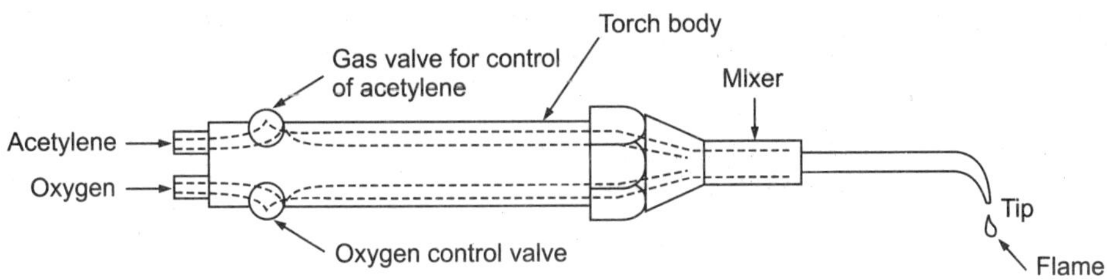

Fig. 9.10 Welding torch

5. **Welding Nozzle or Tips.** It is fitted on the front end of the welding torch. These may be:
   (i) Interchangeable type screw on the head of the blow pipe or
   (ii) Goose neck extension type fitted directly onto the mixer portion of blow pipe.
6. **Pressure Regulator.** These are used to reduce the pressure from the cylinder and maintain it at appropriate value. These are also used in adjusting the pressure of gas to torch.
7. **Gogles.** They are used to protect eyes from harmful heat and radiation of infrared and ultraviolet ray produces during the welding.
8. **Spark Lighter.** It is used for sparking/burning the gases at the torch tip instant and convenient means. It consist of a pointed stone and a rough surfaces to produce spark when rubbed together.
9. **Apron and Gloves.** They are used for the protection of clothes and hand of the operator.

9.2.10 Welding Torch

Welding torch is a device which is used to prepare the mixture of two gases in the desired proportion and burning the mixture at the end of the torch tip. It has a handle to hold it and two inlet connections for two gases at one end such that each inlet has a valve to control the volume of oxygen and acetylene. Welding torch are basically two types:

(i) High pressure
(ii) Low pressure

9.2.11 Gas Welding Technique

They are basically two types:

1. Left-hand (Forehand) welding
2. Right-hand (Backhand) welding

1. Lefthand Welding Technique. It is one of the oldest method of performing gas welding. In this process welding is performed from right hand end and welding torch is held in right hand while filler rod held in left hand. Torch making angle 60–70° with workpieces. Filler rod is held at about 30° to the workpiece. It is important that the cone of the flame should never go outside the puddle which is small pool of molten metal created due to the intense of heat left hand welding. Technique is very suitable for the thickness of workpieces upto 5 mm because as the thickness of workpiece go increasing way good penetration cannot be achieve by this process. For plate thicker than 3 mm beveling of the plate edge to produce V-joint is needed.

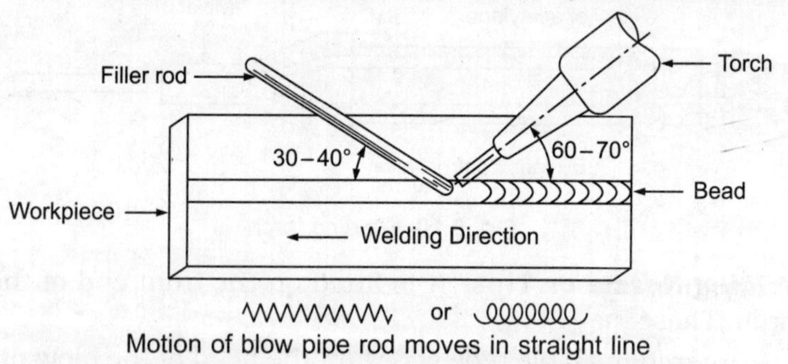

Fig. 9.11 Left ward welding technique

9.2.12 Edge Preparations for Butt Joint with Left-Hand Welding

Job thickness (mm)	***Edge preparation (mm)***	***Filler rod size***
0.8 to 1.3	No root gap, no beveling	No rod
2.4 – 3.0	Root gap 1.5, no beveling	No rod
4.0	Bevel angle 60°–80° roat gap 1.5	2.5 mm
4.8	Bevel angle 70°–80° roat gap 2.4	3.0
6.0	Bevel angle 90°, roat gap 2.4	3.0

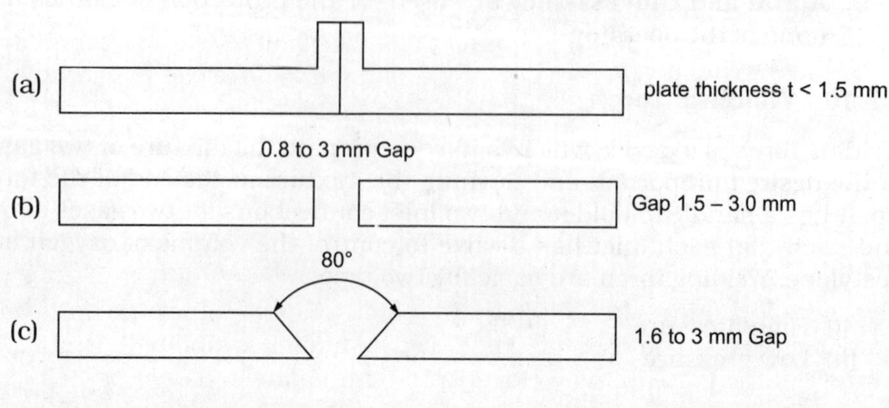

(d) 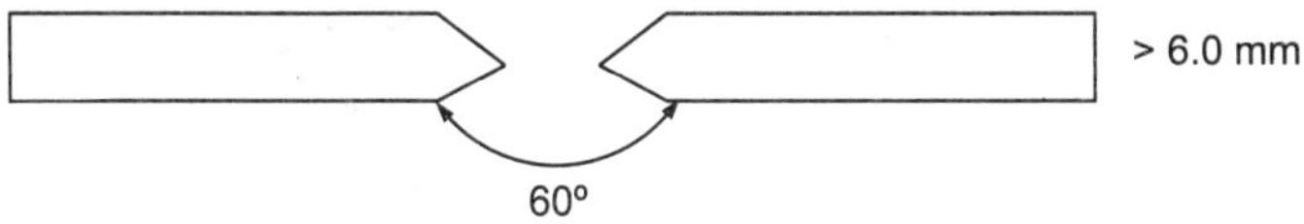

Left ward welding: edge preparation

9.2.13 Right-Hand Welding Technique

In this welding technique torch is held in right hand of the operator while filler rod is held in left hand and welding is performing from left side to right side. This type of welding technique is very suitable when the thickness of workpiece is greater than 5 mm. Weld with generation of approximately 12 mm can be achieved in a single pass.

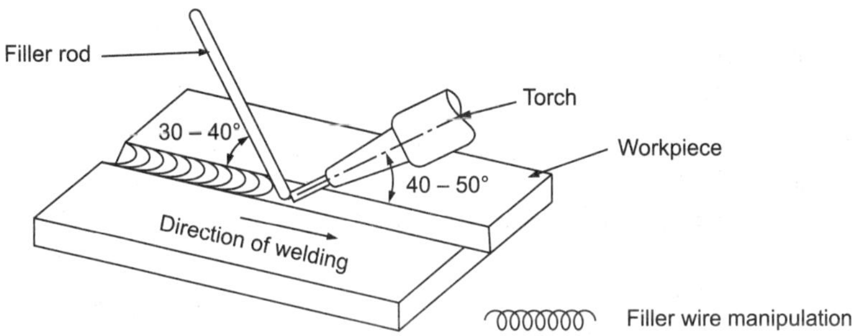

Fig. 9.12 Right ward welding technique

Edge preparation for butt joint with right hand technique:

Job thickness (mm)	*Edge preparation (mm)*	*Filler rod size*
5	Square edge, root gap 2.4	2.4
6.5	Square edge, root gap 3.2	3.2
8.2	Square edge, root gap 4.0	4.0
10.0	Groove angle, 60° V, root gap 3.2	5.0
13.0	Groove angle, 60° V, root gap 3.2	6.5
16.2	Groove angle, 60° V, root gap 3.2	6.5
19.0	Groove angle, 60° V (top) 80° V bottom	6.5
25.0	Groove angle, 60° V (top), 80° V bottom roat gap 4.0	6.5

9.2.13.1 Right-Hand Welding

This method has definite advantages over the left hand welding technique on thick plates but we should be quite aware of its limitation and user rod is given a rotational or circular motion, while the Blow pipe moves in particular a straight line as shown in Fig. 9.12. Angle between blow pipe and rod is greater than that used in the left hand welding method. When using the method good fusion can be obtained with out a V upto 8 mm plate. Above 8 mm plates are prepared with a 60° V. A large blow pipe nozzle is required for a given size plate than in left hand welding as molten pool is controlled by the pipe and rod but the pipe has no side to side motion.

In order to ensure good welds by this method that blow pipe and rod should be held at the correct angle, rod diameter is about half of the thickness of the plate being welded upto 8 mm plate.

9.2.13.2 Advantages of right hand welding technique over left hand welding technique:

(i) Upto 8.2 mm plate thickness no bevel is required. Therefore no edge preparation is required.

(ii) For welding of bigger thickness, where bevelling of plate edge becomes necessary, included angle of υ need be only 60°.

(iii) Better control and higher welding speed.

(iv) Less quantity of deposited metal, as compared to left hand welding reduces shrinkage and distortion.

(v) Quality of welding joint is better than left hand technique.

(vi) Less consumption of filler metal and low cost of welding.

9.2.14 Advantages of Gas Welding

(i) It is one of the most versatile method of welding.

(ii) Rate of heat generation is very less as compared to arc welding process.

(iii) As the source of heat and filler metal are separated, the welder has control over filler-metal deposition rates. Heat can be applied preferentially to the base metal or the filler metal.

(iv) Welding equipment can be transported from one place to another place very easy and the cost of welding equipment are very low.

(v) HAZ is very narrow.

(vi) Maintenance cost of gas welding equipment is low.

(vii) No need of electric current (supply).

(viii) Equipment cost is not so high.

9.2.15 Limitations of Gas Welding

(i) Heavy section can not be joints very efficiently.

(ii) Proper penetration can not be achieve through this process.

(iii) All types of metal cannot be join through this process.

(iv) Flux used in certain welding process produce fumes that is irritating to the eyes, nose, throat and lungs.

(v) More safety is recommended in gas welding process.

(vi) Acetylene and oxygen gases are rather expensive.

(vii) Prolonged heating of the joint in gas welding results in a large heat-affected zone.

(viii) Temperature of flame is less than the temperature of the arc.

(ix) Proper storage need of acetylene cylinder.

(x) Refractory metal (such as ceramics, Tungsten etc.) can not be welded.

9.2.16 Applications of Gas Welding

(i) For joining thin materials.

(ii) For joining material in those case excessively high temperature or rapid heating and cooling of the job would produce unwanted or warmful changes in the metal.

(iii) For welding of ferrous and non-ferrous metal, i.e., **carbon steel**, **alloy steel**, **cast iron**, **aluminium**, **copper**, **nickel**, **magnesium** and its alloy, etc.

(iv) In automotive and aircraft industries.

9.2.17 Oxy-acetylene/Gas-cutting

It is a Chemical Process in the sence that the metal, at the portion where it is to be cut, is actually made to oxides under the action of flame. When thick plates are to cut then oxy-fuel cutting (OFC) is useful with the any fuel gas. Plates upto 2 m thickness can be easily cut by the help of special precaution/method. It is possible to rapidly oxidize (burn) iron and steel when it is heated to a temperature between the 800°C–1000°C when a high pressure oxygen Jet burns the metal and blows it a way causing the cut. This technique was invented in 1887 by Thomas Hatches and extensively used for cutting steel plates of various thickness. Basic advantage of this type process that it is very simple (Equipment) and can be carried any where without handling the heavy steel plates. In oxy-acetylene process of reactions are exothermic all ferrous metal can be cut by means of oxy-acetylene flame. In nature and would provide a good amount of heat to preheat the steel heat generated cause the metal to metal and get blow away by the oxygen pressure.

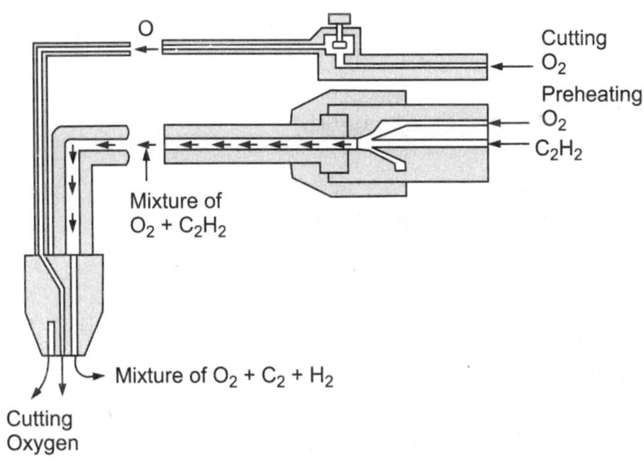

Fig. 9.13 Cross-section through an oxy-acetylene cutting blowpipe [6]

9.2.18 Operating of Cutting Torch

For operating purpose of cutting torch first turned on the Acetylene and Preheating oxygen and establish a neutral flame once the cutting metal heated upto red hot position the cutting oxygen valve is open to make a sharp oxygen jet to strike against the hot metal oxides are formed and melting start. There should be a proper distance between the nozzle tip (usually upto 5 mm) and the upper surface of cutting metal. This could be achieved by fining the cutting torch (blow pipe) on a guide. Speed of blow pipe should not be so faster neither so blow.

9.3 ARC WELDING PROCESS

Definition: Arc welding is a welding process where in coalescence is produced by heating with an electric arc, mostly arc welding is done by without pressure and with or without use of filler metal depending upon the plate (object) thickness.

Electric arc is formed when an electric current pass between two electrodes separated by a short distance from each other. In arc welding one electrode is the welding rod or wire while other is the metal to be welded (work piece), electrode and plate. Arc connected to the supply one to the +ve pole and other to –ve terminal. Arc is started by momenterally touching the electrode on the plate and the with drawing it to about 3 to 4 mm from the plate. When the electrode touches the plates, a current flows and as it is withdrawn from the plate the current continues to flow in the form of a '**spark**' across the very small gap first formed, this cause the air gap to become ionized or made conducting and as a result the current is able to flow across the gap even when it is very wide, in the form of an arc electrode must always be touched on to the plate before the arc can be started.

Arc is generated by electrons flowing from negative (–ve) to positive (+ve) terminal and electrical energy is changed in the arc into heat and light approximately 2/3 of the heat is generated near the positive (+ve) terminal which burns into the form of a creater. Temperature range form **2700°C–5500°C**. While remaining 1/3 is generated near negative (–ve) terminal as electrode connected with positive (+ve) terminal they will burn away 50% faster than if connected to negative (–ve) terminal. Therefore medium coated electrode and bare electrodes are used.

Types of Arc Welding

1. Consumable Electrode Process

- (i) Shielded Metal Arc Welding (SMAW)
- (ii) Submerged Arc Welding (SAW)
- (iii) Gas Metal Arc Welding (GMAW)
- (iv) Flux Cored Arc Welding (FCAW)
- (v) Electrogas Welding (ECW)
- (vi) Electroslag Welding (ESW)
- (vii) Carbon Arc Welding (CAW)

2. Non-consumable Electrode Processes

- (i) Gas Tungsten Arc Welding (GTAW)
- (ii) Atomic Hydrogen Welding (AHW)
- (iii) Plasma Arc Welding (PAW)

1. Flux Shielded Metal Arc Welding or Manual Metal Arc Welding. In this welding process arc coalescence is produced by heating the work-piece with an electric arc setup between a **flux coated electrode** and **workpiece** (object). Flux decomposes due to arc heat and perform several function such as, *arc stability, weldmetal protection* and several more.

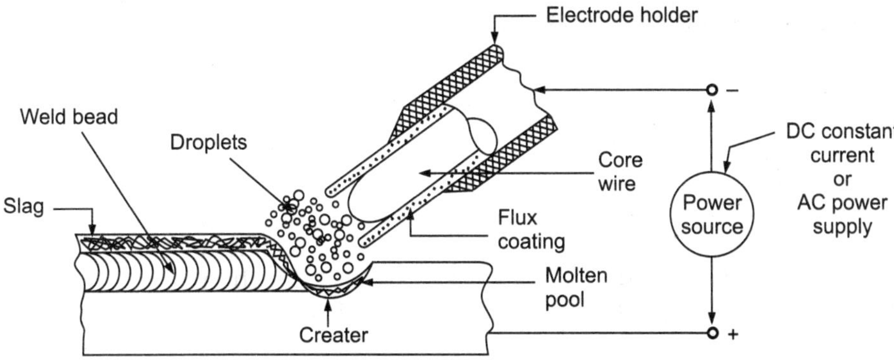

Fig. 9.14 Flux shielded manual metal arc welding

Heat produce during this process by strack between the workpiece and coated electrode. A high current arc with a smaller arc length produces a very intence heat. When arc betwcen the electrode and workpiece. Metal transfer take place in the form of droplets through the arc and arc deposited along the joint to be welded. Shielding is produce as flux coating melts to prevent atmospheric contamination of the molten weld pool.

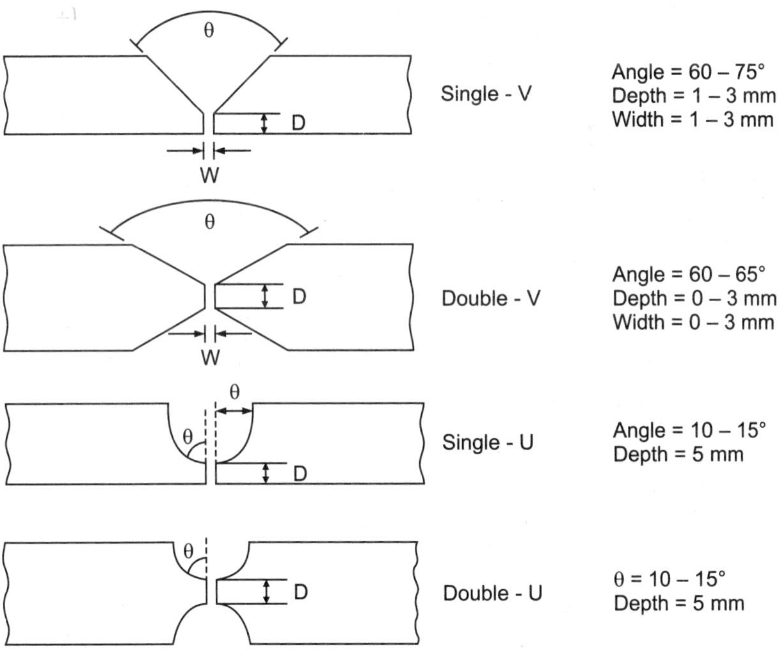

Fig. 9.15 Various type of grooves

Basic limitation or drawback of arc welding that edge preparation is required before welding.

Different types of edge preparation can be used but basically it depends upon the:

(i) thickness of workpiece
(ii) type of process.

(iii) position of welding (1G, 2G, 3G, 4G, 6G).
(iv) access for arc and electrode.
(v) dilution.
(vi) shrinkage and distortion.

9.3.1 Functions of Flux Used for Coating of Electrode

[UPTU-2008-09, 2011-12]

1. Coating provide an atmosphere of inert gas such as carbon dioxide under the arc heat which protect the weld pool from the atmospheric contamination (oxygen, hydrogen and nitrogen).
2. Coating provide flux to the molten metal pool which mixed with oxides and other impurities present in the puddle forms a slag. Slag being lighter, floats on the top of the puddle and protects it against the surrounding air during the weld bead solidification. Slag covering also server the metal to cool slowely to avoid the formation of a brittle weld.
3. Some alloying element added in these coating which is used for stabilizing the arc. Different flux coating is used for AC welding machine and DC welding machine.
4. Some special alloying elements are added in flux coating which Improves the Physical Properties as well as mechanical properties of weld object.
5. As the electrode coating consumed at a slow rate compared to the electrode (bare), coating would be extended beyond the electrode — it helps in concentrating the arc and directing the filler metal in the joint even thermal Losses to the atmosphere from the electrode tip can be reduced. In such a way flux coating also work to improving the metal deposition rate.
6. Flux coating also help to flow of electron in a narrow groove and other different location without any short circuiting problem because flux coating are insulator of electricity.

9.3.2 Edge Preparation

Edge preparation is required during arc welding process. Edge and area adjoining them is cleared of all the **scales**, **rust**, **grease** or any other foreign matter by mechanically or chemically, i.e., my buffing wheel, by head wire brush, by leather device etc.

9.3.3 Striking the Arc

In manual metal arc welding, arc between the electrodes and the workpiece is struck either by just touching the electrode with object or just touching the electrode with workpiece and immediately with drawing from object and maintaining a predetermine distance.

9.3.4 Arc Blow

Arc below is a phenomenon in which a strong magnetic field produced (set up) around the electrode and tends to deflect the arc as through a strong wind were blowing. Therefore termed as '*arc blow*'. Arc may be deflected to the side but it is usually deflected either forward or backward along the direction of travel. Arc blow is noted specially when the electrode is used in a corner or towards the joint end. Arc below is encountered in **DC welding** machine as AC prevents formation of strong magnetic field. In arc blow arc may distort, deflect or rotate.

Several factors are responsible for affecting the arc blow:

Such as:

(a) Magnetic field produced in the workpiece adjacent to welding are due to current flow through arc, presence of bus bars in the neighbourhood of welding place,

(b) Operating multiple welding heads affecting each other,

(c) Magnetic field produced in the workpiece around the earth connection.

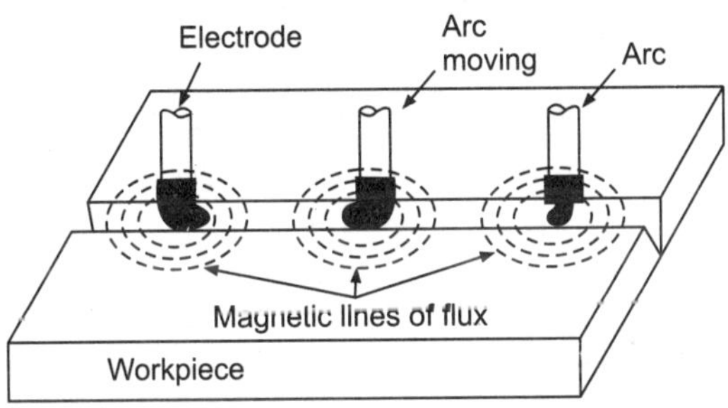

Fig. 9.16 Mechanism of forward and backward arc blow [11]

Increased arc blow results in:

(i) unstable arc,

(ii) poor weld bead,

(iii) lot of undercuts,

(iv) erratic weld deposition,

(v) spattering,

(vi) slag entrapment and porosity of weld.

Arc blow can be decrease or eliminated by:

(i) changing the position of earthing clamp,

(ii) eliminating presence of magnetic material so arounding the workpiece,

(iii) by decreasing arc current,

(iv) putting earthing clamp,

(v) away from the point to be welded using short arc,

(vi) decreasing arc travel speed or pre-heating the workpiece before welding.

9.3.5 Arc Length

Arc length is the distance between the tip of the electrode and the work-piece surface upon which the molten globules are deposited as shown in Fig. 9.17. Short the arc deposits more weld metal with advantage such as **high penetration, maximum strength, slight overlay, minimum porosity, maximum ductility** and better behaviour of alloy steel.

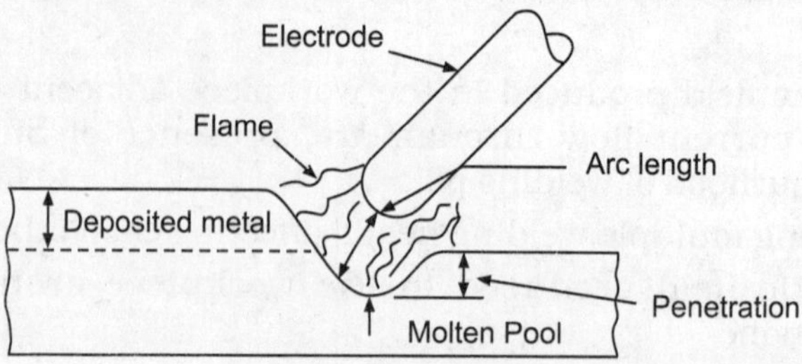

Fig. 9.17 Essentials of an arc

9.3.6 Welding Equipments Required for Arc Welding

(i) AC or DC welding machine (power source)
(ii) Electrode with electrode holder
(iii) Welding leads
(iv) Earthing clamp
(v) Connector
(vi) Face shield
(vii) Chipping hammer.

Mostly AC power source is used for welding. Transformer may be **single phase**, **two phase**, or **three phase**. For low current supply single phase transformer is used therefore they are used for thinner section welding with minor diameter electrode. For high current two phase and three phase supply transformers are used. They are used for thicker welding plates.

Transformer may be oil cooled or air cooled. Big transformer are always oil-cooled. Transformer may be AC or DC generated and generator may be oil/gas operated.

[Currents range (AC welding transformer) — 600 A
Open circuit voltage — 70 to 100 V]

Basically two type of the power source can be used for arc welding purpose, i.e., AC transformer and DC generator rectifier. Power source consist the two cables, electrode cables and earthing clamp; electrode holder stringer and electrode or rod. Both type of power source have certain advantage over each other in AC welding electron flow in both direction. While in case of DC welding direction electron can be changed by simply reversing the cable at the terminals loaded on the generator if electrone is flowing from electrode to workpiece means it is straight polarity (electrode –ve) while workpiece +ve).

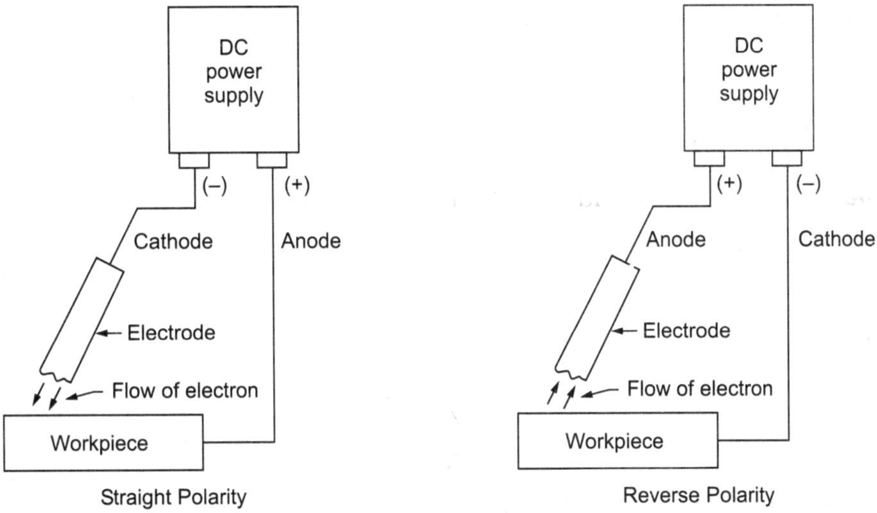

Fig. 9.18 Polarity

If electrone is flowing from workpiece to electrode then it is reverse polarity (electrode +ve while workpiece –ve) if positive ground is used then the penetration will be greater as the heat supplied amount by the electrode force to workpiece. If poles are reversed and there is a negative ground 2/3 of the heat will remain in the tip of the electrode.

Alternating current combines the characteristics of both the shallow the deeper penetration, yielding. A penetration depth that is approximately half way between that achieved by the positive ground and negative ground.

Power source of constant current type having **droping** characteristic are used for **MMAW** process. Power source of constant voltage type are not suitable.

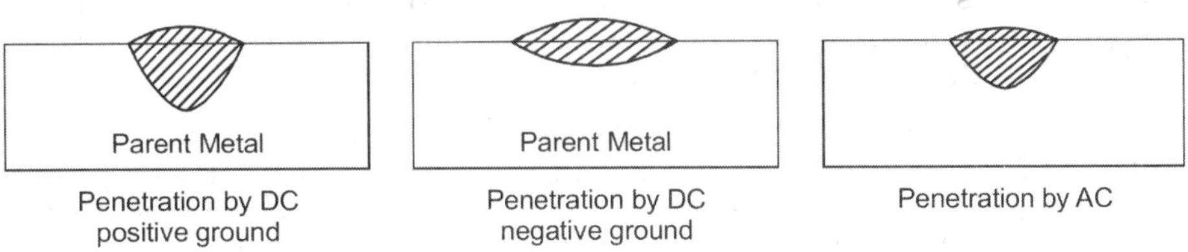

Fig. 9.19 Weld Bead Geometry

Heat developed by the arc is given by —

$$W \text{ (Joule)} = V \text{ (Volt)} \times I \text{ (Amp)} \times t \text{ (sec)}$$

If the arc is travelling at a speed of S mm/min, then **heat input rate (HIR)** of arc will be

$$\text{HIR} = \frac{V \times A \times 60}{S} \text{ Joule/mm length of the joint.}$$

9.3.7 Comparison Between AC and DC Arc Welding

AC arc welding	*DC arc welding*
1. Striking of arc with electrode is relatively difficult maintenance of a short arc is also difficult except with iron powder electrode.	**1.** Developing an arc is easier maintenance of short arc is also easier.
2. No problem of arc blow in AC arc welding. Workpiece do not get magnetized as in DC.	**2.** Arc blow is a severe problem is minimised with the use of proper corrective measure. Workpiece may get magnetized due to current flow in one direction.
3. Arc is never stable.	**3.** Arc is more stable.
4. No polarity change possible and hence no suitable for welding all metals. It is used for welding ferrous metals.	**4.** Polarity (DCSP or DCRP) can be changed and hence suitable for welding both ferrous and non-ferrous metals equally efficient.
5. More suitable for higher current value. It is less suitable for use at low current value with small dia of electrode.	**5.** It is most suitable with lower current value also, for example at low amperage with small dia electrode.
6. Bare electrode cannot be used. Only flux coated electrode with arc stabilizing agent influx can be used.	**6.** Bare and coated electrode can be used.
7. Not suitable for thin sheets or sheet metal work due to difficulty in Stricking the arc.	**7.** It is suitable for welding of sheet metal as striking arc is easier and arc remains steady.
8. Distribution of heat in arc is equal at electrode and job.	**8.** Most of heat (upto 66.67%) is liberated in the positive side of arc, i.e., DC RP.
9. Voltage drop in welding is less and hence welding is suitable for longer distance from welding plant using long welding lead.	**9.** Voltage drop is relatively higher and hence start cables are used to weld only close to the welding plant.
10. Welding transformer have no moving part and working is salient.	**10.** A DC generator set has several moving parts therefore operation is noisy.
11. AC transformer welding set is not costly, simpler in operation maintenance cost is also very low.	**11.** A dc generator set is costly, difficult to operate and very high maintenance cost.

9.3.8 Advantages of Arc Welding

(i) Flux shielded manual metal arc welding is very simple as compared to all arc welding process.

(ii) Cost of equipment is very low.

(iii) Portable equipment is available.

(iv) Various type of metal and their alloys can be welded.
(v) Welding can be carried out in any position, i.e., horizontal, vertical, over head, flat or inclined.
(vi) High deposition of penetration can be achieved.

9.3.9 Limitations of Arc Welding

(i) FSMMA welding is slow process as compared to MIG welding.
(ii) Different material cannot be welded together.
(iii) Due to flux coated electrode, chance of slag entrapment and other defects are more as compared to TIG and MIG.
(iv) Metal transfer is not so clear as composed to MIG welding.

9.3.10 Applications of Arc Welding

(i) Shipbuilding industries
(ii) Pipes and penstocks joining
(iii) Construction work
(iv) Building industries
(v) Automobile and aircraft industries
(vi) Boiler, chemical, fertilizer industries, etc.

9.4 METAL INERT GAS WELDING (MIG)

9.4.1 Introduction *[UPTU-2010-11]*

MIG welding is define as **metal inert gas welding**. It is also one of the type of arc welding. In this process no pressure is applied for welding. In this process of welding wherein coalescence is produced by heat the workpiece with an electric arc establish between a continues feed of metal electrode (copper coated) and the workpiece. No flux is used as used in submerged arc welding (SAW) but a shielding gas (Ar, He, CO_2) is used. It is also known as **gas metal arc welding (GMAW).**

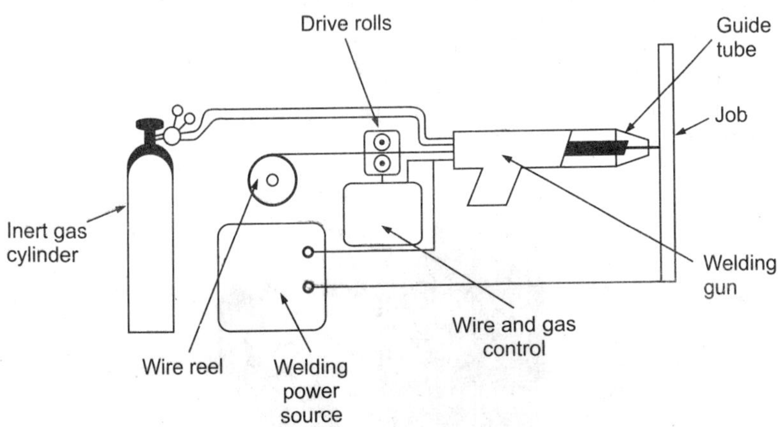

Fig. 9.20 Setup for metal inert gas (MIG) welding

9.4.2 Principle of Operation

Before welding set the current, wirefeed speed and electrical connections. Now arc is struck by one of the two method.

First method current and shielding gas flow is switched on and electrode is scratched against the job as usual practice.

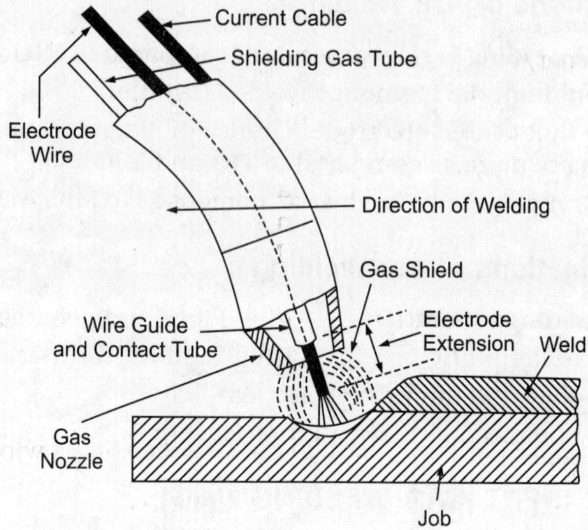

Fig. 9.21 Showing the working of metal inert gas welding (MIG)

For striking the arc by IInd method–electrode is made to touch the job is restructed and moved forward to carry out welding but before striking the arc shielding gas, water and current is switched on during the welding. Torch should be 10 – 12 mm. Away from the workpieces and arc length is kept between 1.5 to 4.0 mm. Arc are basically two types:

(i) Seft adjusted arc
(ii) Self-controlled arc

9.4.3 Equipments Required for GMAW

(i) welding power source with cables.
(ii) welding gun filler wire on a coiled spool.
(iii) shielding gas cylinder, pressure regulator and flow meter.
(iv) control switch.

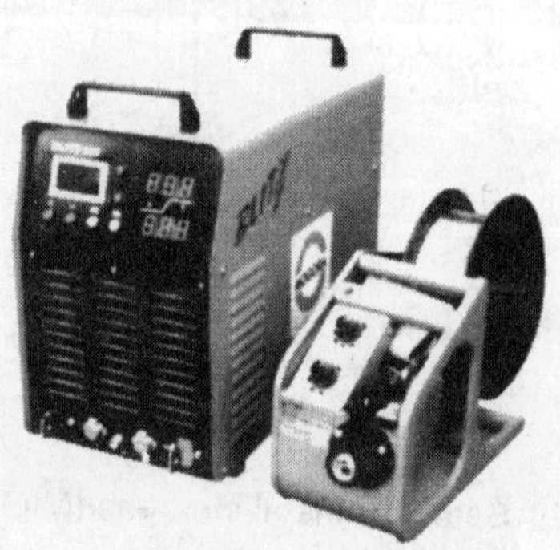

Fig. 9.22 MIG/MAG welding machine [21]

9.4.3.1 Power Source

AC transformer or DC generator with rectifier both type of power source can be used but in usual **dropping** characteristic type power preferred. DC electrode negative produce weld with shallow penetration therefore used for thinner sections DC electrode +ve is preferred when **aluminium** and **magnesium** are welding. DC RP also provide deeper penetration. AC is not recommended because of unequal burn off rater during –ve and +ve half cycles.

Both, transformer/rectifier set and a motor-generator power source can be used in GMAW (MIG). Power source of MIG are generally rated at 60% and 100% duty cycle for semi-automatic or automatic system. For automatic GMAW process usually a rising V-I characteristic power source preferred. A flat type power source for GMAW is very famous as it provide self-regulation of the welding arc. And such can be used. In this type of system welding current is directly related with wire feed rate. The MIG welding machine must produce either a constant arc-voltage or a Rising-Arc Voltage (RAV). The constant voltage amperage (CVA) characteristics gives an operator a great latitude in arc length without any appreciable change in voltage.

9.4.3.2 Welding Gun

Welding gun is device which is used to the copper coated wire for welding purpose.

Push type and pull type of welding gun is available for welding purpose. Welding gun may be aircooled or water cooled gun working up to 200 A should be aircool.

9.4.3.3 Wire feed Mechanism

It may deliver electrode (wire) to the gun at a constant speed. Different diameter of copper coated wire is available for welding such as 0.8 mm, 1.2, 1.6 mm, 2.0 mm etc. according to the material welded. A standard wire spool may have from 1 to 15 kg in weight.

There are basically two types of wirefeed mechanism are available for a standard MIG welding. **In First** a shielding gas is unnecessary in modern flux cored welding as the deoxidizers that take placed in the flux inside a tubular electrode and the **second** one is (type of wirefeed control) for a magnetic flux MIG welding. This type of wirefeed mechanism has some kind of magnetic flux reserviour that feeds the flux directly into the MIG gas.

9.4.3.4 GMAW Variables

GMAW is a semi automatic welding process for this purpose operator requires just only on or off the switch for welding purpose before stabilizing the arc (welding) it is usual to set the open circuit voltage on the panel of the power source not except only setting the open circuit voltage it is also required to set the some other variable (parameters) to control the welding process to obtain the desired results. Several others parameters such as

- wirefeed rate,
- nozzle to plate distance,
- gas flow rate,
- electrode stickout,
- electrode to work angle, etc.

9.4.3.5 *Arc Voltage*

Arc voltage is controlled only by the open circuit voltage (OCV) with a flat characteristic power source.

9.4.3.6 *Shielding Gas*

Basic function of shielding gas to protect the molten metal and the electrode end against atmospheric contamination. Argon is used for **Al, Mg, Cu, Ni, Ti** while **Helium (He)** is used for **Al**, **Mg**, **Cu** and **CO_2** for **mild steel.**

Major shielding gases that are used for MIG processes **argon, helium, carbon dioxide** and oxygen. Argon is the principal inert gas used to weld non-ferrous metal such as **Al, Mg, Cu, Ni, Ti, etc.** on the otherhand helium is used for better contrary of porosity as well as better arc stability because of its greater density and its steel and alloys are calloided then carbon dioxide gas is used as inert gas as CO_2 have an external low cost when compared with Helium, Argon, and it is capable to produce a sound weld.

Compossession chart for different inert gases of different metal

Metal	*Gas*	*Gas effects*
Aluminium	Argon Ar 25% He 75%	Helps to remove surface oxides Oxide removal + Porosity control
Mild steel	Carbon dioxide (CO_2)	Low spatter
Magnesium	Argon	Helps to remove surface oxides
Stainless steel	95–99% He oxygen 1–59%	Oxygen add to arc stability and reduces under cutting
Nickel	Argon Ar 25–80% He 50–75%	Helps to control base metal fluidity and provide good welding bead

9.4.4 Different Type of Materials can be Welded by MIG

Such as:

(i) Carbon and low alloy steel
(ii) Stainless steel
(iii) Heat resistant alloys
(iv) Magnesium alloys
(v) Copper and its alloys
(vi) Aluminium and its alloy
(vii) High strength low alloy steel (HSLA)

9.4.5 Advantages of MIG Welding *[UPTU-2010-11]*

(i) Less number of spatters as compared with MMA welding.
(ii) MIG is very faster process as compared with TIG welding process.
(iii) Deep penetration can be achieved through this process.
(iv) No use of flux during the welding process.
(v) Process can be easily mechanized.
(vi) MIG produces a high quality, weld bead with minimum defeats.
(vii) Large metal deposition rate achieved by MIG welding process.

(viii) High arc travel speed associated with MIG welding reduces distortion considerably.

9.4.5.1 Limitations of MIG Welding *[UPTU-2010-11]*

(i) Welding equipment is more costly and complex as compared to MMA welding.
(ii) Trained operator is required to perform the operation.
(iii) Process is not economically for job shop production.
(iv) All types of material can not be welded.

9.4.5.2 Applications *[UPTU-2010-11]*

(i) For welding of Al, Cu, Mg, Ni and their alloys.
(ii) For welding of aircraft, pressure vessels and ship building industry.
(iii) For manufacturing of refrigerator parts, etc.
(iv) Rail road Industries.
(v) Transport Industries.

9.4.5.3 Features of MIG Welding

(i) It is one of the best way of welding because it can replace TIG, SMAW, SAW, etc.
(ii) Self-adjusting arc is associated in MIG.
(iii) MIG uses constant voltage power source short-circuit current is very-very high in MIG.
(iv) Self adjusting arc (one of the important aspect) voltage is directly proportional to gap between the electrode and the base metal.
(v) Wirefeed rate is very important for the same current if we use the same diameter wire the burn rate will be higher for same current.

9.4.6 Metal Transfer in MIG

(i) Spray transfer
(ii) Global metal transfer
(iii) Dip transfer
(iv) Pluse metal transfer

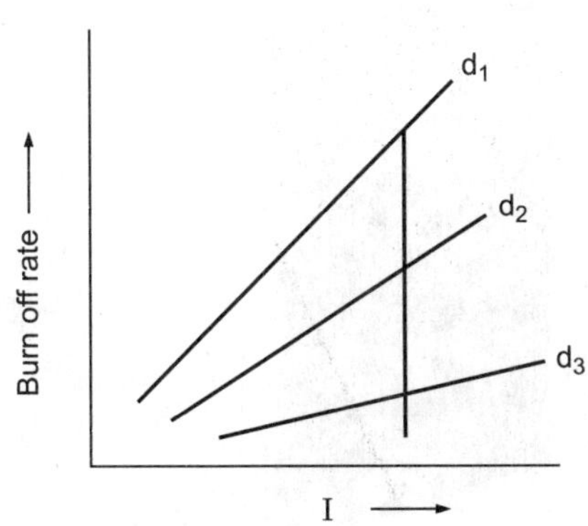

Fig. 9.23 Relation between current and burn off rate [6]

9.5 TUNGSTEN INERT GAS WELDING (TIG)

It is one of the type of arc welding process where in coalescence is produced by heating the job with an electric arc struck between a tungsten electrode and the job. A shielding gas (argon, helium, or both) is used to avoid the atmospheric contamination of the molten metal weld pool. Filler metal may be added if required.

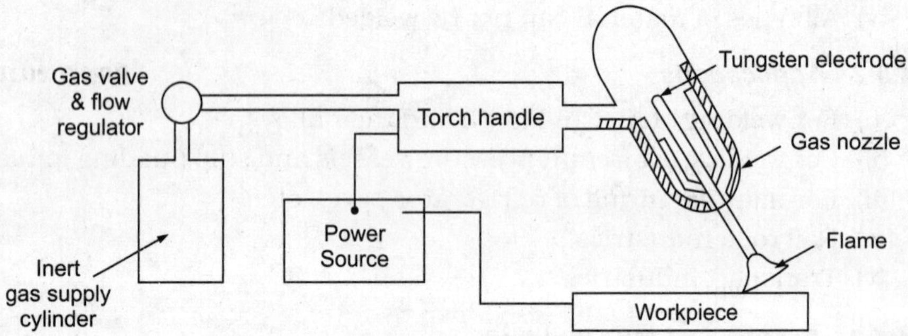

Fig. 9.24 Setup of a typical TIG welding

9.5.1 Principle of Operation

Welding current, inert gas supply, water are turned on. Electric arc is struck between the non-consumable electrode and the workpiece by touching the electrode with workpiece or using a high frequency unit. In Ist method arc is initially struck on a scrap metal piece and then broken by increasing arc length.

In IInd method a high frequency current is super-imposed on the welding current. Welding torch is brought near to the job when electrode tip reaches within a distance of 3 to 2 mm from job. A spark jumps across the air gap between the electrode and job. Air path gets ionized and arc is established.

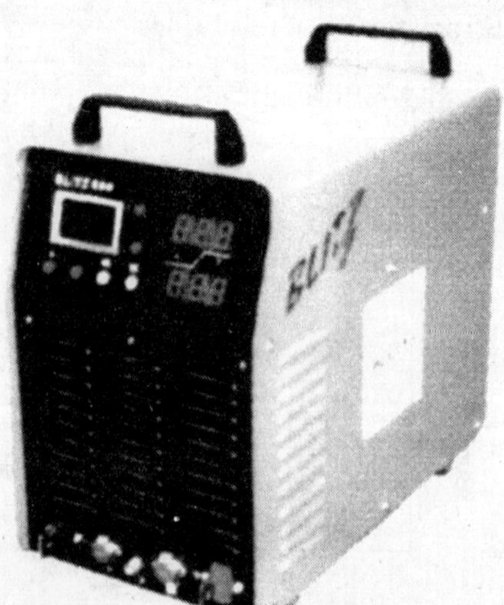

Fig. 9.25 GTAW/TIG welding power source [21]

TIG welding is also known as Gas Tungsten Arc Welding (GTAW) both the **AC** and **DC power** source can be used for GTAW. Electrode employed varies in diameter from 0.5 to 6.5 mm carrying current from 5 A to 650 A; welding torch used for carrying current higher than 1000 A is normally water cooled. GTAW is all position (1G, 2G, 3G, 4G, 6G) welding and gives the highest quantity weld amongst commonly used arc welding process.

DC is prefered over AC for welding of **SS**, **nickel**, **copper** and its alloys.

While AC is used for welding of **Mildsteel**, **Aluminium**, **maginesium** and their alloys.

9.5.1.1 Equipments Used in GTAW

1. Welding torch, tungsten electrode and filler rod.
2. Welding power source, high frequency unit, DC suppressor unit and cables.
3. Inert gas cylinder for shielding purpose, pressure regulator, flow meter
4. Coolings water supply.
5. Gas and water solenoid valve.

9.5.1.2 Welding Torch

Welding torch should be air or watercooled. A tungsten electrode held in a collet and feeds shielding (inert) gas to the weld puddle water cool torch is preferred when welding current go beyond 150 A.

Fig. 9.26 TIG welding torch [21]

9.5.1.3 Inert Gas

Different type of inert gas can be used in TIG.

(i) Argon (Ar)
(ii) Helium (He)
(iii) Ar+He mixture
(iv) Ar–O_2 mixture
(v) Ar–H_2 mixture.

Pressure regulator and flowmeter are used to stepdown the pressure of inert gas from cylinder, i.e., 140 kg/cm^2 to 150 kg/cm^2 and feed the same at definite flow rate for **M.S, Al** and its alloys, **Cu**, **Ni**, and its alloy and **stainless steel** welding generally Ar-He or their mixture is used.

9.5.2 Advantages of GTAW

(i) TIG welding is very suitable for high quality welding of thin material.
(ii) Deeper peneration can be achieve through this process.
(iii) No flux is used so there is no danger of flux entrapment.
(iv) As this process can be seen with the help of goggle therefore operator can exercise a better control on the welding process.
(v) TIG is very suitable for welding of non-ferrous metals and stainless steel.
(vi) TIG welding is suitable for welding in all position (1G, 2G, 3G, 4G, 6G).

9.5.3 Limitations

(i) MIG welding is much faster process as compared to TIG welding under similar application.
(ii) Tungsten if transfer to molten weld pool can contaminate the same. Tungsten inclusion is hard and brittle.
(iii) TIG welding equipment are more costly as compared to other welding equipment.
(iv) Skill or trained operator is required for welding.

9.5.4 Applications of TIG Welding

(i) Basically welding of **aluminium**, **magnesium**, **copper**, **nickel** and their alloys, carbon, alloys or **stainless steel**, **inconel**, **high temperature** and hard surfacing alloys such as **zirconium**, **titanium**, etc.
(ii) Welding sheet metal and thinner section.
(iii) Precision welding in atomic energy, aircraft, chemical and instrument industries.

9.5.5 TIG Welding Bead

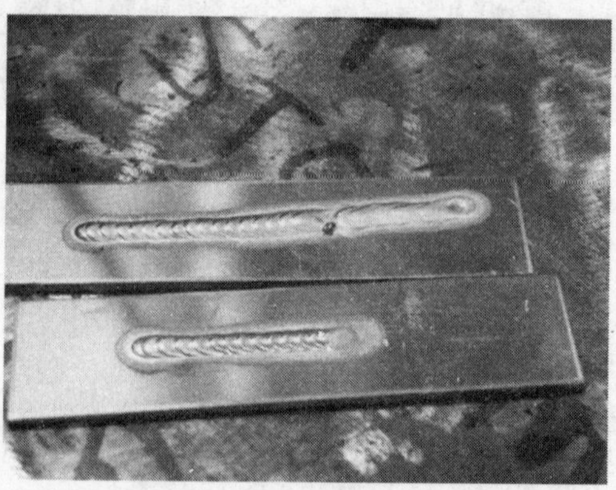

Fig. 9.27 TIG welding bead [21]

9.5.6 Comparision Between TIG (GTAW) and MIG (GMAW) Welding

Tungsten inert gas welding (GTAW)	*Metal inert gas welding (GMAW)*
1. In TIG welding a non-consumable (permanent) tungsten electrode is used.	**1.** MIG uses a copper coated wire of same chemical composition as the parent metal being welded.
2. TIG welding electrode, servers the purpose of producing the arc only filler rod needed if any.	**2.** MIG welding wire servers both the purpose of producing the arc as well as of filler metal.
3. TIG welding process is not so faster as MIG welding process as separately a filler rod is added.	**3.** MIG is faster process as compared to TIG because no filler rod is added separately.
4. TIG welding requires a skill operator.	**4.** Not so much skill operator is needed.
5. TIG welding torch may be watercool.	**5.** MIG welding torch are not water cooled.
6. Penetration is not so much deeper as compared to MIG.	**6.** Deeper penetration can be achieved very easily.
7. TIG welding is not recommended for greater than 6 mm thickness of sheet.	**7.** No such condition applicable in MIG welding.
8. If filler rod is added, operators both hands are engaged. So work must be held in position with clamps or fixture.	**8.** In MIG welding, wire electrode and gases come from same run and thus can be made easily automatic.

9.6 RESISTANCE WELDING *[UPTU 2011-12]*

Resistance welding is one of the type of pressure welding. During the welding, process pressure is applied.

Basic working principle of resistance welding depends upon

$$H = i^2RT$$

where H = heat generated i = current required during welding

T = time in sec R = resistance

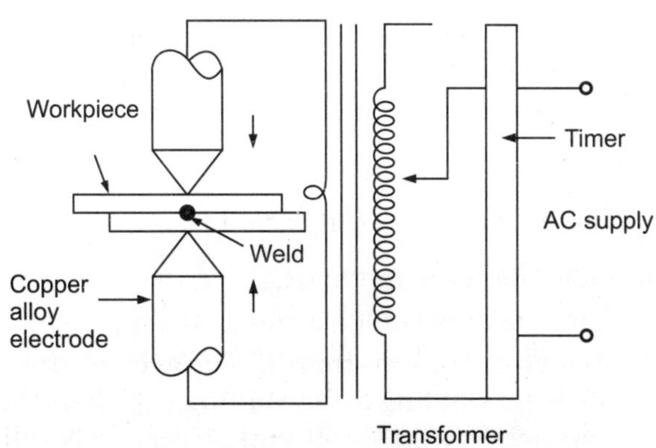

Fig. 9.28 Resistance welding block diagram

Definition: Resistance welding is a one of the type of welding process wherein coalescence is produced electrode by the heat obtained from resistance of the work to the flow of electric current in a circuit of which the work is a part and by the application of pressure. One of the unique feature of resistance welding that no filler rod and no flux is used during the process. Process is very simple.

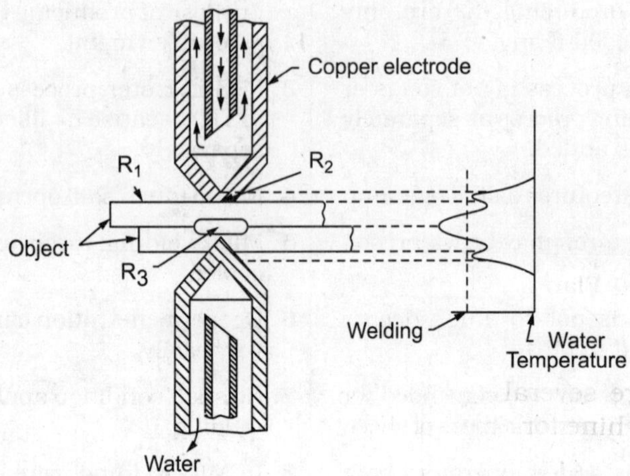

Fig. 9.29 Distribution of temperature in resistance welding [6]

9.6.1 Variables of Resistance Welding *[UPTU-2010-11]*

These are following: (i) Current, (ii) Resistance, (iii) Time and Pressure

9.6.1.1 Current (I)

In resistance welding current is obtained from step-down transformer, magnitude of current may be controlled through taps on the primary of the transformer. Low current does not provide proper fusion where as if welding current is too high the entire thickness of the work metal between the electrode is heated to the plastic state by the time.

AC or DC system are used for supply purpose. High frequency resistance welding is used for application of continuous seam or butt seam welding. But single phase AC of mains frequency, i.e., 50 Hz are used.

9.6.1.2 Resistance

Here resistance is define as total resistance, i.e.,

(i) Resistance of workpieces metal (R_1).
(ii) Contact resistance between electrodes and workpiece (R_2).
(iii) Resistance between the two mating surfaces of the workpiece (R_3) for getting a sound welding and avoiding over heating the electrode value of resistance R_1 and R_2 should be relatively much less.

Shape and size of the electrodes and pressure exerted on them also matter in reducing R_2. Electrode should be made from highly conductive material such as copper-cadmium or Cu-Cr alloys. Back resistance is define as the sum of all resist.

Shape of Electrodes

(i) Flat.
(ii) Truncated.
(iii) Spherical.

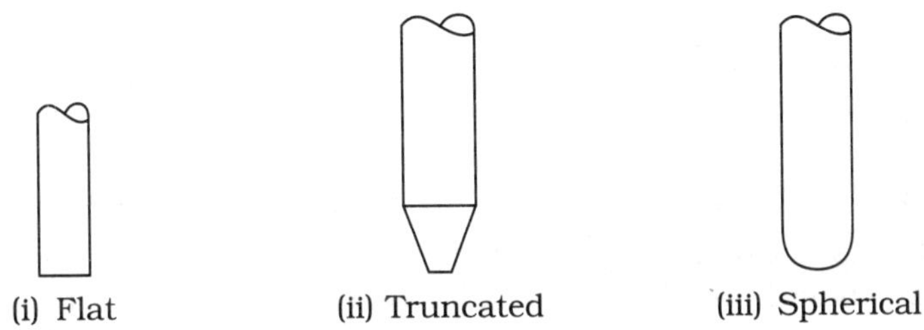

Fig. 9.30 Shapes of electrodes

There are several segments (period) of timing setup on a resistance welding machine for example a spot welding machine

(i) squeeze time
(ii) weld time
(iii) hold time
(iv) off time.

9.6.1.3 Advantages of Resistance Welding

(i) No filler rod/electrode is required.
(ii) High rate of production.
(iii) Semi-skilled operator can perform welding very easy.
(iv) Semi-automatic equipment.
(v) No part warping or distortion.
(vi) No need of edge preparation.
(vii) Less chance of workpiece burn through.
(viii) Less time consuming process.
(ix) Suitable for mass production.
(x) Metallurgy of workpiece is not so much change.

9.6.1.4 Limitations of Resistance Welding

(i) All types of metal cannot be welded.
(ii) Complex shapes of part cannot be join.
(iii) Initial cost of equipment is high.
(iv) Welding joint is not so strong.
(v) Bigger job thickness cannot be welded.

9.6.1.5 *Applications of Resistance Welding*

Resistance welding is used for joining **sheets, tubes**, **brass** for making furniture, **aircraft** and **automobile parts, fuel tank**, **wire fabrics**, **grills** and **containers**.

9.6.2 Types of Resistance Welding

Resistance welding are different types:

(i) Spot welding
(ii) Seam welding
(iii) Projection welding
(iv) Percussion welding
(v) Resistance butt welding
(vi) Flash butt welding
(vii) Stud welding

9.6.2.1 *Spot Welding*

Spot welding is one of the type of resistance welding. It is one of the easy and simplest way of welding. In spot welding two or more pieces of metal are held in between two copper electrode applying pressure then passing sufficient current to make the weld. Pressure and weld time are the fundamental variable of spot welding.

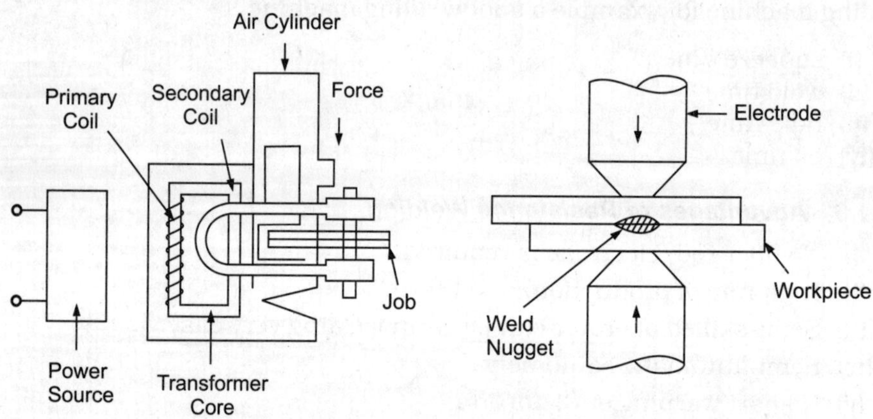

Fig. 9.31 Spot welding

9.6.2.2 *Advantages of Spot Welding*

(i) Cost of equipment is not high.
(ii) Untrained operator can perform welding easily.
(iii) High speed of production.
(iv) Less chance of warpage and distortion.
(v) No edge preparations is required.
(vi) Dependability.
(vii) No skill operator is required to welding.
(viii) Operation may be automatic or semi-automatic.
(ix) No metallurgical change.

9.6.2.3 *Applications of Spot Welding*

(i) Welding of low carbon steel, high speed steel bits of steel shanks and steel plates (up to 12 mm) are spot welded satisfactorily.
(ii) Different type of stainless steel like utensil and cutlery are spot welded.
(iii) Aluminium, Al-Mg alloys, copper and its alloys, Ni and its alloys, monel metals are also welded.
(vi) Spot welding frequently used in automobile and aircraft industries.

9.6.2.4 *Types of Spot Welding Machines*

Several types of spot welding machine is available:

(i) Rocker-arm type machine.
(ii) Portable machine type.
(iii) Press type machine.
(iv) Multiple electrode type spot welding machine.

(i) Squeeze Time. It is the time between the initial application of the electrode pressure on the workpieces and initial application of current to make weld. Welding current applied at end of squeeze time.

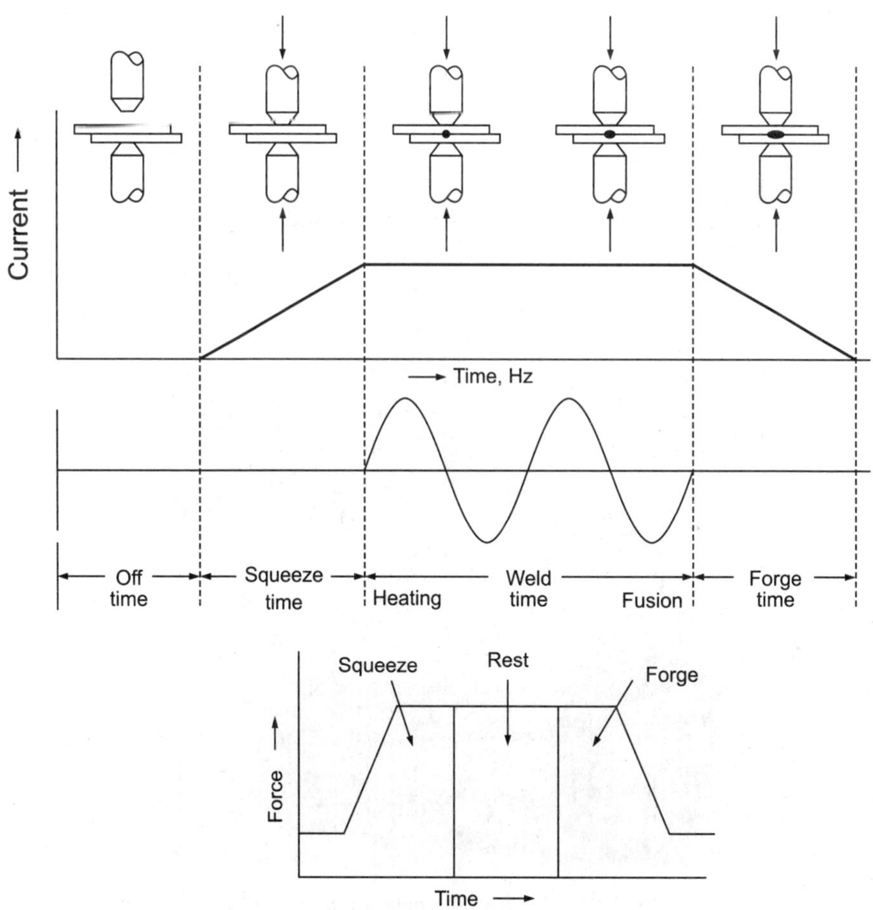

Fig. 9.32 Typical stages of spot welding [6]

(ii) Weld Time. During this stage current flows through the circuit from electrode to workpieces.

(iii) Hold Time. It is the time during which forces acts at the point of welding after the last impulse of welding current ceases electrode pressure is maintained until the metal has some what cooled.

(iv) Off Time. Once joint is formed it is the time when electrode will be with draw from the job.

9.6.3 Seam Welding

Seam welding is another type of resistance welding process wherein coalescence at the flying surface is produced by the heat obtained from resistance to electric current through workpiece held together under pressure by electrode roller. Seam welding is also known as **continuous spot welding** as the spot produced in workpiece continuously.

9.6.3.1 Principle of Operation

In continuous spot welding or seam welding workpieces passed through two copper roller under pressure continuously a current impulse is applied through the roller to the workpiece in contact with them heat generated thus makes the workpiece plastic and pressure from the electrode complete the weld. A step-down transformer is used for supply.

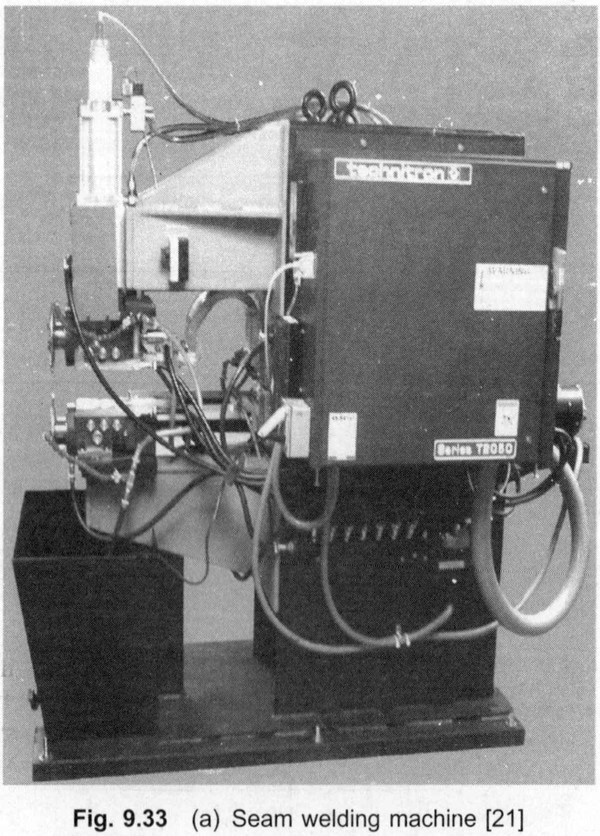

Fig. 9.33 (a) Seam welding machine [21]

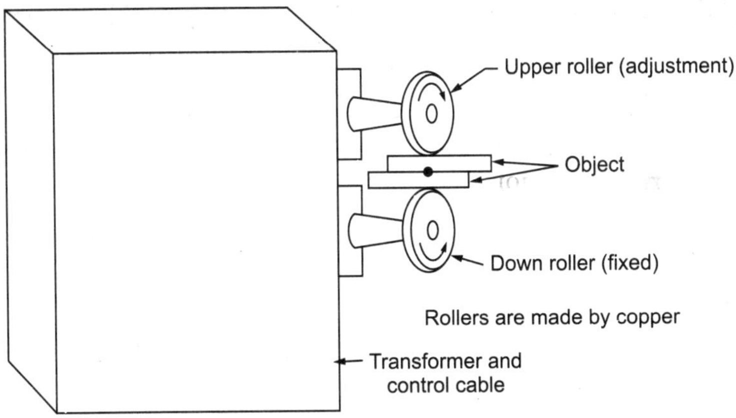

Fig. 9.33 (b) Seam welding machine (Block diagram)

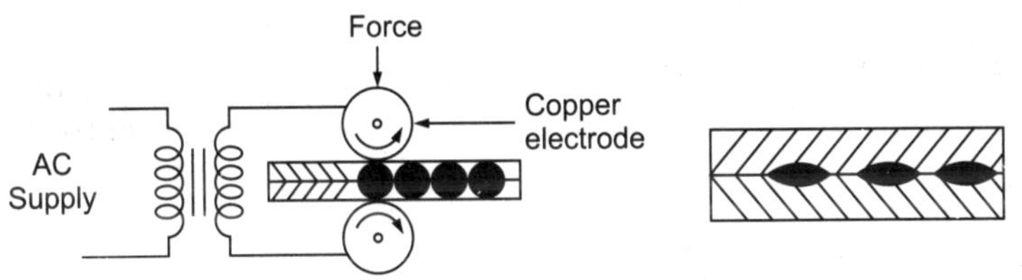

(i) Stitch seam welding

(ii) Roll-spot seam welding

Fig. 9.34

9.6.3.2 Seam Welding Equipment

Basic construction of seam welding is same as spot welding machine. Except that the electrode are mechanically driven rotating discs. Generally seam welding operated by single phase AC through, some are designed to operate on three-phase also.

(i) Power Source. A step down transformer is used for power source which decrease the voltage and increases the ampear (I).

(ii) Electrode. A good conductor of conductivity material is used as electrode material such as copper.

(iii) Control Pannel. It is used for regulating, timing and sequencing the application of welding current and force and the rate of movement of the work between the welding electrode.

9.6.3.3 Types of Seam Welding Machine

(i) Circular. This is very common type of seam welding machine in which face of the electrode wheels are parallel to the throat of the machine and throat depth is typically 30 cm to 90 cm.

(ii) Longitudinal. Longitudinal type of machine are used for welding short seam in containers.

(iii) **Universal.** Seam welding replaces spot welding for producing continuous leak proof joint for use in sheet metal item (containers) which are frequently used in automobile industries. Figure 9.32 shows a seam welding machine which consist the two roller upper and lower both roller are made by copper and its alloys. One (upper) roller is adjustable while down roller is fixed. Wheel/roller may be aligned either in line with throat or transverse to it. When two roller/wheel are in line they are known **longitudinal seam welding machine.**

9.6.3.4 Advantages of Seam Welding

(i) It is capable to produce gas tight or liquid-tight joint.
(ii) Semi-skill person can perform welding very easy.
(iii) Overlap can be less than for spot or projection weld.
(iv) Process is suitable for mass production.

9.6.3.5 Limitations

(i) Process is limited only lap joint.
(ii) Complex shape joint cannot be welded.
(iii) Very high thickness of plate (> 3 mm) cannot be welded.

9.6.3.6 Applications

Seam welding is used for welding:

- carbon steels
- alloys steels
- stainless steel
- alloys of aluminium
- nickel and
- magnesium.

Seam welding produce **water**, **air tight** joint. Copper and its alloys cannot be welded by seam welding it is limited only to lap-joint. Seams of cylinders and cabinet to circular seams for welding bottoms in end of cylindrical tanks.

9.6.4 Projection Welding

Projection welding is a resistance welding process wherein coalescence is produced by the heat obtained from resistance to electric current flow through the workpiece held together under pressure by electrodes. Projection welding is another variation of spot welding. Small projections are raised on one side of the sheet or plate with a punch and die. Projection act to localize the heat of welding circuit during welding process. Projection collapse owing to heat and pressure, and part to be joined are brought in close contact.

One of the most common application of projection welding is for attaching small fastness, nuts, special bolts and similar part to large component.

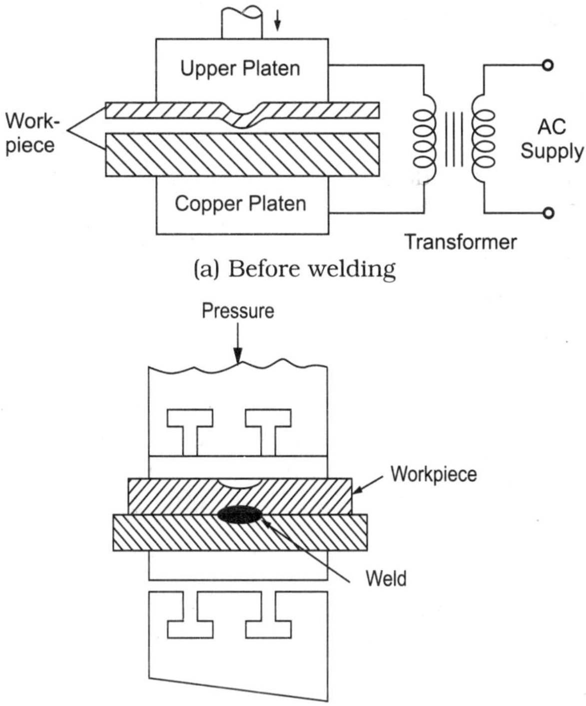

(a) Before welding

(b) After welding

Fig. 9.35 Working principle of projection welding

9.6.4.1 Advantages of Projection Welding

(i) A large number of joint can be weld at a time.
(ii) Scale, rusted, oil, metal coated can be joint easily.
(iii) Projection welding requires less pressure, less amount of current for welding purpose.
(iv) Projection welding locates the weld at certain desired points.
(v) Projection welding electrode possess longer life.
(vi) Weld can be made in a fraction of seconds.

9.6.4.2 Limitations of Projection Welding

(i) Very thick sheet cannot be weld very efficiently.
(ii) Metals those are not stronger enough to support projections cannot be projection welding satisfactorily.
(iii) For same strength of joint projection (all) should be same height.
(iv) Very difficult to weld two different metal.

9.6.4.3 Materials that can be Join by Projection Welding

Low carbon steels, coated metals such as:

- Galvanized steel
- Tin plate
- Stainless steel
- Titanium alloys
- Naval brass, etc.

9.6.5 Percussion Welding

In this resistance welding, heat obtained for welding purpose from an arc produced by a rapid discharge of stored electric energy. Pressure is applied rapidly during and immediately following the electric discharge. For storing the electrical energy a electrostatic capacitor is used.

9.6.5.1 Procedure for Welding

(i) First of all workpiece to be joined by C clamp in moving platens.

(ii) Now two workpiece are brought near to each other by applying a appropriate pressure.

(iii) Arc produced between the workpiece by carrying the workpiece in light contact in such a manner current start to flow after this gap is maintained between them to produce the arc. Sometimes nib is formed on the one workpiece surface of the arc where joint is needed. Current start to flow through this nib which result melting and arc is produced due to gap.

(iv) At this stage, more pressure is applied which extinguisher the arc.

(v) Pressure released as the workpiece cooled.

(vi) After this workpieces unclamped.

9.6.5.2 Advantages of Percussion Welding

(i) Arc temperatures is high as compared to flash butt welding.

(ii) Strong joint are produced.

(iii) Upsetting is very minimum. In arc it is upto nil.

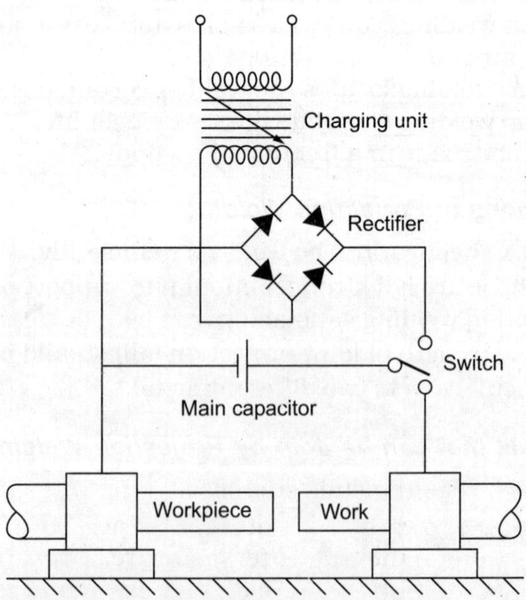

Fig. 9.36 Setup for percussion welding

9.6.5.3 *Disadvantages of Percussion Welding*

(i) This process is only limited up to butt joint only.
(ii) Equipment is very costly as compared to spot welding.
(iii) Equipment must be extremely rigged and provided with accurate holding fixtures and requires sensitive timing devices, etc.

9.6.5.4 *Applications of Process*

(i) This process is applicable for welding of stellite tips to tools, silver contact tips to copper, copper to aluminium, etc.
(ii) It is used in telephone industries.
(iii) This process is used for welding fine wire leads to filaments in lamps, etc.

9.6.6 Flash Butt Welding

Flash butt welding has largely replaced upset butt welding.

Definition: Flesh butt welding is a resistance welding. In flash butt welding, the ends of the stock are clamped with a slight separation before welding as the current is turned on, it jumps the gap and creates of flash with a great deal of heat.

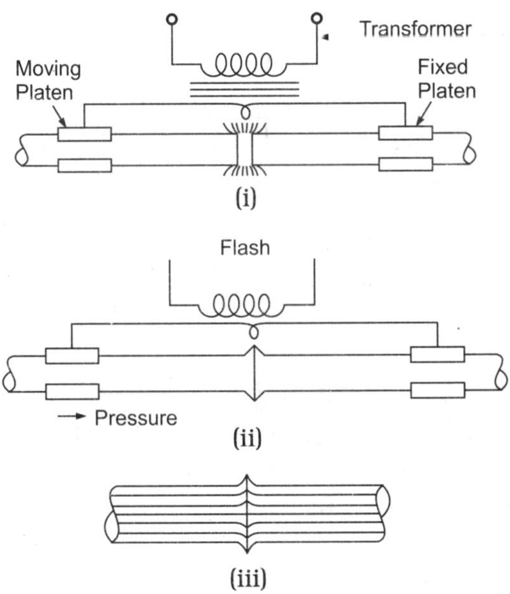

Fig. 9.37 Steps in flash butt welding.

9.6.6.1 *Working Principle*

Two pieces are moved towards each other in an accelerated motion (under pressure) and they reach the proper temperature. They are forced together under high pressure and current is cut off. Process produced a homogeneous weld between two sheets, wire or bars without overlapping and without addition of filler metal. Most of flash welding machine are semi-automatic or automatic type there are two plates in machine, stationary and moving and later is controlled either mechanically or by hydraulically. One workpiece held in stationary platen and other held in movable platen high voltage high current is turned flashing is established between the two abutting ends of the workpiece.

9.6.6.2 *Metal Welded by Flash Welding*

Low carbon steel, magnesium alloys, **Tool steel**, **Stainless steel**, **Aluminium** and its alloys **copper alloys,** etc.

9.6.6.3 *Advantages of Flash Welding*

(i) A lot of dissimilar metal can join.
(ii) Strength factors up to 100% produce up flash welding.
(iii) No edge preparation is required before welding.
(iv) It is fast welding process than other.
(v) Cost of joint is low.
(vi) An account of only a small portion of the metal be ignited current consumption is less as compared to butt welding.
(vii) Length of alignment of workpiece is maintained to a high degree of accuracy.

9.6.6.4 *Limitations*

(i) Metal lost during flashing and upsetting.
(ii) Straitness of workpiece difficult to maintain.
(iii) Operator need protection from flying particles as process is fire hazard.
(iv) Cast iron, lead and its alloys etc. cannot be welded by this process.
(v) Opening between the Jaws of the gripping clamps also limits the size of the workpiece.
(vi) Process is suitable only for lap joint
(vii) Very heavy shapes can not be welded.

9.6.6.5 *Metals can be Welded by Flash Welding*

As a general rule it can be said that any metal which can be forged can be also flash melted easily. Also it is possible to weld.

Metal commonly welded by flash welding are:

- low carbon steel
- low alloys steel
- tool steel
- stainless steel
- copper alloys
- aluminium alloys
- nickel alloys
- molybdenum alloys
- magnesium alloys and
- titanium allays.

9.6.6.6 Some Applications

This process of welding used in to mild steel or alloys steel shanaz welding of pipes and many other similar jobs.

- automobile industries
- welding of tubular and
- solid structural assemblies, etc. in
- aircraft, industries
- welding of band saw blade
- welding of tool steel
- drills, reamers
- taps.

9.6.6.7 Resistance Butt Welding

This type of the welding is also known as upset welding process. Basically upset welding process is used to join the metal piecess end to end. In this process the metal piece usually *bar* and *rods* of the same cross-section are held in suitable clamp or vice with their previously squared ends abilting against each other. Projecting length of the piece between the clamp are adjusted according to their cross-section and the corresponding materials of which they are made. A longer projection will offer more resistance and as short length will provide less resistance abuilting ends of the piece having cqual cross section should be kept exactly in the middle of clamp. Endeavour should be to adjust the projected length in such a way that the resistance affected by them to the flow of electric current is sufficiently enough to generate the desired amount of heat at the respective ends. Clamp holding the piece either from the electrode themselves or are fitted with separate electrode in them one of the clamp is rigidly fixed and other one is movable clamp. After building the ends together the current is switched on and the contacting surface heated to the fusion point at this moment additional mechanical pressure is applied by means of the hand lever or by spring attachment and in such a way weld completed. This pressure should be maintained for a few second, actual time depending upon the cross-section of the piece, to allow the metal piece to joint together.

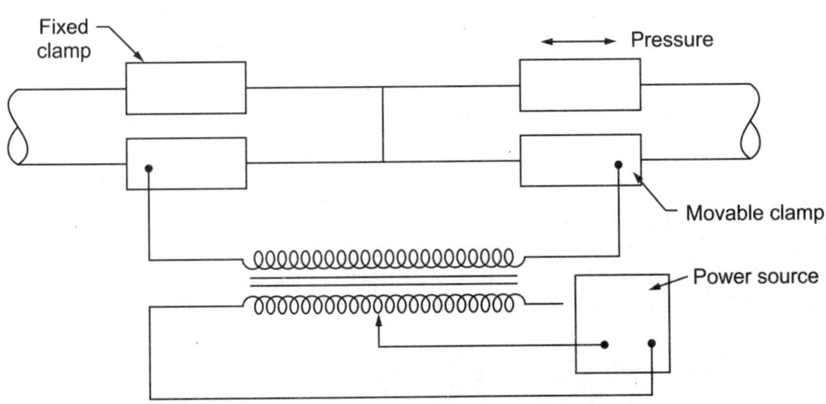

Fig. 9.38 Principle of resistance butt welding (UPSET)

In hydraulic type machine the hand lever is replaced by a hydraulic plunger. The *Butt welding* method is very suitable for joining end to end the items like **brass, rods, tubes wires,** etc.

For large sizes of stock, a better practice is to start the weld with a lower pressure etc. localise the heat at the joint. This can be accomplished by employing a handwheel or lever is place of the spring. When molten metal is seen flowing from the outer surface, the weld is consolidated by application of higher pressure, preferably hydraulically or pneumatically. Enough care should be taken in gripping very thin wires to ensure that the grip is at the edge of the clamp.

9.6.6.8 *Difference Between Flash Welding and Upset Butt Welding*

(i) No arcing (flashing) take place in upset butt welding which take place in flash welding.

(ii) Less current is consumed in flash welding but welding time is more.

(iii) Heat application in flash welding precedes pressure application whereas in upset flash welding, constant pressure is applied during heat processes particularly to eliminate flashing only.

(iv) Dissimilar metal can weld very efficiently.

(v) Faster process.

(vi) Flash welding can be used to weld the high alloy steels.

9.7 ATOMIC HYDROGEN WELDING

9.7.1 Introduction

In atomic hydrogen welding, welding heat is obtained from an electric arc maintained between two electrodes of tungsten in an atmosphere of shielding gas (hydrogen). Filler rod and pressure may or may not be applied it depends upon the need.

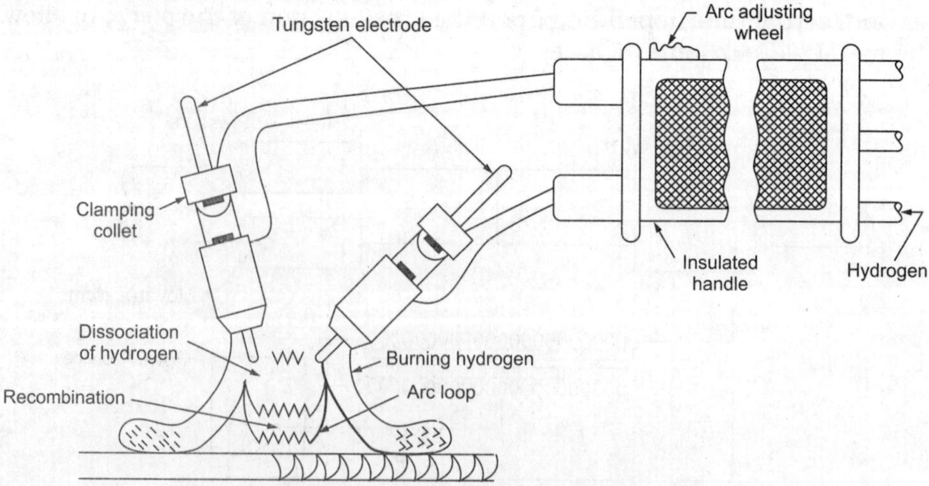

Fig. 9.39 Principal of atomic hydrogen welding

This welding process actually a combination process of electric arc welding and gas welding. Electrode holder consists two tungsten electrode which are arrange in an inclined position and hydrogen nozzle is in between the tips of these electrodes to eject the hydrogen, hydrogen molecule break into atomic hydrogen as electric arc strikes. This atomic hydrogen when touch the relatively cold metal, recombines into molecular hydrogen thus consider table heat generate, which melts the metal to be welded formation of atomic hydrogen proceeds with the absorption of heat:

$$H_2 \rightleftharpoons 2H - 421.2 \text{ kJ/mol.}$$

This atomic hydrogen recombines to form molecular hydrogen outside the arc, particularly on the relatively cold surface of the work being welded releasing the heat gained previously

$$2H \rightleftharpoons H_2 + 421.2 \text{ kJ/mol.}$$

metal is heated by the indirect arc and by the heat evolved above producing a temperature of about 3700°C of the flame.

9.7.2 Procedure

Supply the hydrogen gas and switch on the current:

- Arc is struck by touching the electrode momentarily and then with drawing immediately a proper distance (arc length approximate 1.5 mm).
- Current value can be adjusted as per the thickness of the workpiece.
- Atomic hydrogen welding, arc is held over the job to be welded till a molten pool forms.
- Depending upon the thickness, filler metal may or may not be added.

9.7.3 Equipments

1. Welding torch and tungsten electrode.
2. Hydrogen cylinder, pressure regulator, hose pipes etc.
3. AC power source.

9.7.4 Advantages of Atomic Hydrogen Welding

(i) High heat concentration is available.

(ii) Hydrogen gas itself acts as shielding gas.

(iii) Job is separate from the electric circuit.

(iv) Smooth, uniform, strong and ductile welds can be obtained.

(v) Hydrogen gas also cools the electrodes.

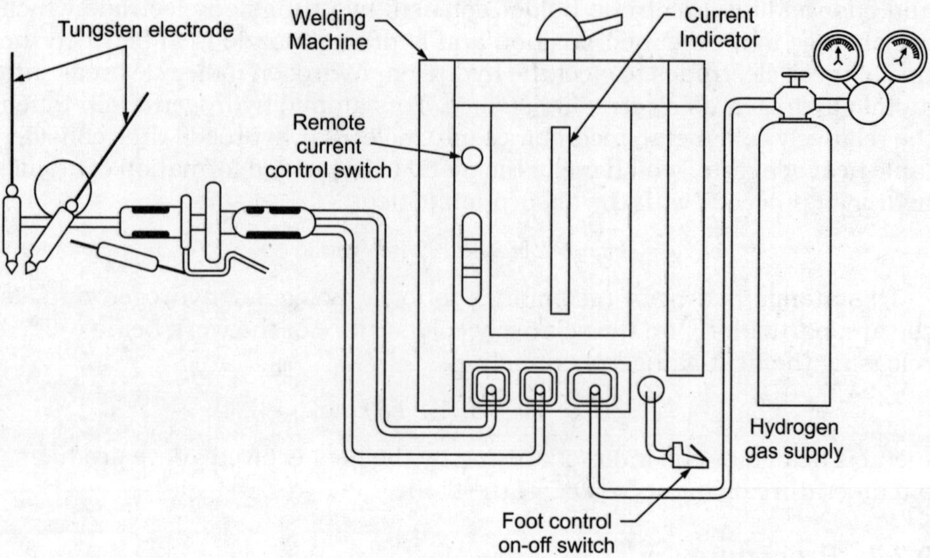

Fig. 9.40 Setup for atomic hydrogen welding

9.7.5 Disadvantages of Atomic Hydrogen Welding

(i) Welding speed is not high
(ii) Operating cost is very high
(iii) Skill person is needed for operating (welding).

9.7.6 Applications

(i) It is used in die repairs, welding of stainless steels, aluminium, copper etc.
(ii) It is also used to apply carbides and many hard facing alloys on soft metals.

9.7.7 Submerged Arc Welding *[UPTU 2010-11, 2011-12]*

It is one of the arc welding process wherein coalescences is produced by heating with an electric arc setup between the bare electrode (copper coated) and workpiece under a blanket of **granular flux** metal electrode continuously feed and act as a filler metal. No pressure is applied during welding process one of the important variable of SAW that welding is completed without usual **spark**, **spatter** and **smoke**. Manual metal arc welding is replaced by submerged arc welding.

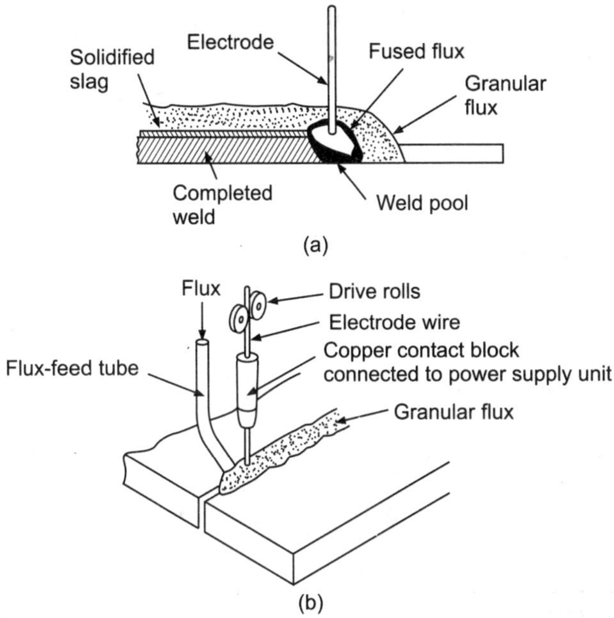

Fig. 9.41 Schematic diagram of submerged arc welding

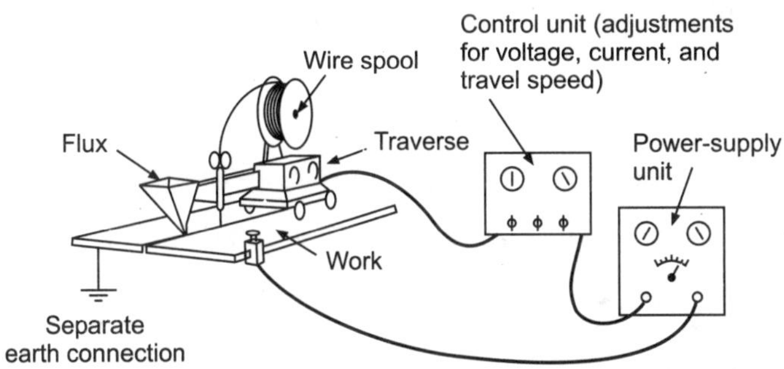

Fig. 9.42 Equipment for submerged arc welding (note that other types of traverse mechanism can be used)

9.7.7.1 Working Principle

Hopper is full with flux, as trigger is pulled and the flux starts depositing on the joint to be welded, arc may be struck either by touching electrode with job or by placing steel wool between electrode and workpiece before switching on the welding current or by using high frequency unit arc length is kept constant by using the **principle of self** adjusting arc. Backing strip or plate may be used to control penetration and to support large amount of molten metal associated with the process.

9.7.7.2 Aspects of SAW

(i) Current is very-very high.

(ii) Deposition of metal is very high.

(iii) It is used for 3 to 8 mm thickness sheet.

(iv) Heat affected zone (HAZ) is very large due to high current.
(v) Constant potential power source is used.
(vi) SAW is replaced by MIG welding as more metal deposition in MIG.

9.7.7.3 *Welding Equipments*

(i) Welding head.
(ii) Flux hopper.
(iii) Welding power source.
(iv) Flux.
(v) Bare electrode.

Fig. 9.43 Welding head and power source

9.7.7.4 *Advantages of Submerged Arc Welding (SAW)* *[UPTU-2010-11]*

(i) Deep penetration can be achieve by SAW.
(ii) Weld bead is very smooth.
(iii) Less probability of spark, smoke, flash or spatters.
(iv) High rate of metal deposition.
(v) Faster welding speed.
(vi) Strength of welded joint upto 100%.
(vii) No helmat is required.
(viii) Thicker plate can also weld easily.
(ix) No wastage of electrode.

9.7.7.5 *Limitations of Submerged Arc Welding (SAW)* *[UPTU-2010-11]*

(i) As welding is under a blanket therefore operator cannot see the weld performance.
(ii) Flux cannot be reused.
(iii) Skill operator is required.
(iv) Equipment is costly.
(v) Large **HAZ** which affect the mechanical and metallurgical properties of parent metal.

(vi) Edge preparation is must.
(vii) Cast Iron, Al alloys, Mg alloys, Pb and Zn cannot be welded by this process.
(viii) If backing strip is not used more chance burn off the weld.

9.7.7.6 *Applications*

SAW welding extensively used for welding of:

- pipes
- penstocks
- pressure vessels
- structural shape
- railroad, building industries
- shipbuilding
- nuclear power industries
- railroad industries.

9.8 ELECTROSLAG WELDING

Electroslag welding is also another type of the arc welding. Electroslag welding was discovered in Russia (Ukrane) and it is a welding process where in coal sense is produced by molten slag which melts the filler metal and the surface of the work to be welded, electroslag welding is quite similar to vertical submerged arc welding.

9.8.1 Working Principle

In electroslag welding process a granular flux is placed in the gap between the plate being welded and as the current is turned on, welding takes place in a watercooled copper shoes that bridge the gap of the joint as the flux melts, a slag blanket from 25.4 to 38.1 mm thickness is formed, high resistance of the slag causes most of the heating for the reminder of the weld thus electroslag welding is a progressive process of melting and solidification from the bottom upward.

Electroslag welding is used to weld metal from 3.8 mm to 38 mm thickness and the maximum thickness that can be weld by this process is upto

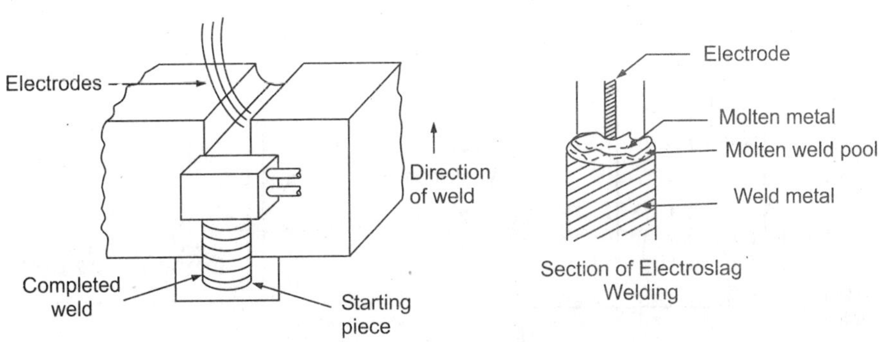

Fig. 9.44 Electroslag welding

100 mm. Molten metal and slag are retained in the joint by means of copper shoes that automatically move upward as the weld progresses by means of a temperature sensitive mechanism.

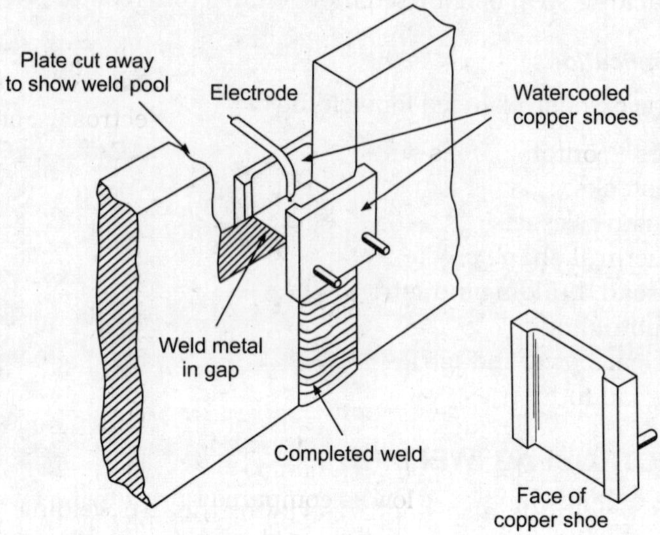

Fig. 9.45 Schematic diagram of the arrangement for electroslag welding

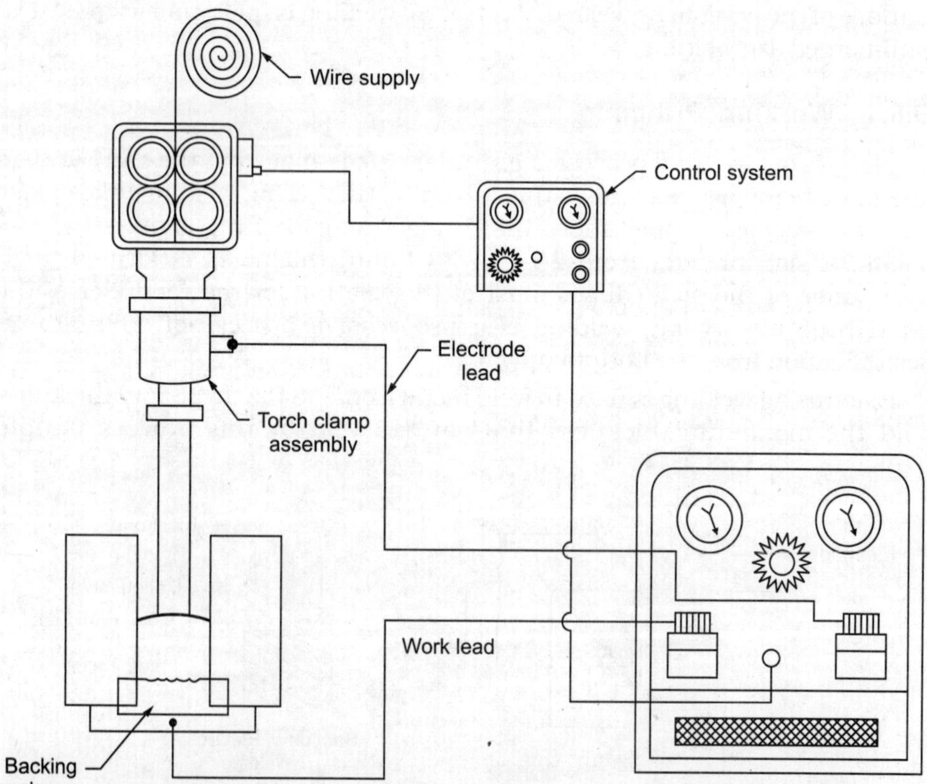

Fig. 9.46 Schematic diagram of the arrangement for electroslag welding

9.8.1.1 *Equipments Used in Electroslag Welding*

1. **Power Source.** DC (constant potential) feeding two electrode wire in parallel.
2. **Welding head.**
3. **Electrode Wire.** Both solid and tubular electrode wire can be used as the filler metal in both system of electroslag welding.
4. **Flux.** Molten flux is the slag that gives the electroslag process its name flux contain oxides such as – $\mathbf{SiO_2}$, $\mathbf{Al_2O_3}$, $\mathbf{CaF_2}$, **CaO**, etc.

9.8.2 Advantages

(i) Joint preparation is quite simple as compared to other welding process.

(ii) Very high thickness plate can be very easily weld in a single pass with more economically.

(iii) It gives extremely high deposition rate.

(iv) Distortion and thermal stresses are in very low percentage.

(v) Flux consumption is very low as compared in submerged arc welding.

(vi) No spark, no flash, no spattering during the electroslag welding process.

9.8.3 Limitations

(i) It is more economical than electroslag welding for joint blow 60 mm.

(ii) More chance of hot cracking and notch sensitivity in the heat affected zone (HAZ).

(iii) Complex shape joint cannot be welded by this process.

(iv) Electroslag welding tends to produce rather large grain size.

(v) Process is only limited to vertical position.

9.8.4 Electrogas Welding

Electrogas welding is a method of gas-metal-arc welding or flux cored arc welding wherein an external (shielding) gas is supplied and molding shoes, confine the molten weld metal for vertical position welding. Concept of electrogas welding is quite similar to electroslag welding the **basic difference is that an inert gas is used in electrogas welding** to shield the molten metal. Process is used on thinner material than those associated with electroslag welding.

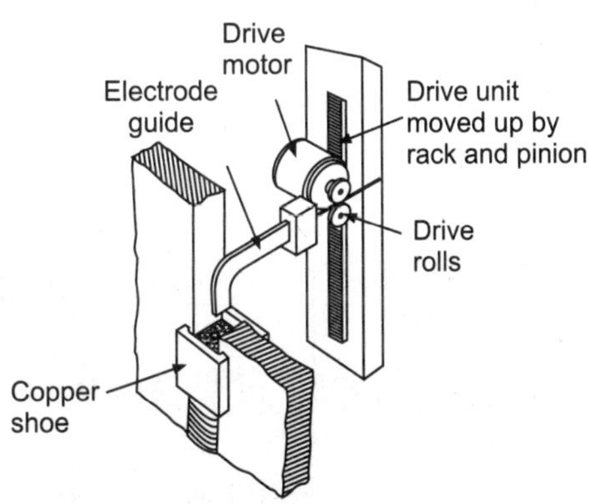

Fig. 9.47 Electroslag wirefeed unit

9.8.4.1 *Difference between Electroslag and Electrogas Welding*

Electroslag welding	*Electrogas welding*
(i) Energy required for welding is generated by electrical resistance of a molten slag bath which convert electrical energy into thermal energy.	(i) Heat required for welding purpose is obtained by maintaining a continuous electric arc between the electrode and the molten metal (weld) pool.
(ii) Weldpool is shielded by molten slag.	(ii) An inert gas (CO_2) is used for shielding purpose.
(iii) Either AC or DC can be used for electroslag welding.	(iii) Only DC can be used for electrogas welding.
(iv) In electroslag welding, arc is not continuous.	(iv) In electrogas welding, arc is continuous.
(v) Chance of flux entrapment is more.	(v) In this as no flux is used therefore no chance of flux entrapment (inertgas is used).

9.8.4.2 *Principle of Operation*

(a) Solid electrode (b) flux cored electrode.

(i) In solid electrode process gun (welding) is installed over the joint between the two plates to be welded, CO_2 is used as shielding purpose and electrode is fed through the welding gun and arc is produce between the starting plate at the bottom of the joint and electrode no flux is present on the top of the molten metal pool by this process upto 75 mm thick plates can be welded.

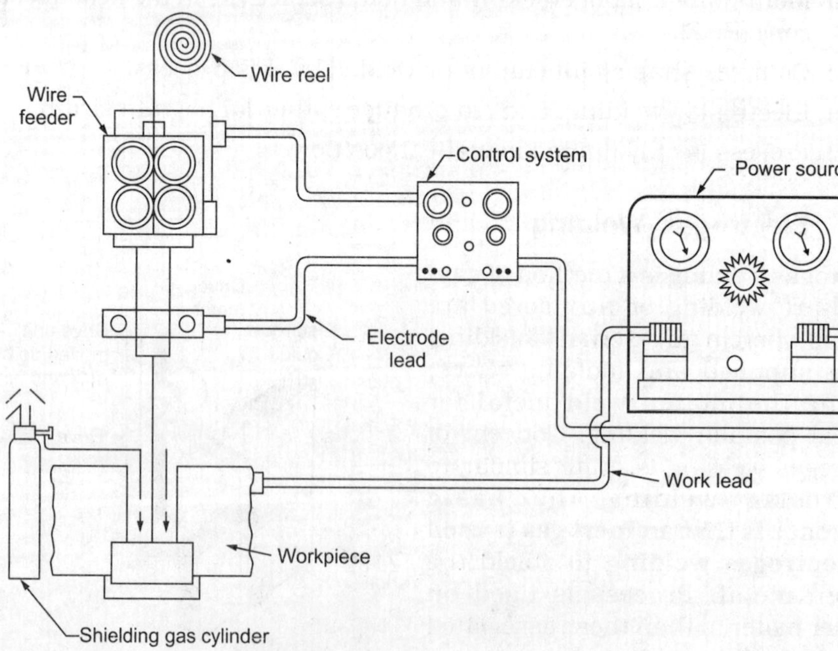

Fig. 9.48 Setup of electrogas welding

9.9 SOLID STATE WELDING

Solid state welding is also one of the important type of welding process in which the heat required for welding purpose is produced by rubbing or by applying hammering operation on two workpiece. Solid state welding are different type such as **friction welding, ultrasonic welding, diffusion welding, explosive welding,** etc., pressure is also applied.

9.9.1 Friction Welding

Without no doubt, the most successful development in the field of pressure welding has been that of friction welding machine used for this process looks some what like a large lathe fitted with two chuck one driven by a motor and other fixed, two parts to be joined are clamped in the chuck and one part is rotated this rotating component must be round in cross-section, but the part held in the fixed chuck can be a matching section or flat when the rotating chuck reaches the welding speed, parts come into contact under a light axial load, as the force rub two parts friction then generate heat and localised heat plastic zones are produced, with the end load maintained, heat continuous to be generated until the whole interface has reached a uniform temperature at the same time, the plastic metal starts to flow outwards towards the periphery, carrying with it any oxides present at the joint face. When sufficient heat produced, the relative rotation of the parts is stopped rapidly and the end load may be increased, weld time are short, being around 20 to 100 sec.

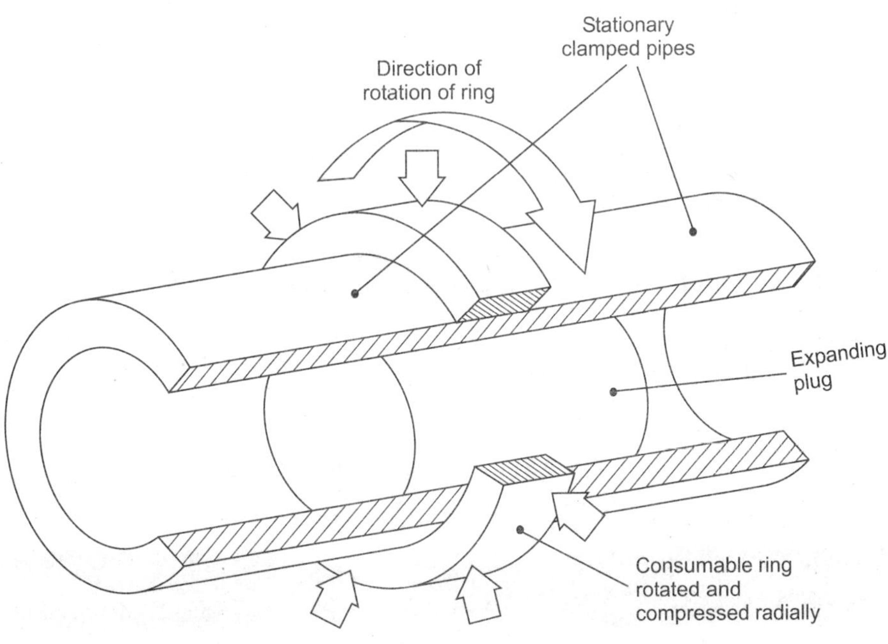

Fig. 9.49 Different forces for radial friction welding

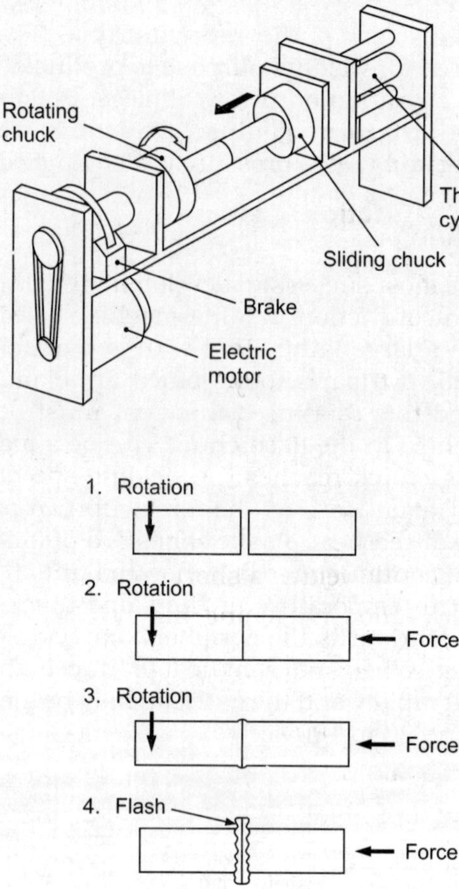

Fig. 9.50 Steps of friction welding [6]

Friction Welding

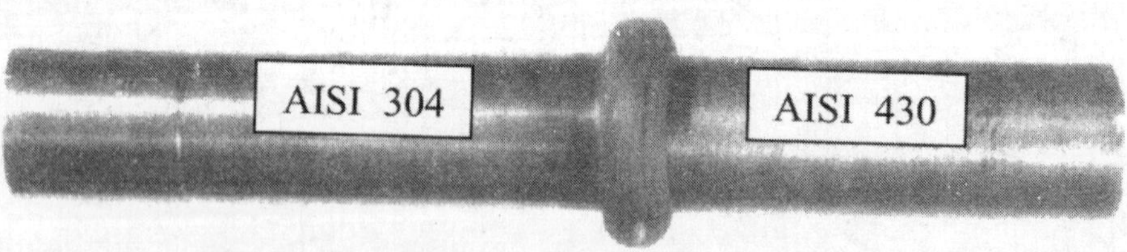

Fig. 9.51 Friction welded joint

Fig. 9.52 Friction welded tensile test specimen after test [6]

9.9.2 Welding Parameters (Variable)

(i) Relative speed.
(ii) Friction pressure.
(iii) Duration of heating.
(iv) Forge pressure.

9.9.3 Advantages of Friction Welding

(i) Operation is quite simple.
(ii) Low magnitude power is required.
(iii) Welding takes few sec once parameter is set.
(iv) Heat Affected Zone (HAZ) is very narrow.
(v) No flux, gas filler metal or slag present to cause imperfection in weld no smoke too.
(vi) Process suitable for mass production.
(vii) Good mechanical properties of joint.
(viii) Produce high quality weld in a short cycle time.
(ix) Cost effective method of joining highly dissimilar material and geometries.

9.9.4 Limitations of Friction Welding

(i) All type of metal cannot be joint.
(ii) Process is limited only butt welding, complicated or complex shape cannot be welded.
(iii) Process is limited only joining of small pieces in the form of bar stock.
(iv) Occasionally heavy flash is forced.
(v) Process is not suitable for short run production.
(vi) Tooling and equipment cost is very high.
(vii) Process is limited to joint pre-design joint.

9.9.5 Materials that Can be Welded

(i) Aluminium and its alloys.
(ii) Alloys steel.
(iii) Law carbon steel.
(iv) Carbon-steel.
(v) Stainless steel.
(vi) Tantalum.
(vii) Stainless steel to carbon steel.
(viii) Steel, alloys steel to carbon steel.
(ix) Aluminium to stainless steel.
(x) Copper to aluminium, etc.

9.10 FLUX CORED ARC WELDING (FCAW)

The flux cored arc welding process introduced in early 1950 is a modification of MIG/CO_2 welding process, in which solid wire is replaced by a flux cored electrode wire, i.e., a tubular wire filled inside with flux and alloy additions.

This process is growing in popularity. It is being used for more than 20% of arc welding. Some FCAW still uses CO_2 shielding, but the use of flux cored wire alone is increasing. In many cases, the flux-cored wire alone produces welds equal to or better than the original metal and its use eliminates the need for the gas shield equipment and cost of the gas.

9.10.1 Definition and Concept

The FCAW is a process in which coalescence is produced by heating with an electric arc between a continuous tubular consumable electrode and the work. The electrode is flux cored, i.e., the flux is contained within the electrode which is hollow. In addition to flux, mineral and ferro alloys in the core can provide additional protection and composition control.

The flux cored electrode is coiled and supplied to the arc as a continuous wire as in CO_2 welding. The flux inside the wire provides the necessary shielding of the weld pool. Additional shielding may (or may not) be obtained from an externally supplied gas (e.g., CO_2) or gas mixture.

9.10.2 Principle of Operation

As explained above, FCAW utilizes the heat of an arc between a continuously fed consumable flux cored electrode and the work. The heat of the arc melts the surface of the base metal and the end of the electrode. The metal melted off the electrode is transferred through the arc to the workpiece where it becomes the deposited weld metal. Shielding is obtained from the disintegration of ingredients contained within the flux cored electrode. Additional shielding may be obtained from an envelope of gas supplied through a nozzle to the arc area. Ingredients within the electrode produce gas for shielding and also provide deoxidizers, ionizers, purifying agents and in some cases alloying elements (for composition control). These ingredients form a glasslike slag, which is lighter in weight than the deposited weld metal and which floats on the surface of the weld as a protective cover. The flux cored electrode is fed into the arc automatically from a coil. The arc is maintained automatically and arc travel can be manual or by machine.

9.10.3 Welding Equipments

(i) A variable speed motor and motor control to power feed rolls which drive the electrode at a preset and uniform rate.

(ii) (a) A gun which houses a trigger to initiate and stop the electrode feed and flow of gas, electrical current to the arc, and, if used, water for cooling the torch,

(b) a nozzle which directs the shielding gas to the arc and weld pool (except for self shielded FCAW),

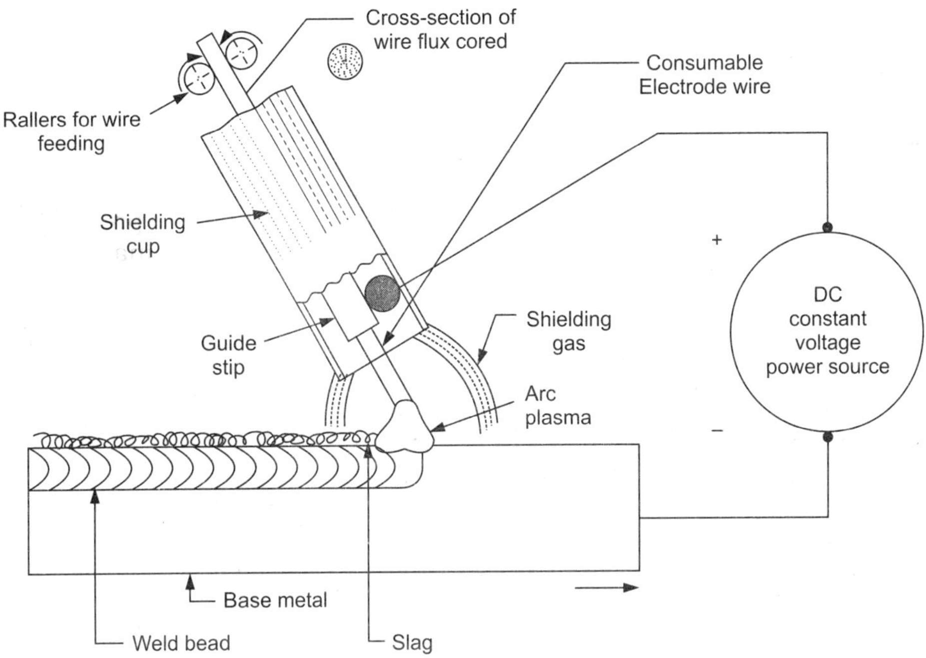

Fig. 9.53 Setup flux cored arc welding

(c) a contact tube at the axis of the nozzle to transfer welding current to the electrode and

(d) a system of cables, hoses, electrical connections, and casings to direct the gas, electrode, power and water if used.

(iii) A mount for the spooled or coiled electrode.

(iv) A control station containing the relays, solenoids and timers needed to integrate the system.

(v) A source of shielding gas, if needed, and a device for metering the flow rates of the gas.

(vi) A power supply to provide an appropriate amount and type of current.

(vii) A water supply for cooling if necessary.

The equipment described above can be adapted for use in semi automatic welding or mounted on fixtures for automatic or machine welding. As compared to GMAW, in FCAW higher current power sources and larger welding guns or torches are used. When FCAW without additional shielding is employed, the entire gas supply system is eliminated. This removes the gas cylinders, the regulator and flow meter, the hoses, the solenoid valve and the nozzle on the welding gun. In view of the amount of smoke produced by FCAW, it is becoming necessary to include smoke suction nozzles surrounding the gun nozzle to reduce smoke and fumes in the shop atmosphere.

9.10.4 Weldable Metals

- Low to medium carbon steels.
- Low alloy high strength steels.
- Quenched and tempered steels.
- Cast iron.
- (Certain) stainless steels.

FCAW with CO_2 gas shielding gives deeper weld penetration and can weld thicknesses from 1.6 mm to 13 mm with no edge preparation. When CO_2 is not used, the maximum thickness welded is only about 6 mm.

9.10.5 FCAW Advantages and Applications

- FCAW may be an "all-position" process with the right filler metals (the consumable electrode).
- No shielding gas needed making it suitable for outdoor welding and/or windy conditions.
- A high-deposition rate process (speed at which the filler metal is applied) in the 1G/1F/2F.
- Some "high-speed" (e.g., automotive applications).
- Less precleaning of metal required.
- Metallurgical benefits from the flux such as the weld metal being protected initially from external factors until the flux is chipped away.

Used on the following alloys

- Mild and low alloy steels
- Stainless steels
- Some high nickel alloys
- Some wearfacing/surfacing alloys

9.10.6 FCAW Disadvantages

Of course, all of the usual issues that occur in welding can occur in FCAW such as incomplete fusion between base metals, slag inclusion (non-metallic inclusions), and cracks in the welds, etc . . . But there are a few concerns that come up with FCAW that are worth taking special note of:

- Melted Contact Tip – Happens when the electrode actually contacts the base metal, thereby fusing the two.
- Irregular wire feed – Typically a mechanical problem.
- Porosity – The gases (specifically those from the flux-core) don't escape the welded area before the metal hardens, leaving holes in the welded metal.
- More costly filler material/wire as compared to GMAW.
- Less suitable for applications that require painting, such as automotive body works.

9.11 SOLDERING AND BRAZING

Soldering and brazing are the joining process as welding process. By the using of heat and a filler metal (in case of solder) whose melting point is lower than the melting point of metals to be joined. In *soldering* the melting point of solder (filler rod) is usually less than 427°C whereas in ***Brazing*** the melting point of filler metal is higher than 427°C. In soldering and brazing, there is no direct melting of the **parent metal** of workplace being joined, rather the solder or filler metal flows between the surface to be joined through capillary action. Most common use of soldering and brazing is in joining two dissimilar metals. **diffusion bonding** refers to the metallurgical joining of metal surface by the application of heat/pressure to cause comingling of atoms at the joint interface. Interface surface should be very clean and free of contamination. Diffusion bonding in most instance is accomplished entirely in solid state.

9.11.1 Soldering

Soldering is define as a metal joining process wherein coalescence is produced by heating the surface to be joint to a suitable temperature and melting the filler metal which is a fusible alloy called **solder (*melting point usually less than 427°C*).** So that it may be distributed between properly filled surface of the joint through capillary action. Soldering operation is performed by bringing molten solder in contact with the preheated surfaces (being joined) and heating the joint area to a good wetting temperature (about 55 to 80°C above the melting point of soldering alloy). Solder is then left to cool and freeze as quickly as possible to avoid development of internal cracks in the joint. Principle underlying soldering is that when the surface to be jointed are cleaned off well from oxides they can be joined together using molten solder that may adhere easily with the workpiece. Surface due to molecular attraction molecules of solder enter with the parent metal molecules and from a strong bond.

9.11.1.1 Metals Joined by Soldering

Efficiency of soldering joint basically depends upon the characteristics of oxides of metal being soldered and mutual solubility of solder and base metal therefore, all alloys cannot be equally welded by solders:

(a) all type of carbon steel are soldered easily but the wetability decreases with increase in carbon contents in steel.

(b) **Cast irons** are very **difficult** to solder as graphite carbon flanks resist wetting.

(c) **Stainless steel** are also **difficult** to solder as the chromium oxides on stainless steel rougher the surface, better the steel can be soldered.

(d) Copper and its alloys (brass, bronze) are easily soldered.

(e) Phosphor bronze are easily soldered

(f) Aluminium or silicon bronze are not easy to solder due to the presence of oxides.

(g) Aluminium and its alloys are soldered with difficulty.

(h) Nickel alloys are easy to solder.

9.11.1.2 Soldering Joint

In soldering, butt joint are avoided and lap joints are preferred as we know that soldering joints are weaker in strength as compared to brazing joint hence the former needs to be designed such that in service, the soldered joint does not take load, the base structure should take the load; joint strength can however be enhanced further up to some extent by designing proper interlocking system at the joint.

Different type of soldering joint is shown in Fig. 9.54. Simple butt joint is not prefer as no strength.

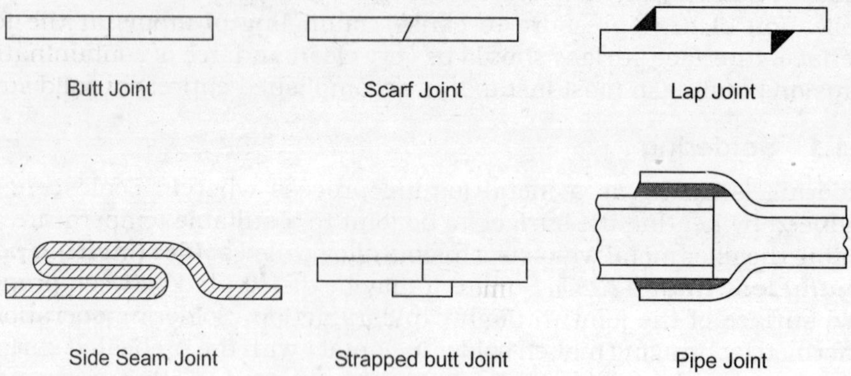

Fig. 9.54 Different soldering joint

A lap joint is most widely used where overlapping may be three times thickness of the workpiece metal. Lap joint fits more easily and gets good strength due to extended overlap joint gap or clearance is necessary for the capillary action of the molten solder it is usually kept 0.075 to 0.25 mm.

9.11.1.3 Types of Solders

Solders are basically two type:

(i) soft solder
(ii) hard solder

Soft solders melt usually below 427°C whereas hard solder melt at temperature above 600°C.

Soft solder are primary alloys of lead (Pb) and tin (Sn). This solder melts at a temperature which is lower than the melting point of alloys element. For example, Pb melts at 327°C, Sn at 232°C but their solder melts at 205°C. Occasionally **antimony silver**, **bismuth** and **cadmium** are added to the soft solder.

Hard solder are basically the alloys of copper and zinc (Zn) to which silver or phosphorus are sometimes added hard soldering is employed when a stronger joint is need hard.

Soldered joint can stand higher service temperature, hard solder are used in the brazing process wherein melting point of filler rod is greater than 600°C.

9.11.1.4 Soldering Alloys

They are:

(i) Tin-lead (Sn-Pb) solders
(ii) Tin-Antimoney lead solder
(iii) Tin-Zinc (Sn-Zn) solder
(iv) Lead-silver (Pb-Ag) solder
(v) Cadmium-silver solder
(vi) Cadmium-zinc (Cd-Zn) solder
(vii) Zinc-Aluminium (Zn-Al) solder.

9.11.1.5 Steps Involved in Soldering Process

(i) Surface to be joined by soldering must be thoroughly cleaned to any **dirt**, **grease** or **oxides** by pickling (acid cleaning) process wipe out the joint clean.
(ii) Joint surface are heated with a soft oxy-acetylene flame.
(iii) Soldering flux is later applied onto the heated surface of the joint it is heated further sometimes.
(iv) Solder is now applied with the help of soldering iron and it is allowed to melt only by heat present in the heated surface of the workpiece.
(v) Soon after the soldering operation is over, soldcring flux salted be cleaned from the joint with a wire brush or by filing smooth.

9.11.1.6 Method of Soldering

Method of soldering can be classified according to the method of applying heat to melt the solder and also for preheating the surface to be soldered these include:

(i) Soldering iron method
(ii) Torch method
(iii) Dip bath method
(iv) Resistance method
(v) Ultrasonic method
(vi) Induction method
(vii) Wave method soldering, etc.

9.11.1.7 Advantages of Soldering

(i) Low investment of cost.
(ii) Simplicity and cheapness of the equipment.
(iii) Properties of parent metal are not affected due to low operating temperature.
(iv) Good and effective sealing in fabrication as compared to other process like riveting spot, welding and bolts.
(v) Process is simple.

9.11.1.8 *Applications of Soldering*

(i) Connections in wireless set (radio), TV sets, etc.
(ii) Wiring joints in electrical connections, battery and other terminals.
(iii) Drain water gutters and pipes.
(iv) Radiator brass tube for motor car.
(v) Copper tubing carrying liquid fuel, gas or air used on engine.
(vi) Brass halved bearing are jointed with solder when relined with white metal and bored on lathe.
(vii) When halved bearings are to be lined with babbitt metal.
(viii) It is sometimes used to repair utensils.

9.11.2 Brazing

It is also one of the type of welding process which is used for joining the two similar or dissimilar metal piece togather by heating the surface and by using a non-ferrous filler metal having its **melting point above 427°C** but below the melting point of metal to be brazed. The molten filler metal is distributed between the joint surface by the capillary action which on cooling results in a sound joint.

The main advantage of brazing process is to joining of dissimilar metals and thin section. Process is mostly used to joining pipes and other **fittings, carbide tipping** on the tool shanks, **electrical parts, radiator**, repair of cast iron and **heat exchanger.**

In brazing, bond is produced either by the formation of solid state selection (diffusion bonding) or intermetallic compound of the parent metal (Job) and one of the metals in the filler material (Brazing alloy). The strength of the brazed joint is provided by metallic bonding. Brazing is one the oldest method of joining metal pieces.

It is used for brazing to:

- cast iron
- wrought iron
- carbon steel
- alloy steel, high speed steel
- stainless steel
- alloys of copper
- aluminium and
- magnesium.

In certain cases brazing is *preferred over welding process* for preserving specific metallurgical characteristics even after making the joint. Brazing is very effective technique for joining cast metals to wrought metals, non-metals to metals, dissimilar metals etc. thin walled tubes and light gauge sheet metal assemblies are better joint by brazing as *welding results in burning of the base metal* and joint is not formed, complex and heavy structure can be easily joint by brazing without thermal stresses in the joint. Brazing maintain precision production tolerances in making the joint. It has the ability to preserve protective metal coating/cladding. Brazing is one such technique which makes possible to join pieces of greatly different thickness. Brazed joints are pressure tight.

Brazing as a production process has certain limitation in its application. It calls for a properly machined fitting in making parts to be joint for proper capillary action. **Limitation** on the size of components to be brazed is there as in the process, outer area of the joint is to be heated. Large components therefore cannot be heated properly to brazing temperature **brazed joint develop corrosion if flux is not properly removed**. Brazing required certain degree of skill and experience on the part of welder for handling special brazing jobs. Brazing is also used for **brazing tool bits** (small piece of high speed steel/tungsten carbide) on high **carbon steel shanks** for making tipped cutting tools.

9.11.2.1 Brazing Operation

Brazing operation is carried out in the some manner as soldering. The surface to be brazed should be clean and heated to a temperature enough to melt the flux spread over the surfaces to be joined as the flux melts at a temperature which is lower than the filler material it wets the joint surfaces and remove the oxide film, etc. Giving a clean surface since the capillary action between the base metal and the filler metal is several times higher than that between the base metal and the flows the molten filler metal replaces the flux and flows in between the mating surfaces of the joint by capillary attraction. **Lesser the joint clearance**, better will be the capillary action upon cooling the joint is found filled with filler metal and the solidified flux is found on the joint periphery.

9.11.2.2 Brazing Joint

Brazing joint needed certain requirement such as a good joint allows enough clearance for filling to the filler metal into the joint area when the filler metal placed at one side is sucked through the joint by capillary action as the brazing alloys should not be subjected to direct stress, the **lap joint** is considered best because this makes the brazing alloy subjected to shearing which is better situation. In case of brazing, closer fit between the surfaces to be joined is essential therefore setting up of proper clearance (to permit flow of filler metal) between mating surfaces is an important factor. Gap or clearances depends on fluidity of filler metal, its wettability and susceptibility to slug (oxides) entrapment.

For different type of filler metals joint clearance is specified as follows:

(i) Al-Si alloys - 0.15 – 0.25 mm (lap less than 6 mm)
(ii) Cu-P alloys – 0.25 – 0.62 mm (lap greater than 6 mm)
(iii) Cu-Zn alloys – 0.05 – 0.125 mm
(iv) Silver-brazing alloy – 0.05 – 0.125 mm
(v) Magnesium – 0.10 – 0.25 mm
(vi) Nickel – 0.05 – 0.125 mm.

9.11.2.3 *Different Type of Brazing Joints*

These are lap, butt, scarf, flanged bottom, butt lap and modified lap.

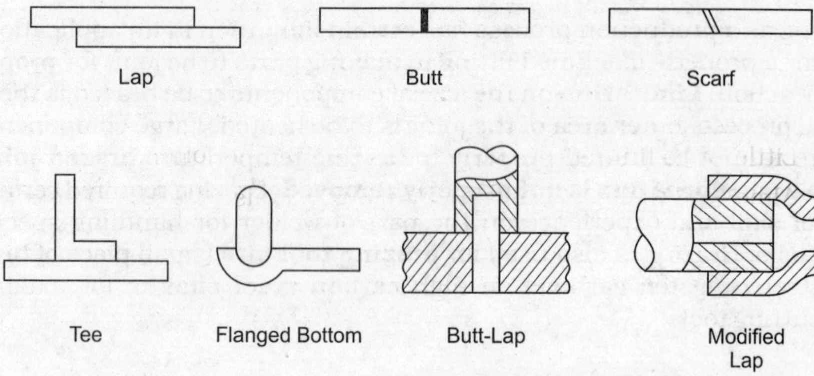

Fig. 9.55 Brazing joint

9.11.2.4 *Metals that can be Joint by Brazing*

They are following type:

1. **Low Carbon Steels:** They can be easily brazed. Low alloys steel tend to harden during brazing and hence there should be brazed carefully allowing then to cool down slowly.
2. **Stainless Steels:** They posses some problem in wetting due to chromium and need special flux to be used.
3. **Cast Irons:** Cast iron, gray, malleable, ductile cast iron are easily brazed with silver brazing filler rods.
4. **Aluminium and its Alloy:** With copper silicon and magnesium are easily brazed. However, higher contents of magnesium in Al-alloys are difficult to braze.
5. **Magnesium and its Alloys:** Containing aluminium, zinc, manganese are easily brazed.
6. **Copper Zinc Alloys:** Need to be brazed carefully with copper application of flux to reduce loss of zinc during brazing. **Copper silicon alloys** need to be stress-relieved before brazing to avoid intergranular penetration of filler metal resulting in reduced strength of the joint **copper Aluminium Alloys** call for special fluxes to take care of refractory oxides formed during brazing *copper-nickel alloys* are readily brazed with brazed rod *copper-tin alloys* (phosphor bronze) with low Sn contents are brazed with silver copper phosphorus, or copper zinc filler rods.
7. **Other Material that can be Brazed Include:** High carbon steels, high speed steels, carbide, ceramics and dissimilar metals.

9.11.2.5 *Advantages of Brazing*

(i) Cast and wrought material can be joined easily.

(ii) Metallurgical properties of the base material are not seriously disturbed.

(iii) Assemblies can be brazed in a stress free condition.
(iv) Dissimilar metal can be joined.
(v) Non-metal can be joined to metal, when the non-metal is coated.
(vi) Material of different thickness can be joined easily.
(vii) Complex assemblies can be brazed in several steps by using filler metal with progressively lower melting temperature.
(viii) Little or no finishing is required by the brazed joints.
(ix) Almost all the common engineering materials may be statisfactorily braze welded.

9.11.2.6 Disadvantages of Brazing

(i) It requires tightly mating parts.
(ii) It requires proper clearing.
(iii) Size of the job are limited.
(iv) Joint are not successful at elevated temperature.
(v) Colour of the filler metal may not match that of the base metal.
(vi) High degree of skill required.

9.11.2.7 Applications of Brazing

(i) Parts of bicycle such as frames and rims
(ii) Pipe joints subjected to vibrations
(iii) Exhaust pipe in motor engine
(iv) Band saws
(v) Tipped tool
(vi) Nipples and unions mild steel and copper tubing.

9.11.2.8 Comparison Among Welding, Brazing and Soldering

[UPTU 2008-09] V. Imp.

Welding	Soldering	Brazing
1. In welding process fusion is obtained by heat and/or pressure.	In soldering joint is obtained by means of filler material whose melting point is less than 450°C and less than the melting point of base metal.	In brazing the joint is obtained by means of filler material whose melting point is above the 450°C and less than the melting temperature of base metal.
2. The strength of the joint is highest comparing to soldering and brazing.	The soldered joints are weakest among the welding and brazing.	The brazed joints are stronger than soldered joint but weaker than welded joints.
3. The joint strength all times may be equal to even greater than the strength of the base metals.	The joint strength depends upon the adhesive qualities of filler material and can never reach the strength of base metal.	The joint strength depends upon the adhesive qualities of filler material and can never reach the strength of base metal.

(Contd...)

Welding	Soldering	Brazing
4. The filler material may or may not be used.	The filler metal is essentially used because joint is made by filler material only.	The filler material is essentially used because joint is made by filler material only.
5. The composition of filler metal is normally same as that of base metal (the metal to be welded).	Essentially filler metals are alloys of lead and tin.	The filler metals are alloys of copper (e.g., copper zinc and copper silver alloys).
6. Welded joints can withstand high temperature.	Soldered joints are not suitable for high temperature service because of lower melting point of the filler metal.	Brazed joints are also not suitable for high temperature service because of lower melting point of filler metal.
7. The process cannot joint dissimilar metals which are insoluble in other.	The process can join dissimilar metals which are insoluble in each other.	The process can joint dissimilar metals which are insoluble in each other.
8. The parent metals are heated to the melting temperature, so welding is likely to cause the metallurgical damage and change in the properties.	Heating of parent metal is negligible to cause any change in their structure and properties.	Heating of parent metal is negligible to cause any change in their structure and properties.
9. Welding has wide applications in construction and maintenance of boilers, pressure vessels, tanks bridges building constructions, cutting tools dies, and furnaces.	Soldering is widely used in joining small assemblies, electric and electronics parts.	Brazing is used for joining electrical parts, joining carbide tips to tools, heat exchangers, radiators, etc.

9.11.2.9 Different Type of Brazing Operations

Various brazing methods are based on the ways and means employed for heating the workpiece and melting of filler metal, for example:

(i) Torch brazing

(ii) Furnace brazing

(iii) Dip brazing

(iv) Induction brazing

(v) Infrared brazing etc.

9.11.2.10 *Difference between Soldering and Brazing* (UPTU 2011-12)

Soldering	*Brazing*
(*i*) Melting point of solder/filler rod is lower than 427°C.	(*i*) Melting point of filler rod is greater than 427°C but lower than the melting point of parent metal.
(*ii*) Soldered joint do not resist corrosion.	(*ii*) Brazing joints resist to corrosion.
(*iii*) Soldering joint weaker in strength. They cannot stand at high temperature.	(*iii*) Where as brazing joint are stronger in strength and can withstand high temperature.
(*iv*) Diffusion bonding is the secondary importance.	(*iv*) In brazing bonding condition are set up to allow large amount of diffusion to take place along surface being joint.

QUESTIONS

1. Define the welding process. ***[UPTU]***
2. On what basies we can classified the welding process. ***[UPTU 2008-09]***
3. Describe the advantage and Limitation of Welding Process.
4. Enlist the application of welding process in detail.
5. Enlist the material that can be join by welding process.
6. Compare the welding process with other joining process like riveting and bolting.
7. Write about differences and similarities of welding and Casting Process. ***[UPTU 2009]***
8. Define gas welding in detail with figure.
9. Weather gas welding process is used for the thick plate or thin plate (sheet).
10. Define gas welding equipment and accessories with the help of figure. ***[UPTU 2003-04]***
11. What are the major component of oxyacetylene torch?
12. What are the advantage of two-stages regulator over a single stage regulator?
13. What are the two major types of welding blow pipes? What are their differences?
14. Advantage of the gas welding.
15. Write short notes of gas welding and gas cutting. ***[UPTU 2004-07]***
16. Differentiate between flash back and back fire. ***[UPTU 2006-07]***
17. What are the colour of oxygen and acetylene cylinder?
18. What are the safety devices and how do they work for the acetylene cylinder and the oxygen cylinder?
19. Why are all fuel line threaded left handed?
20. Enlist the different equipment used in gas welding.
21. Define resistance welding and advantage of resistance welding.
22. What are the limitation of high frequency resistance welding?

23. What is the two cycles in resistance welding?

24. Define the seam welding.

25. Why is projection spot welding a higher production output method than a standard spot welding?

26. Differentiate flash welding and upset (resistance butt) welding.

27. Write a brief note on flash welding.

28. What are the specific advantage and disadvantages of the resistance welding process? ***[UPTU 2007-08 MFS]***

29. What do you understand by the term 'heat balance'? Explain its relevance?

30. What are parameter to be controlled in the resistance welding?

31. Comparison between AC and DC arc welding machine.

32. What is the effect of alloying elements of a flux?

33. Explain the difference between constant current and constant voltage characteristics of the welding power sources for arc welding. List the different type of power source for arc welding available.

34. Write about modes of metal transfer in arc welding.

35. Write the short notes on arc welding. ***[UPTU 2007-08 MFS]***

36. Define polarity in arc welding.

37. What is an arc column?

38. Define the arc welding in detail.

39. What is arc blow? ***[UPTU 2006-07]***

40. What do you mean polarity in DC arc welding? How can their be advantageously utilized?

41. Define MIG welding with the help of a neat sketch.

42. List of the material that can be welded by MIG process.

43. What types of power source is used for MIG welding?

44. How does the power supply compensate for difference in arc length?

45. What is the basic carbon dioxide (CO_2) MIG welding process?

46. Why oxygen is used with argon as a shielding gas?

47. Advantage and limitation of MIG welding.

48. Define the TIG welding in detail with the help of neat sketch.

49. Define the advantage and limitation of TIG welding process.

50. List of the material that can be welded by TIG welding process.

51. Differentiate between the TIG welding and MIG welding process. ***[UPTU 2002-03, 2006-09]***

52. Differentiate between the TIG welding and plasma arc welding (PAW) process.

53. Define the submerged arc welding process in detail.

54. Write the advantage and limitation of the SAW process.

55. What types of flux used in submerged arc welding?

56. Write the uses of submerged arc welding (SAW).

57. Write the principle of SAW.

58. What is the difference between electroslag and electrogas welding? ***[UPTU 2003-04]***

59. What is the function of the molten slag pool?

60. Define the friction welding in detail.

61. Define the advantage and limitation of friction welding.

62. Enlist the material that can be welded by friction welding.

63. Define the soldering in detail.

64. Differentiate between the soldering and brazing process. ***[UPTU 2008-09]***

65. At what circumstances soldering and brazing process preferred over welding? ***[UPTU 2007-08 MFS]***

66. Enlist the material that can be joint by soldering process.

67. Enlist soldering joint.

68. Define different type of solders and soldering alloys.

69. Function of soldering flux and types of soldering flux.

70. Different type of soldering method.

71. Define the Brazing (joining process) in detail.

72. Differentiate between welding, soldering and brazing.

73. Enlist the material that can be join by brazing process.

74. Enlist the brazing joint.

75. How many types of filler material used in brazing?

76. Enlist the advantage of brazing process.

77. Define the different type of brazing operation.

78. Define the dip brazing and vacuum brazing.

79. Why direction kept at a minimum in electroslag welding?

80. What are the power source requirements for electrogas and electroslag welding process?

81. What is the thickness limit that electroslag can successfully weld?

82. What are the factors which decide weld quality? ***[UPTU 2008-09]***

83. What is the difference between submerged arc welding and electroslag welding? ***[UPTU 2007-08 MFS]***

84. Write short notes on spot welding. ***[UPTU 2007-08 MFS]***

85. Why sometimes weld joins fails during use? Give precaution to enhance life of weld joint. ***[UPTU 2008-09 MFS]***

86. Write the short notes on hard facing. ***[UPTU 2008-09]***

87. Explain the process of electroslag welding with figure and what are the properties of the flux used in the process?

❑❑❑

UNIT-V

Unconventional Machining Processes and Welding

Chapter 10

Metallurgical Aspect in Welding

10.1 METALLURGY OF FUSION WELD

As we know that for welding purpose we have required a very high temperature due to high temperature as the electrode strike with the workpiece. Workpiece also heat up in such a way, that the metallurgy of parent metal near by the weld joint completely change (due to high temperature). Basically a weldment is divided into three zone, *i.e.*, **parent metal (PM)** or **Base metal (BM)**, **weld metal (WM)** forming the weld bead and the last one is heat affected zone **(HAZ)**. For the welding of low carbon steel, which is used widely throughout the world, three zone of a weldment can be drawn on the iron-carbon equilibrium diagram (Fe_3C). Which is shown in the Fig. 10.1 mention above out of then weld metal region consist weld bead and is a cast structure. The HAZ is, in a way, heat treated portion of weldment,

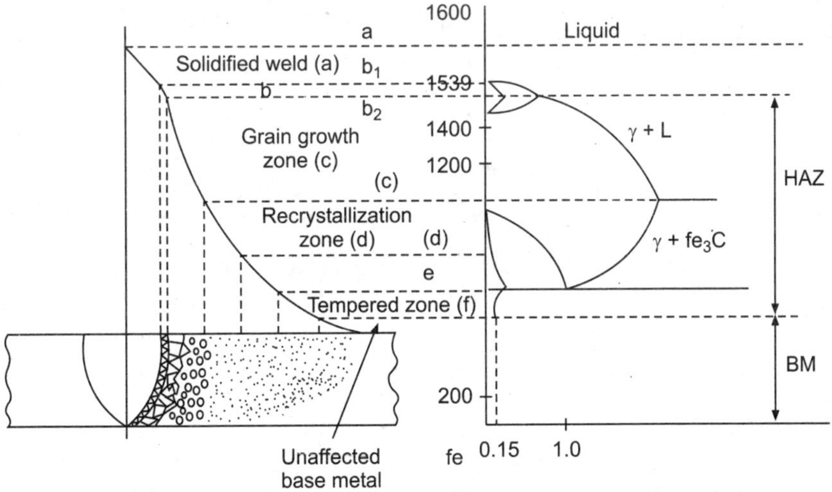

a = partially molten metal

b_1 = under bead zone

b_2 = grain growth zone (1150–1450°C)

d = partially transformed zone

e = zone of sphereodised carbide (350–A1)°C.

a = Solidified weld

b = Solid-liquid transition zone

c = Grain growth zone

d = Recrystallization zone

e = Partially transformed zone

f = Unaffected base metal.

Fig. 10.1 Different zone of a steel weldment as represented on Fe_3C diagram

on the other hand. Base metal/unaffected metal is the original work material with a small region which has been heated up to 650°C and hence has undergone a minor metallurgical change in its grain size and results change in mechanical properties. Different type of microstructures produces due to welding speed, material composition, and the amount of heat input by fusion welding.

10.1.1 Heat Affected Zone (HAZ) *[UPTU 2011-12]*

As the molten metal weld starts to cool, only a small amount of heat escapes through the surface of the pool. The majority of heat flows through the parent metal on each side of the joint as a result the parent metal is subjected to a heating and cooling cycle. Two critical components of the cycle are the highest temperature reached and subsequent rate of cooling. Peak temperature depends on the distance from the weld parent or base metal near the fusion boundary is heated to just below the melting point. The region of the parent metal which has undergone a metallurgical change as a result of thermal cycle is known as heat affected zone (HAZ).

HAZ (heat affected zone) in steel can be sub-divided into the following zone, on the basics peak temperature reached it is starting from weld metal side.

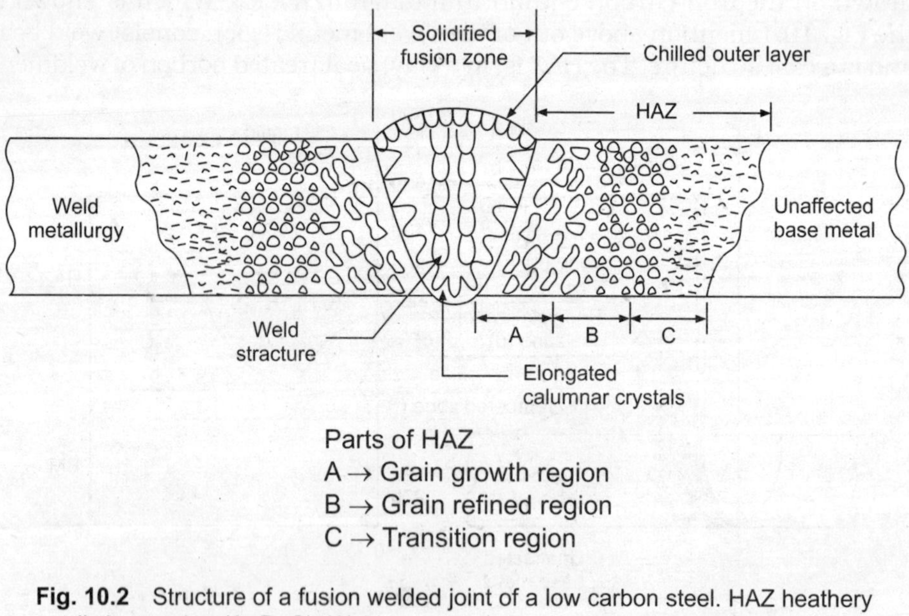

Fig. 10.2 Structure of a fusion welded joint of a low carbon steel. HAZ heathery distinct region (A, B, C). Weld structure may comprises of coarse pearlite

1. Under bead zone, *i.e.*, part of HAZ which is heated to beyond the critical temperature of grain growth and extends upto fusion boundary zone
2. Grain growth zone beyond 1111°C to peritectic temperature.
3. Grain refined zone, temperature 950-1150°C, *i.e.*, beyond A_3 utpo grain refined temperature.

4. Partially transformed zone, 750-950°C, *i.e.*, temperature between A_1 to A_3. Zone of spherodized carbide, *i.e.*, 550° – 750°C, *i.e.*, below A_1.
5. Zone of unchanged parent metal upto 550°C.

Above figure of fusion weld in steel, presenting proper terminology for the various region and interfaces part of the heat affected zone has been heated above transformation temperature because of altered structure the HAZ may no longer passes the desirable properties of the parent metal and since it was not molten it cannot assume the properties of solidified weld metal.

10.1.2 Effect of Heat on Welding and Heat Affect Zone (HAZ)

Two inherent problems in a welding process should be anticipated by the weldment designer:

(a) effect of localised heating and cooling on microstructure and properties of the base metal.
(b) effect of residual stresses locked upon the weldment as a result of uneven cooling than correctly designed casting and forgings.

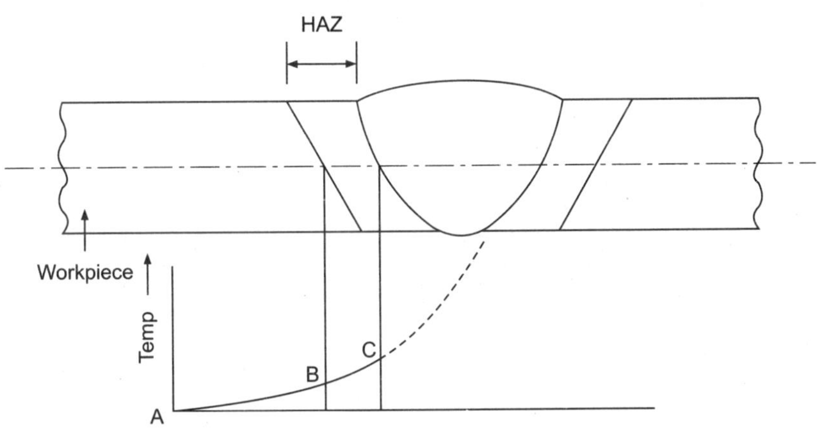

A - Original temperature of base metal
B - Temperature AC1, 723°C
C - Melting point of base metal
BC - Heat affected zone (HAZ)
AB - Region of low temperature with no change in microstructure

Fig. 10.3

In welding process, a thermal cycle is introduced during heating and fusion of metal. Base metal is heated over a range of temperature over melting of fusion figure showing thermal gradient during a fusion welding process. Fusion zone is the weld metal pool zone in which metal are in molten state. HAZ is the heat affected zone which could not melt but the undergone change in microstructure.

Metal at a far off distance from weld will be simply warmed up but as weld area is approached, progressively higher temperature are obtained, Fig. 10.36 shown the thermal gradient during a welding process. Slope of thermal gradient depends upon the *welding process*, *metal being welded* and the way by which heat is supplied/unit volume of metal/unit time besides the thermal conductivity of the base metal, flow of heat in the weld zone is highly directional towards the adjacent cold metal and this result on cooling in the formation of columnar grains at right angle to the fusion face and going towards the centre of fusion zone.

10.1.2.1 Weld Metal Zone

This zone produce as the weld solidifies from molten state. Ratio of parent metal and filler metal in this zone depends on **welding technique**, **type of joint**, **thickness of workpiece**, etc. microstructure of this zone reflects the cooling rate in the weld, based on chemical composition and cooling rate, there may be martensite (due to fast cooling) and the fine pearlite and coarse pearlite showing comparatively slower rate of cooling. It may be noted that columnar (long elongated) crystals are formed near fusion face due to directional cooling of weld towards the centre.

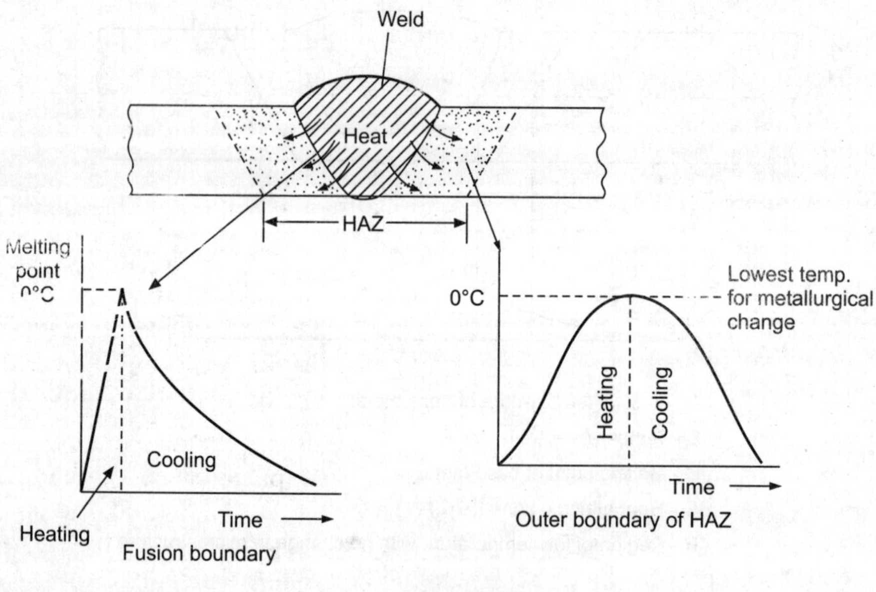

Fig. 10.4 HAZ boundary

Consequently, this is often the weakest area in the as weld joint. Except where there are various defects in the weld deposit most welding failure originate in the heat affected zone. The HAZ terminates where the base metal has experienced too little heat to be affected or altered by the welding process.

10.1.2.2 Effect of Cooling Rate on Formation of HAZ

Controlling of the cooling rate play an important (major) role in producing metallurgical suitable welds specially in heat treatable steel during the welding operation; near the weld bead (metal) zone is the heat affected zone that is composed of parent metal that did not melt but heated to a high enough temperature for a predetermined time that grain growth occurred. In all welding processes except cold welding heating and cooling are essential and integral parts of the process. In many fusion welding process high degree of superheat can be obtained in the weld metal. Maximum temperature exceeds various critical temperature at which phase transformation occurs in the metal involved in welding. Thermal gradient's slope and extent is determine by the heat supplied by welding process/unit volume of metal/ unit time and thermal conductivity of the parent metal.

Study of welding metallurgy requires consideration of the following metallurgical phenomena that play an important role in fusion welding.

1. Electrode and base metal melt and then solidify as an alloying element in the weld metal zone.
2. Parent metal (that is unmelted) is subjected to a complex heat treatment in the form of a temperature gradient extending from the melting range to room temperature and followed by a cooling cycle which is induced by the near by cold metal.

Phase change result due to welding depend on:

(i) process for welding (arc, gas spot, etc.).
(ii) composition of filler rod and composition of flux used for coating.
(iii) welding joint is made under present of which atmosphere, *i.e.* oxygen, nitrogen, etc.
(iv) base metal's chemical composition.

10.1.2.3 Weld Zone

Weld metal zone is formed as the weld metal solidifies from molten metal.

Weld zone is the mixture of parent metal and electrode. Ratio depends upon the welding process used, plate thickness (parent metal thickness), type of welded joint, etc.

Weld metal is a cast metal of the particular composition of the mixture that has cooled. Its microstructure reflects the cooling rate and cooling rate depending upon the chemical composition. A martensite structure in the weld indicates a very fast cooling rate, fine pearlite and coarse pearlite showing comparatively slower rates of cooling respectively.

From the molten weld pool molten metal solidity grows epitaxially upon the solid grains of the unmelted parent metal. Weld solidify in a cellular or dendritic growth node due to the composition of parent metal and rate of solidification.

Both mode cause segregation of alloying element and consequently. Weld metal is less homogeneous on the micro level than the parent metal and

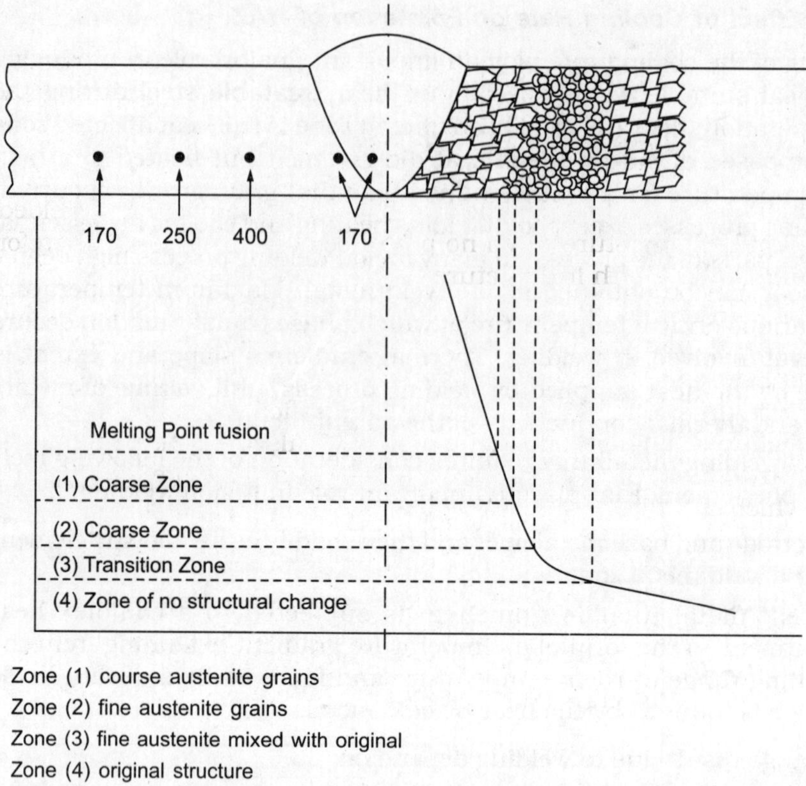

Fig. 10.5 Heat affected zone

therefore cannot be expected to have same properties as the wrought parent metal. In less, the filler metal has the same deposited condition properties equal to the parent metal.

Heat affected zone is subjected to a complex thermal cycle (sudden heating followed by rapid cooling) in which all temperature from melting range of the steel down to comparatively much lower temperature are involved and HAZ there fore consists of a series of graded, structure ringing the weld bead.

HAZ consists a variety of microstructure example in plain carbon steel the microstructure may range from very narrow region of hard martensite to cause pearlite. This render the HAZ, the weakest area in a weld.

Heat affected zone is some time define by the response of the welded joint to hardness or etching effect test with of HAZ. Varies according to the welding process and welding technique, in arc welding it extends only upto few mm from fusion boundary but in oxy-acetylene and electroslag welding it is some time very wide.

10.2 WELDING STRESSES

During the welding due to high temperature (fusion) in SMAW, GMAW, ESW, SAW, GTAW, etc., results in complex thermal stresses and strain which finally lead to generate the residual stresses and distortion in the weld structure. Residual stress are responsible to reduce the service life of a welding object or even cause cut as tropic, failure. Distortion usually due to by misalignment with consequential difficulties in assembly as well as poor appearance finish welding structure.

10.2.1 Residual Stresses

Residual stresses are defined as the stress which remain within a welded/ structure/casting structure when no presence of external load. Therefore they are self balance with in structure itself.

As residual stresses exist without interference, therefore the resultant moment produce by stresses are zero residual stresses arise from permanent changes in the shape of structure. Residual stresses developed during the welding operation which hamper the functional efficiency of the component leading to failure of engineering components. It is also responsible for brittle fracture. Stresses left in the joint after welding are referred as residual stresses which are major responsible for crack. Finally they may lead to cracking. A common property of residual stresses is that they are purely elastic in nature even if they result from plastic deformation. Residual stresses can be reduces by reducing the elastic deformation.

10.2.2 Method for Reducing/Relieving the Stresses

Following method are used to reduce the stress:

(i) Weld form fixed end to free end.
(ii) Peening.
(iii) Heat treatment.

10.2.3 Chief Factors Responsible for Residual Stress

(i) Parent metal quality, filler rod or electrode quality.
(ii) Size and shape of weldment.
(iii) Comparative weight of parent metal and weld metal.
(iv) Joint nature and method of welding (backing strip, tracking, etc.).
(v) Heat input into weldment, i.e., value of current, diameter of electrode and speed of welding in arc welding.
(vi) Nature of structure and near by joints.
(vii) Expansion and contraction of structure after welding.
(viii) Cooling rate — High / Low
(ix) Stress present in parent metal.

Residual stresses in the weldments can cause series specified problem such as hydrogen induced cracking, yield and stress corrosion cracking. To avoid the residual stresses some steel are preheat before welding.

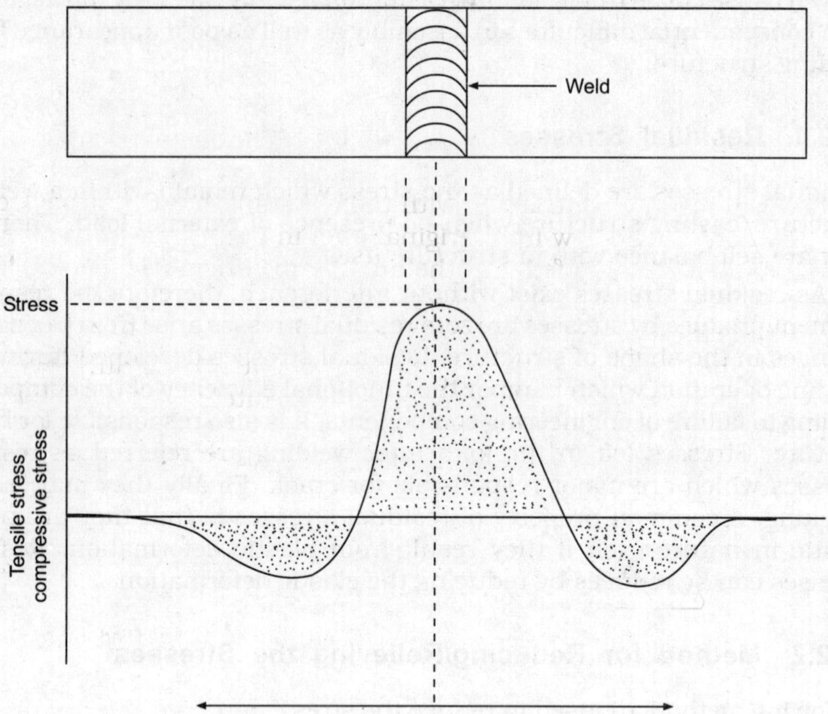

Fig. 10.6 Distance from weld centre line distribution of residual stresses in a butt welded joint.

10.2.4 Measurement and Determination of Residual Stresses in Weldments

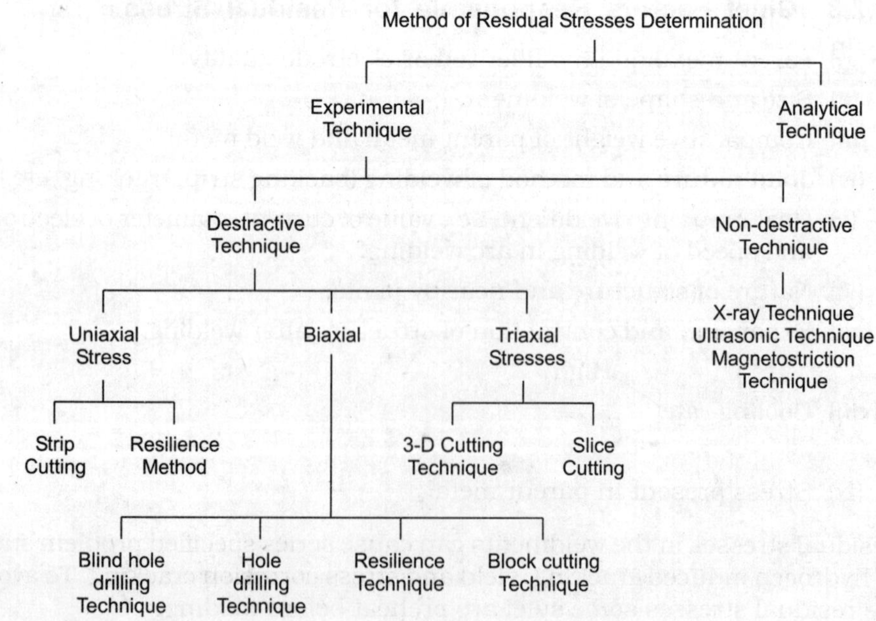

Fig. 10.7

10.2.5 Solidification Shrinkage

It is very difficult to separate shrinkage during solidification. The contraction of liquid metal and change of volume on solidification as melting and solidification in welding are at the same time, as the solidification front progresses upwards towards the centre line. Solid metal occupies a very smaller space than the liquid metal it replaces, *i.e.*, its density is higher. Molten metal is also contracting with the resulting that the surface of the pool should recede below from original level in between at the same time further molten metal is fed into the area by melting of the parent metal. Therefore a appreciable shrinkage is producing during cooling and solidification from the liquid state. At the end of weld run gives an interesting example of the importance of weld-metal feeding but, since no further melting is taking place level of the surface falls and a creater is produced.

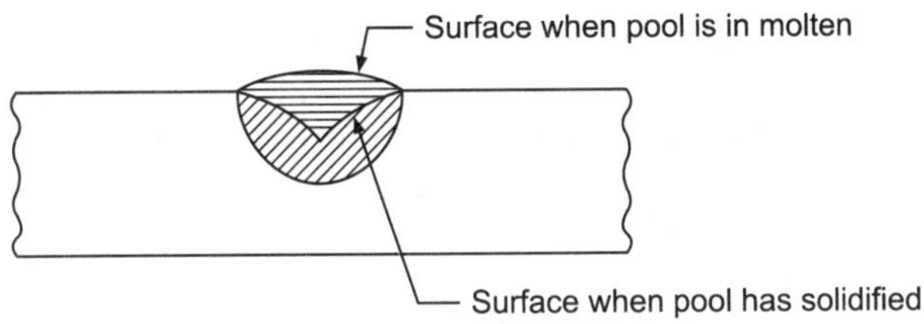

Fig. 10.8 Shrinkage

10.2.6 Contraction of Solid Metal

After welding, weld metal is still hot and heat continues to flow into the parent metal until the whole joint has cooled to room temperature during this period of cooling, weld metal contracts both along the length of the joint and at right angle to it. This means that its volume should become smaller, but in fact this is prevented from happening by the restraining influence of the parent metal to which it is bonded.

10.2.7 Longitudinal Shrinkage

Decreasing in length of the weld can be obtained by using the equation

$$L_1 = l_0(1 - \alpha\Delta\theta)$$

or

$$l_1 = l_0 - l_0\alpha\Delta\theta$$

where L_0 = original length

L_1 = length after cooling through $\Delta\theta$°C

α = co-efficient of linear expansion

$\Delta\theta$ = temperature change °C contraction

Linear shrinkage is therefore equal to $l_0 \propto \Delta\theta$ for low-carbon steel.

The following data would apply

$$\alpha = 14.3 \times 10^{-6}/°C$$

$$\Delta\theta = \text{Melting temperature} - \text{Room temperature}$$

$$= 1500 - 20 = 1480°C$$

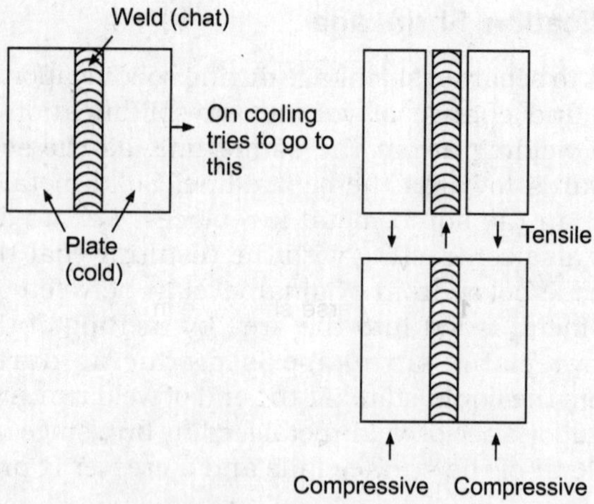

Fig. 10.9 Deformation of weld metal element during cooling

As the value of α does not remain fixed over the temperature range but it indicate that the contraction along the length of the weld should be applicable base plate may be considered as more or less rigid. So that when the weld metal tries to contract. It is plastically deformed by being held to the original length. It shows that the tensile forces set up in the weld region, reacted by compressive forces in the plate after cooling cycle completed weld joint became same length as base plate if plate were perfectly rigid.

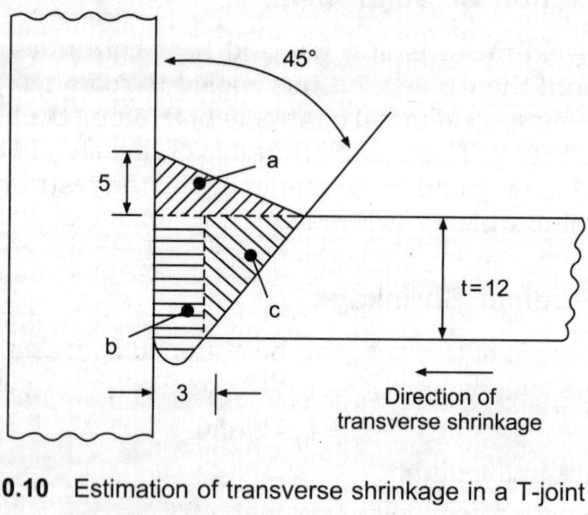

Fig. 10.10 Estimation of transverse shrinkage in a T-joint (butt)

10.2.8 Transverse Shrinkage

Transverse shrinkage = 0.1 × A

$$A = a + b + c$$

$$= \frac{1}{2} \times 5 \times (12+3) + 3 \times (12) + \left(\frac{1}{2} \times 12 \times 12\right) = 145.5 \text{ mm}$$

$$\text{Transverse shrinkage} = \frac{0.1 \times 145.5}{12} = 1.213$$

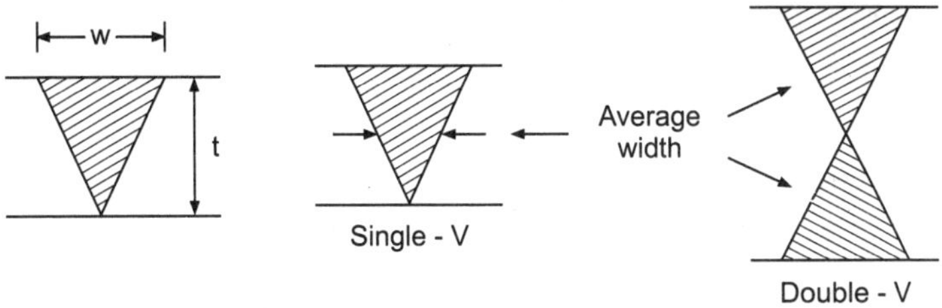

Fig. 10.11 Transverse shrinkage in V-b butt weld

$$\text{Area of weld } A = \frac{1}{2} \times w \times t$$

$$\text{Transverse shrinkage} = 0.1 \times A$$

$$= 0.1 \times \frac{\frac{1}{2} \times w \times t}{t}$$

$$= 0.1 \times \frac{w}{2} = 0.1 \times \text{average width}$$

In shrinkage transverse to the weld where contracting weld metal tries to pull the plate towards the centre line of the joint and due to this the whole joint area is in transverse tension.

10.3 MAIN CAUSES FOR DISTORTION

As welding involves highly localised heating of joint edges to fuse the material, non-uniform stresses are set-up in the component because of expansion and contraction of the heated material. Initially, compressive stresses are created in the surrounding cold parent metal when the weld pool is formed due to the thermal expansion of the hot metal (heat affected zone) adjacent to the weld pool. However, tensile stresses occur on cooling when the contraction of the weld metal and the immediate heat affected zone is resisted by the bulk of the cold parent metal.

The magnitude of thermal stresses induced into the material can be seen by the volume change in the weld area on solidification and subsequent cooling to room temperature. For example, when welding C–Mn steel, the molten weld metal volume will be reduced by approximately 3% on solidification and the volume of the solidified weld metal/heat affected zone (HAZ) will be reduced by a further 7% as its temperature falls from the melting point of steel to room temperature.

If the stresses generated from thermal expansion/contraction exceed the yield strength of the parent metal, localised plastic deformation of the metal occurs. Plastic deformation causes a permanent reduction in the component dimensions and distorts the structure.

10.3.1 Main Types of Distortion

Distortion occurs in six main forms:

- Longitudinal shrinkage

- Transverse shrinkage
- Angular distortion
- Bowing and dishing
- Buckling
- Twisting

The principal features of the more common forms a distortion for butt and fillet welds are shown in Fig. 10.12.

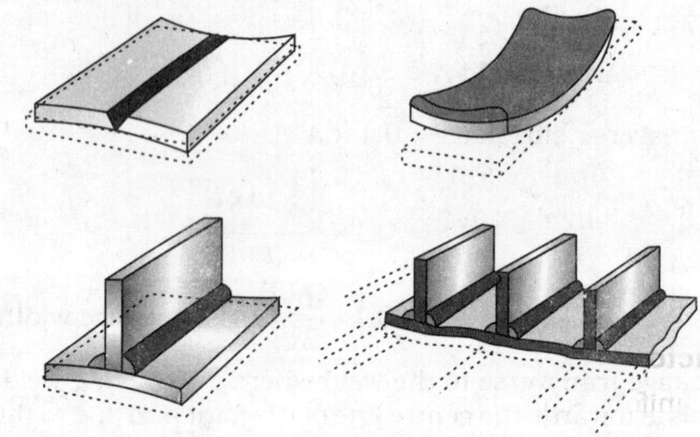

Fig. 10.12

Contraction of the weld area on cooling results in both transverse and longitudinal shrinkage.

Non-uniform contraction (through thickness) produces angular distortion in addition to longitudinal and transverse shrinkage.

For example, in a single V butt weld, the first weld run produces longitudinal and transverse shrinkage and rotation. The second run causes the plates to rotate using the first weld deposit as a function. Hence, balanced welding in a double side V butt joint can be used to produce uniform contraction and prevent angular distortion.

Similarly, in a single side fillet weld, non-uniform contraction produces angular distortion of the upstanding leg. Double side fillet welds can therefore be used to control distortion in the upstanding fillet but because the weld is only deposited on one side of the base plate, angular distortion will now be produced in the plate.

Longitudinal bowing in welded plates happens when the weld centre is not coincident with the neutral axis of the section so that longitudinal shrinkage in the welds bends the section into a curved shape. Clad plate tends to bow in two directions due to longitudinal and transverse shrinkage of the cladding; this produces a dished shape. Dishing is also produced in stiffened plating. Plates usually dish inwards between the stiffeners, because of angular distortion at the stiffener attachment welds.

Twisting in a box section is caused by shear deformation at the corner joints. This is caused by unequal longitudinal thermal expansion of the abutting edges. Increasing the number of tack welds to prevent shear deformation often reduces the amount of twisting.

10.3.2 How Much Shall We Allow for Weld Shrinkage?

It is almost impossible to predict accurately the amount of shrinking. Nevertheless, a 'rule of thumb' has been composed based on the size of the weld deposit. When welding steel, the following allowances should be made to cover shrinkage at the assembly stage.

10.3.3 Transverse Shrinkage

Fillet Welds: 0.8 mm per weld where the leg length does not exceed 3/4 plate thickness.

Butt Weld: 1.5 to 3 mm per weld for 600 V joint, depending on number of runs.

10.3.4 Longitudinal Shrinkage

Fillet Welds: 0.8 mm per 3 m of weld.
Butt Welds: 3 mm per 3 m of weld.
Increasing the leg length of fillet welds, in particular, increases shrinkage.

10.3.5 Factors Affecting the Distortion

If a metal is uniformly heated and cooled there would be almost no distortion. However because the material is locally heated and restrained by the surrounding cold metal, stresses are generated higher than the material yield stress causing permanent distortion. The principal factors affecting the type and degree of distortion, are:

- Parent material properties
- Amount of restraint
- Joint design
- Part fit-up
- Welding procedure

10.4 WELDING DEFECT

Performance of welding structure/component during their functioning period depends upon their quality of fabrication, which is turn is based on the presence/absence of defects in weld joint. If a welding structure is not fulfilling the need as per their standards is defined there is some part of defect is present. A welding structure should be free from defect otherwise chance of welding structure failure are more. Defect impair the strength of weld joint and may result in the failure of a whole structure when load is applied.

Several types of welding defects can be produced during the process.

These are following:

(i) Cracks (Internal/External)
(ii) Pin holes
(iii) Blow holes
(iv) Creater
(v) Incomplete penetration
(vi) Improper fusion
(vii) Porosity
(viii) Undercut

(ix) Poor welding bead
(x) Excess of spatters
(xi) Distortion
(xii) Overlapping
(xiii) Burn through
(xiv) Surface oxidation.

10.4.1 General Sources of Welding Defects

(i) Improper selection of process.

(ii) Wrongly applying welding method for a particular application, i.e., incorrect voltage, amperes, shielding gas, etc.

(iii) Undesirable and size of weld bead due to overfill/poor profile.

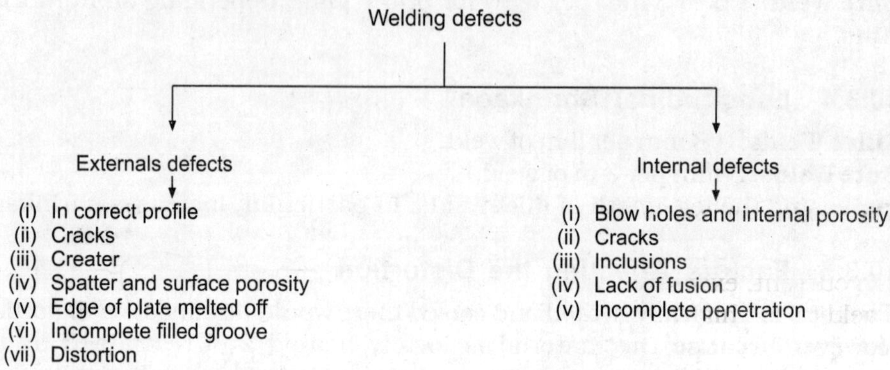

Fig. 10.13 Block diagram of welding defect

10.4.2 Excessive Penetration Bead

Remedies	Causes
(i) joint should prepared properly (ii) are length should be proper (iii) current should be appropriate	(i) incorrect joint preparation (ii) high current (iii) large arc length

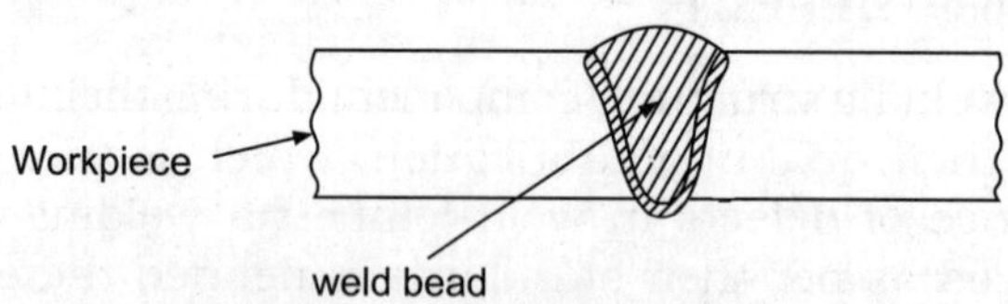

Fig. 10.14 Excessive penetration

10.4.3 Lack of Penetration

When penetration is not proper upto root of welded joint then it is define as the lack of penetration.

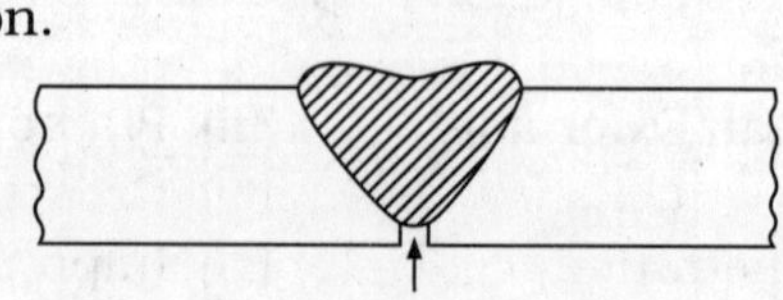

Fig. 10.15 Lack of penetration

Causes for Incomplete Penetration

Remedies	Causes
(i) arc travel speed should be uniform	(i) faster arc travel speed
(ii) root gap should be proper	(ii) root gap is very small
(iii) angle should be proper	(iii) bevel angle is also very small
(iv) arc current should be appropriate	(iv) very less arc current
(v) diameter of electrode should not be very large.	(v) Very large diameter of electrode
	(vi) Longer arc length
	(vii) Improper joint.

10.4.4 Undercut

Undercut is also one of the type of welding defect. Undercut basically occurs in the parent metal nearby the weld bead and the basic cause of under bead is due to very high current, excessive arc length and speed of welding too high. Improper manipulation of electrode and dampness in electrode and arc blow are also responsible for undercut.

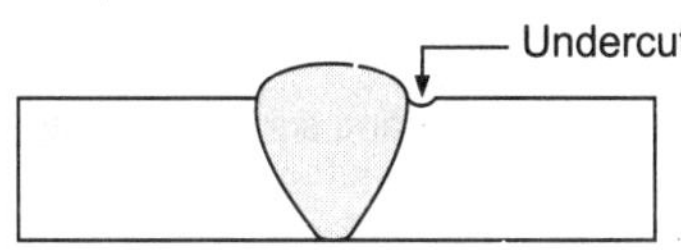

Fig. 10.16 Undercut

Causes	*Remedies*
(i) High current range.	(i) Use proper current.
(ii) Too large electrode diameter.	(ii) Use proper electrode.
(iii) Excessive weaving.	(iii) Weaving should be proper and as per need.
(iv) Faster arc travel speed.	(iv) Maintain proper arc traval speed.

10.4.5 Cracks

Cracks are also one of the type of welding defect. Cracks may be different type such as:

(i) surface cracks
(ii) hot cracks
(iii) cold cracks
(iv) stress corrosion cracks, transverse crack
(v) longitudinal cracks in butt welding.

One of the important characteristic of cracks is that they have sharp notches and notch propagate the weld joint. Which will be responsible for failure of welding joint if there are lot of carbon contents in electrode (1% - 2%) then there are more chance of cracks. Cracks also due to presence of hydrogen contents.

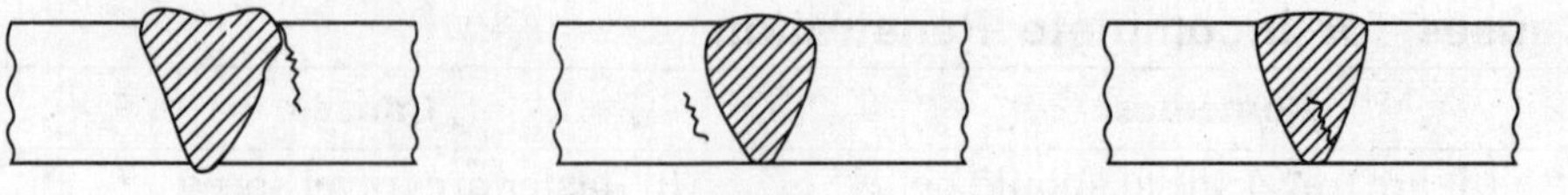

Fig. 10.17 Different types of cracks

Causes	*Remedies*
(i) Cooling rate is very high.	(i) Cooling should be reduced.
(ii) Low ductility.	(ii) Preheat, anneal the parent metal before welding.
(iii) Residual stresses inside the material.	(iii) Remove the residual stress by preheat or a applying heat treatment process.
(iv) Small weld bead.	(iv) Increase electrode size, increase current, reduce welding speed.
(v) Hydrogen content produced as damp coating or rusty and wet workpiece.	(v) Always use fresh (new) electrode or bake electrode before welding.

10.4.6 Hot Cracking

Hot cracking occurs during the welding process means when welded portion is on hot position, if there is high restraint then hot cracking. It occurs mostly in weld metal.

Cracks and Their Remedies

1. **Hot Cracking:** It generally occurs at high temperature and size can be very small to be visible. Crack in most parts is irregular and its magnitude depends upon the strain involved in solidification. They are most likely to be formed during the root pass when mass of base metal is very change compared to be weld metal deposit.

Remedies

(i) It can be prevented by preheating the base metal.

(ii) Increases the cross-sectional area of root bead.

(iii) By changing the contour or composition of the weld bead.

10.4.7 Factors Promoting Hot Cracking

Several factors are responsible for hot cracking such as:

(i) High current density.

(ii) Weld metal composition having wide freezing range.

(iii) Distribution of heat and thermal stresses in the weld metal itself.

(iv) Impurities such as sulphur and high carbon or nickel content and those which form low freezing point liquid films.

(v) Joint restraints and high thermal sensitivity.
(vi) Electrode crack sensitivity.
(vii) Dilution of weld metal.
(viii) Welding parameters which stimulate cause dendritic growth.

10.4.8 Cold Cracking

Cold cracking occurs after 2 or 3 days of welding in weld metal. Presence of hydrogen is most important parameter for responsible for cold cracking.

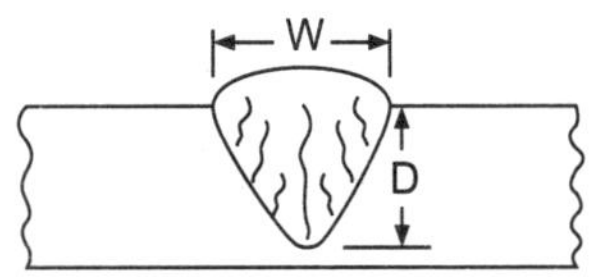

Fig. 10.18

$$\frac{D}{W} > 1.5$$

Centre line cracking occurs in metal after welding due to shrinkage of metal.

10.4.9 Cold Cracking—Causes and Remedies

It generally occurs at room temperature after the weld is completely cooled this can be generally seen in the heat affected zone (HAZ).

Causes	***Remedies***
(i) Excessive restraint of the joint, which induces very high residual stresses.	(i) Stress releaving the weldment immediately would help in relieving the residual stress and potential for cold cracking reduces.
(ii) Martensitic transformation making the metal very hard as a result of rapid cooling.	(ii) Pre- and post-heating of the weldment helps in reducing the cooling rates and the consequent of the stress.
	(iii) By reducing the cooling rate of welded first.

10.4.10 Factor Responsible for Cold Cracking

(i) Hydrogen contents in the weld ment.
(ii) Welding joint restraint and high thermal sensitivity.
(iii) Presence of impurities.
(iv) Embrittlement of the heat affected zone in low-alloy steel.
(v) Weld of insufficient sectional area.
(vi) High welding speed and low current density.

10.4.11 Distortion

Distortion is define as the change in the shape of the workpiece after welding distortion occurs due to the temperature gradient that exits in different sections of the workpiece during welding. More number passes of welding bead, slow arc travel speed, high residual stresses in plates. Responsibles for welding distortion.

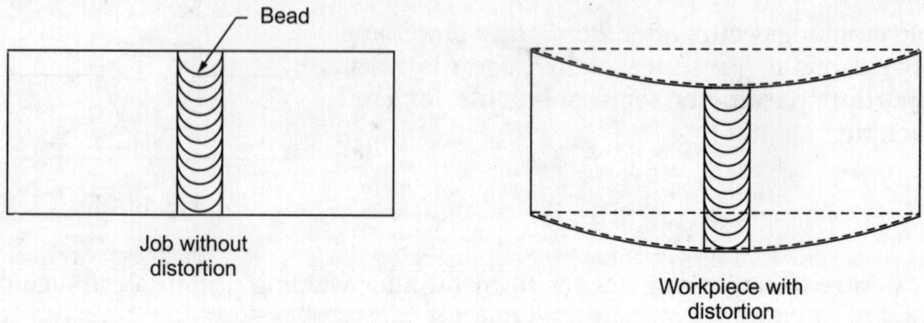

Fig. 10.19 Welding distortion

Causes	*Remedies*
(i) Cooling rate very high.	(i) Cooling rate should be reduced.
(ii) Small diameter of electrode.	(ii) Use appropriate diameter of electrode.
(iii) Residual stresses in base metal.	(iii) Remove/reduced the residual stresses.
(iv) Clamping and fixing of parent metal is not proper.	(iv) Use proper clampings jigs and fixture use for welding purpose.
(v) Arc travel speed very slow.	(v) Maintain proper arc travel speed.

10.4.12 Creater

Creater occurs in a welded structural due to end of a weld run and also due to improper welding technique when finishing a weld, the welder should not draw away the arc quickly, rather maintain the arc at the end of weld with no travel movement to give chance the creater to fill up.

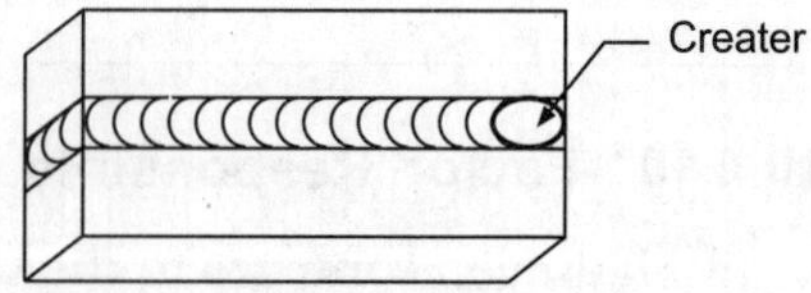

Fig. 10.20

10.4.13 Porosity

Porosity is also another type of welding defects they are define as a small voids and produce in weld bead due to entrapement of gases. Parent metal melt under arc and absorbs gases such as H_2, CO, N_2 and oxygen. If they are near by molten metal weld pool these gases get spread

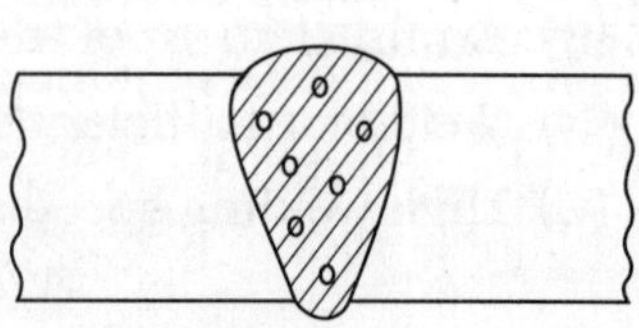

Fig. 10.21

throughout the molten pool by diffusion and convection phenomena. **Improper electrode, longer arc length, faster arc travel speed, welding with dump** and **damage electrode**. Improper base metal composition are responsible for this type welding defect.

Causes	*Remedies*
(i) High welding speed.	(i) Reduced welding speed to allow time for gases to escape.
(ii) Dirty base metal.	(ii) Clean base metal properly.
(iii) Electrode with damp and damaged coating.	(iii) Proper coated electrode should be used.
(iv) Improper arc length, welding current and welding method.	(iv) Use proper arc length, welding current and welding method.
(v) More hydrogen, nitrogen or oxygen content in welding atmosphere.	(v) Use filler metal having high contents of deoxidizers.

10.4.14 Slag Inclusion

Flux used in welding react with impurities and forms a slag. If this slag is mixed with weld bead that defect called as slage inclusion.

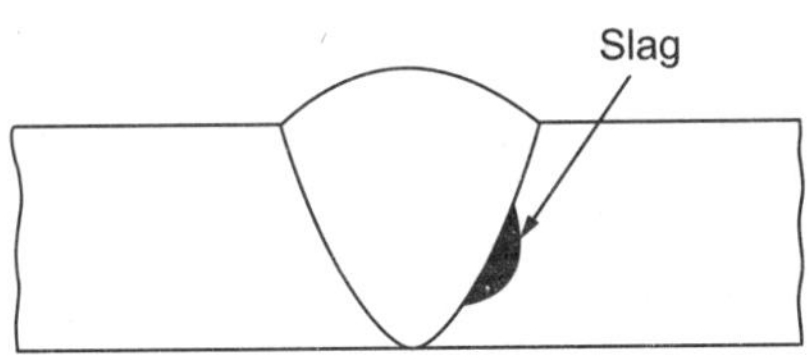

Fig. 10.22 Slag inclusion

Causes	*Remedies*
(i) No proper slag removal.	(i) Remove the slag properly if any.
(ii) Arc length is very long.	(ii) Use proper arc length.
(iii) Less electrode angle, slow speed and very large electrode.	(iii) Use larger electrode angle, more speed and small electrode.
(iv) Oxide inclusion.	(iv) Provide proper gas shielding.
(v) Entraped pieces of electrode covering.	(v) Use proper and fresh electrode.
(vi) Joint design is not proper.	(vi) Joint design should be proper.

10.4.15 Incomplete Fusion and Penetration

If molten do not mix well with the parent metal that defect is known as in complete fusion. This is responsible for poor weld bead.

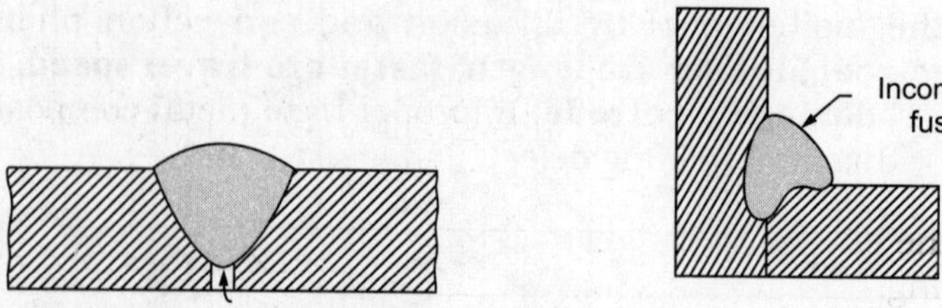

Fig. 10.23

Incomplete penetration produces when the welding joint bead is insufficient (distance from base metal plate top surface to the maximum extent of the weld snuggest).

Causes	*Remedies*
(i) Less arc current.	(i) Use proper arc current.
(ii) Higher arc travel speed.	(ii) Use proper arc travel speed.
(iii) Longer arc length.	(iii) Use shorter arc length.
(iv) Very larger electrode diameter.	(iv) Use small electrode diameter.
(v) Electrode position is not proper.	(v) Use proper electrode position.
(vi) Improper joint design.	(vi) Redesign or change the joint.

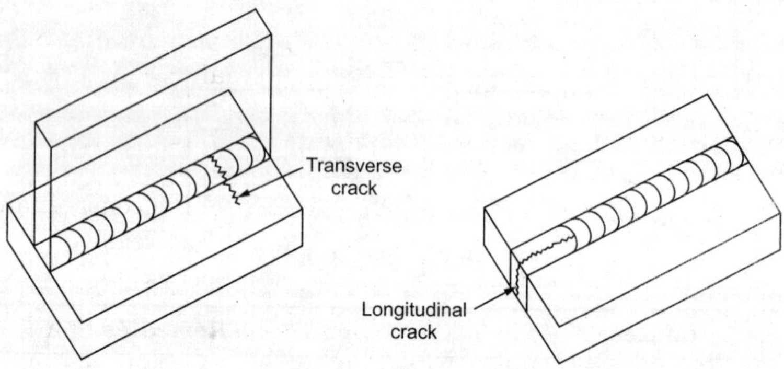

Fig. 10.24 Type of cracks

10.4.16 Inclusions

Inclusion is also another welding defect. It is define as slag or any other foreign material present in weld bead. Inclusions is responsible for lower the strength of joint and make it weaker. Several factors are responsible for entrapment of inclusion such as

(i) longer arc length,
(ii) high or low arc current
(iii) very larger electrode diameter
(iv) insufficient chipping

and cleaning of multiple pass, undercut, etc.

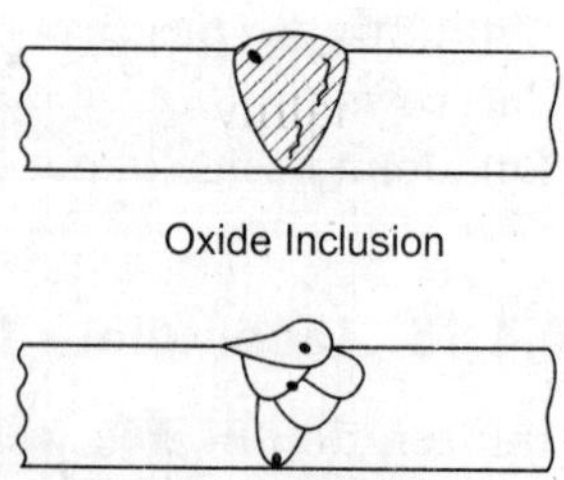

Fig. 10.25

10.4.17 Spatters

Spatters are also welding defect (externally). They are define as the small spherical particles produced around the weld bead due to metal particles expelled out from the molten metal pool under the effect of harsh arc during welding. Spatters are not the part of weld but freeze out in small tiny balls along the weld bead. Excessive current, a longer arc and damp electrode are responsible for spatters.

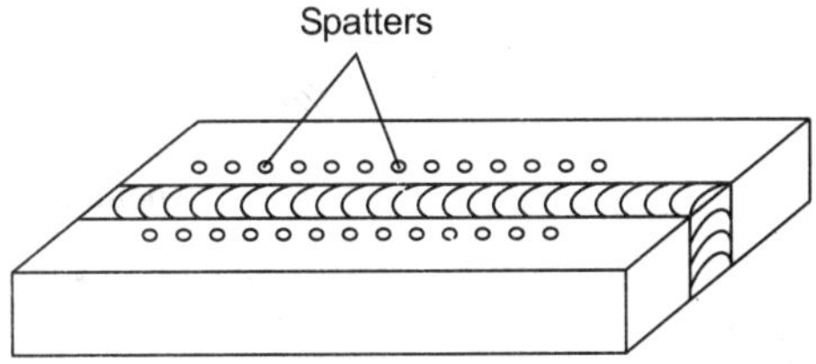

Fig. 10.26 Spatter

Causes	*Remedies*
(i) Excessive current for welding. (ii) By using too long arc length or too high arc voltage. (iii) Use of damp electrode. (iv) Arc blow. (v) Presence of hydrogen contents.	(i) Use proper welding current. (ii) Use proper arc length and proper arc voltage. (iii) Use fresh or baked electrode. (iv) Reduce arc blow (use AC power). (v) Reduce hydrogen contents.

10.4.18 Poor Weld Bead Appearance

If the weld bead deposited is not straight, width of bead is not constant, ripples are not consistent, then the weld bead is called poor bead.

Causes	*Remedies*
(i) Arc length is not constant.	(i) Maintain the constant arc length throughout welding.
(ii) Improve welding technique.	(ii) Use proper welding technique.
(iii) Damaged electrode coating.	(iii) Use fresh electrode.
(iv) Poor earthing clamp and connector.	(iv) Use proper earthing clamp and connector.

10.5 WELD DECAY IN STEEL HAZ OF STAINLESS STEEL WELD

When stainless (18/8 type) are heated within the range of 600-850°C or cooled slowly through that range, the carbon is precipitated from solid solution mainly at grain boundary initiating with Crome to form Crome rich carbides. In other words, Cr-Carbide is formed and precipitated at grain boundaries. The area around weld thus improverished in cromium have a much lesser resistance to corrosion than the remainder of the metal and in this zone may occurs, this is called ***weld decay***. Weld decay is one of the most widely effect in heat affected zone.

10.5.1 Preventing of Weld Decay

(a) Decreases the carbon contents of steel to an amount insufficient to form carbides.
(b) Add to both parent plates and weld filler rod material alloying element such as titanium and molybdenum to form carbides preferentially at the boundaries and preserve the chromium content.
(c) Add molybdenum to reduce precipitation tendency.
(d) Solution treat the steel at 1050°C after welding to take carbide phase back into solution.

10.5.2 Work-Hardened Metals

Work-hardened metal are characterised by increasing hardness and tensile strength occurs due to deforming (working) at room temperature. Typical examples of work-hardened materials used in welded fabrications are pure **aluminium and its alloys containing Mn and Mg and steel** with thickness of 0.7 to 0.9 mm. Which is used for manufacture of refrigerators, work hardening of the steel occurs due to rolling process which is used to reduce the thickness of metal.

A material's weldability, or jointability, can refer to whether it can be welded together without cracks, and in this case a weldable material is a material that can be welded together creating a crack-free weld. Weldability can also refer to more qualitative characteristics of how easy or difficult it is to achieve a functional weld. This kind of weldability is not easily quantified, and can depend on a variety of factors, including the welding method, the physical specifics of the joint created, and what the welded structure is to be used for. There are several welding methods, and knowledge of a material's weldability is important in order to choose the proper welding process.

According TWI -to many people 'weldability' is simply a measure of how easy it is to make a weld in a particular material without cracks. If it is easy to avoid cracking, the material is deemed 'weldable'. For a weld to be truly successful, however, it is also necessary for it to have adequate mechanical properties, and to be able to withstand degradation in service (e.g., corrosion damage). Thus, weldability is a measure of how easy it is to.

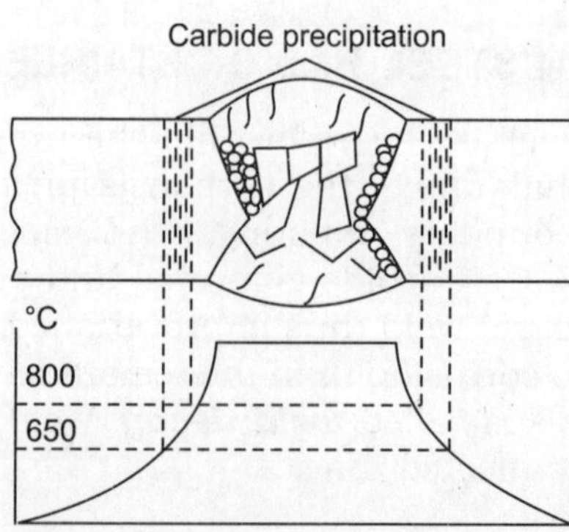

Fig. 10.27 Weld decay

10.6 FACTORS THAT AFFECT THE WELDABILITY

There are several factors that affect that weldability of the metal such as

(i) **Service:** It refer as the ability of the process material combination to tern out a weld that will stand up to thermal and mechanical stresses, corrosion and other service requirement.

(ii) **Fabrication:** Fabrication is define as the combination of materials and process to convert raw material into desired shape by minimizing the porosity absence of cracks, proper weld geometry.

(iii) **Melting Point:** Melting point is also affect the weldability of any material. Melting point of any material refers as the point at which the material will start to melt. Melting point is different for different material, for example, **aluminium** have **low** melting point while **lead** have very **high** melting point.

(iv) **Thermal Conductivity and Thermal Expansion:** It is also an important factor that affect the weldability of material thermal conductivity and thermal expansion of copper have higher thermal conductivity and they are different to bring to the fusion temperature. In adition the rapid expansion and contraction of the heat affected zone (HAZ) causes stresses and distortion.

(v) **Surface Condition (Texture):** Surface (material or workpiece) that to be welded should be free from any foreign matter such as dirt, grease, paint, scale and rusting.

(vi) **Alloying Element:** It is very important factor that affect the weldability of material as increasing or decreasing the alloying element hard enability of HAZ. The element that have greatest effect on hard enability of steel are C, Mn, Mo, Vd, Cr, tungsten, Ni, etc.

10.6.1 Weldability Testing

Weldability testing is carried out to determine the correct welding condition, weldability test provide the following information such as:

(i) preheat requirement
(ii) selection of correct welding process
(iii) required energy input
(iv) joint design, etc.

Weldability testing may be classified as:

(i) Simulated test.
(ii) Their fiscal test.
(iii) Actual welding test.

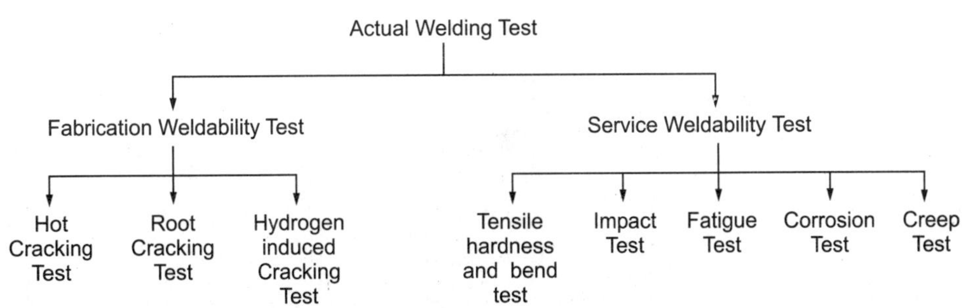

Fig. 10.28 Block diagram of weldability test

10.7 WELDING JOINTS

Different types of welding joints are shown in the Fig. 10.29(*a*) and 10.29(*b*). Full open corner joints are recommended for any thickness of the plate. Whereas half open corner joints and close corner joints are used for plate of smaller thickness.

Joint Name	***Joint***	***Symbol***
Square Butt		
Single-V Butt		
Double-V Butt		
Single-U Butt		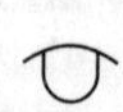
Double-U Butt		
Single Bevel Butt	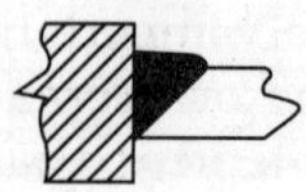	
Double Bevel Butt	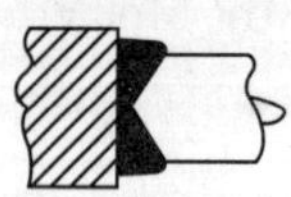	
Single-J Butt	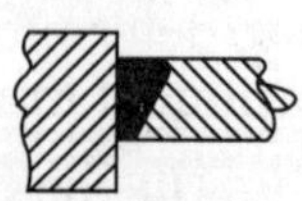	
Double-J Butt		
Stud		⊥
Fillet		
Flash		N

Fig. 10.29 (a) Schedule of basic symbols for welding (according to BIS)

Joint Name	***Joint***	***Symbol***
Butt Resistance	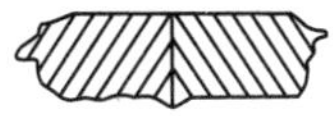	
Fillet		
Bead		
Spot		
Seam		
Mashed seam	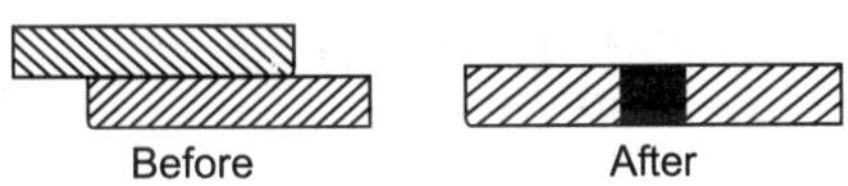	
Backing strip		
Projection	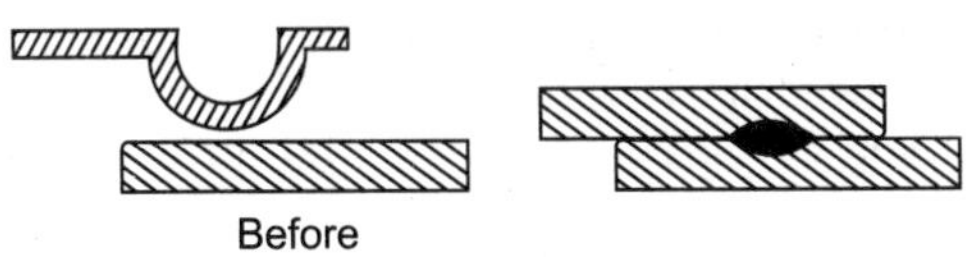	
Stitch	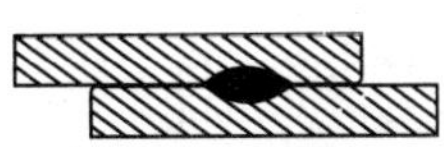	

Fig. 10.29 (b) Schedule of basic symbols for welding (according to BIS)

OBJECTIVE TYPE QUESTIONS

1. Which one of the following statement is correct?
(a) AC can be used for GTAW process
(b) No flux is used in gas welding of mild steel
(c) Borax is the commonly used flux coating on welding electrode
(d) Laser beam welding employs a vacuum chamber and thus avoides use of shielding method ***[IES-2000]***

2. Consider the following process:
(a) Gas welding (b) Thermit welding
(c) Arc welding (d) Resistance welding

The correct sequence of these processes in increasing order of their welding temperature.
1. 1, 2, 3, 4 2. 1, 3, 4, 2
3. 4, 1, 3, 2 4. 4, 3, 1, 2 ***[IES-2000]***

3. The maximum heat in resistance welding is at the
(a) Interface between the two plates
(b) Tip of the positive electrode
(c) Tip of negative electrode
(d) Top surface of the plate at the time of electric contact with electrode ***[IES-2001]***

4. Arc blow is more common in
(a) AC welding with base electrode
(b) DC welding with base electrode
(c) AC welding
(d) DC welding with straight polarity ***[IES-2001]***

5. Pinch effect in welding is the result of
(a) Expansion of gases in the arc
(b) Surface tension of the molten metal
(c) Electro-magnetic forces
(d) Electric forced ***[IES-2001]***

6. In arc welding DC reverse polarity is used to bear greater advantage in
(a) flat welding of lap point (b) flat welding of butt joint
(c) overhead welding (d) edge welding ***[IES-2001]***

7. In manual metal arc welding (SMAW), the equipment should have dropping characteristics in order to maintain
(a) constant voltage when arc length change
(b) temperature in the arc constant
(c) constant current when arc length change
(d) weld pool red-hot ***[IES-2001]***

8. Oxyactylene reducing flame is used while carrying out the welding on
(a) alloy steel (b) high carbon steel
(c) mild steel (d) grey cast iron ***[IES-2001]***

9. In which one of the following welding technique is vacuum environment required
(a) Electron beam welding (b) Ultrasonic welding
(c) Laser beam welding (d) Plasma arc welding ***[IES-2002]***

10. In resistance welding, heat is generated due to the resistance between
(a) Electrode and workpiece
(b) Two dissimilar metal being in contact
(c) As peritics between touching plates
(d) Interatomic force *[IES-2003]*

11. High speed electron beam of electron beam welding is focussed on the weld spot using
(a) Vaccum lens (b) Optical lens
(c) Inert gas lens (d) Magnetic lens *[IES-2003]*

12. Match List I (welding defect) with List II (causes) and select the correct answer using the codes given below the list

	List I		**List II**
A.	Spatter	1.	Damped electrode
B.	Distortion	2.	Arc below
C.	Slag inclusion	3.	Improper cleaning in multi pass
D.	Porosity	4.	Poor Joint Selection

(a) A B C D
4 2 3 1

(b) A B C D
2 4 3 1

(c) A B C D
4 2 1 3

(d) A B C D
2 4 1 3

13. Match List I (welding problem) with List II (causes) and select the correct answer using the codes given below the List: *[IES-2004]*

	List I (Welding problem)		**List II (Causes)**
A.	Cracking of weld metal	1.	Excersive Stresses
B.	Cracking of base metal	2.	High joint rigidity
C.	Porosity	3.	Failure to remove slag from previous deposit
D.	Inclusion	4.	Oxidation
		5.	Excersive H_2, O_2, N_2 in the welding atmosphere

(a) A B C D
3 1 4 2

(b) A B C D
2 1 5 3

(c) A B C D
2 4 5 3

(d) A B C D
3 4 2 1

14. Consider the following statement
The size of the heat affected zone (HAZ) will increases with,
1. increased starting temperature
2. increased welding speed 3 incremental thickness.

Which of the following statement given above are correct:

(a) 1, 2 and 3 (b) 2 and 3
(c) 1 and 3 (d) 1 and 4

15. Consider the following statement:
1. In arc welding 65% to 75% heat is generated at the anode
2. Duty cycle in case of arc welding is the cycle of complete welding of work piece from the begining
3. Arc below is more common with DC welding.

Which of the statement given above arc correct?

(a) 1 and 2 (b) 2 and 3
(c) 1 and 3 (d) 1, 2 and 3 *[IES-2005]*

16. In atomic hydrogen welding, hydrogen acts as
(a) one of the gases to generate the flame
(b) A heating agent
(c) A lubricant to increase the flow characteristic of weld metal
(d) An effective shielding gas protecting the weld. ***[IES-2005]***

17. Which of the following are the major characteristics of submerged arc welding
1. high speed welding
2. High deposition rates
3. Low penetration
4. Low clean lines.

Select the correct answer using the code given below:
(a) 2 and 3
(b) 3 and 4
(c) 1 and 2
(d) 1, 2 and 3

18. Fabrication weld ability test is used to determine
(a) Mechanical properties required for satisfactory performance of weld joint
(b) appropriate machining process
(c) susceptibility of welded joint for cracking
(d) suitability for joint design ***[IES-2006]***

19. In which of the following welding processes, flux is used in the form of granules
(a) AC arc welding
(b) DC arc welding
(c) sub-merged arc welding
(d) argon arc welding ***[IES-2006]***

20. Which one of the following is not a fusion welding process
(a) Gas welding
(b) Brazing
(c) Arc welding
(d) Resistance welding ***[IES-2006]***

21. Which one of the following welding processes consists of minimum heat affected zone (HAZ)?
(a) Ultrasonic welding (USW)
(b) Laser beam welding (LBW)
(c) Metal inert gas welding (MIG)
(d) Shielded metal arc welding (SMAW) ***[IES-2006]***

22. In metal arc welding, the transfer of metal from the electrode is due to
(a) molecular attraction
(b) surface tension and attraction
(c) gravitational force
(d) ionisation of the space between the electrode and workpiece
(e) (a) and (b) above.

23. Filler metal is used in
(a) electric spot welding
(b) seam welding
(c) projection welding
(d) continuous welding
(e) none of the above.

24. In MIG welding, helium or organ is used in order to
(a) provide cooling effect
(b) act as flux
(c) act as shielding medium
(d) facilitate welding process
(e) protect electrode.

25. In which of the following welding techniques, the weld pool is surrounded by an inert gas
(a) arc welding
(b) carbon arc
(c) MIG
(d) submerged arc
(e) electroslag welding.

26. The main advantage of MIG welding over TIG welding is that
(a) former can be used to weld hard to weld metal
(b) former permits use of large currents thereby allowing higher deposition
(c) welding rate is very fast
(d) welding is completely automatic
(e) all of the above.

27. The striking voltage as compared to voltage during welding in arc welding is
(a) same (b) more
(c) less (d) unpredictable
(e) none of the above.

28. The arc length in electric arc welding is the distance between the tip of the electrode and the
(a) workpiece (b) centre of crater
(c) bottom of crater
(d) workpiece top surface minus electrode diameter
(e) none of the above.

29. In inert gas arc welding, following is used for welding carbon steel
(a) non-combustible electrode in combination with helium and D.C. current
(b) combustible electrodes and argon in combination with A.C. current
(c) straight polarity D.C. current
(d) carbon dioxide, because of its excellent penetration and high speed
(e) none of the above.

30. Which of the following metals can be suitably welded by TIG welding?
(a) aluminium (b) stainless steel
(c) magnesium (d) pure titanium
(e) all of the above.

31. The following gas is used in tungsten-inert-gas welding process
(a) acetylene (b) oxygen
(c) hydrogen (d) argon
(e) helium.

32. The arc in argon arc welding takes place between the
(a) parent metals
(b) consumable tungsten electrode and workpiece
(c) non-consumable tungsten electrode and filler rod
(d) non-consumable tungsten electrode and workpiece
(e) consumable tungsten electrode and filler rod.

33. In resistance welding the electrode material is made of
(a) carbon steel (b) stainless steel
(c) copper (d) high speed steel
(e) graphite.

34. Seam welding is
(a) arc welding (b) multi spot welding
(c) continuous spot welding (d) used for forming sound bars
(e) gas welding.

35. Which of the following materials is best weldable with itself?
(a) stainless steel (b) copper
(c) aluminium (d) mild steel
(e) cast iron.

36. Flash butt welding belongs to the following category of welding
(a) gas welding
(b) arc welding with straight polarity
(c) arc welding with reverse polarity
(d) resistance welding
(e) thermit welding.

37. Submerged arc welding uses following type of electrode
(a) bare rods
(b) coated electrodes
(c) core wires
(d) copper electrodes
(e) stainless steel rods.

38. The flux in submerged arc welding is in the form of
(a) coating on the electrodes
(b) core wires
(c) granules
(d) paste.

39. Thermit welding is a form of
(a) resistance welding
(b) gas welding
(c) fusion welding
(d) forge welding
(e) arc welding.

40. Oxy-acetylene flame cuts metal by its
(a) evaporation
(b) oxidation
(c) burning
(d) rusting
(e) intensive oxidation.

41. Neutral flame has
(a) 1 zone
(b) 2 zones
(c) 3 zones
(d) 4 zones
(e) no zone.

42. Oxygen to acetylene ratio in case of carburising flame is
(a) 0.5 : 1
(b) 0.9 : 1
(c) 1 : 1
(d) 1 : 1.2
(e) 2 : 1.

43. In gas welding the combustion takes place by mixing oxygen with
(a) hydrogen
(b) fuel gas
(c) CO
(d) CO_2
(e) mixture of several gases.

44. Carburising flame has
(a) 1 zone
(b) 2 zones
(c) 3 zones
(d) 4 zones
(e) no zone.

45. Carburising flame is used to weld metals like
(a) steel
(b) copper and brass
(c) aluminium, stainless steel, zinc diecastings, nickel, monel etc.
(d) carburised steel
(e) none of the above.

46. Which of the following materials is best cut by oxy-cutting process?
(a) brass
(b) copper
(c) bronze
(d) stainless steel
(e) mild steel.

47. Weld spatter refers to *[UPPSC-2010]*
(a) welding electrode
(b) flux
(c) filler material
(d) welding defect
(e) shielsd.

48. Grey cast iron is best welded by
(a) TIG
(b) electric arc welding
(c) MIG
(d) submerged arc
(e) oxy-acetylene.

49. Grey cast iron is usually welded by
(a) gas welding
(b) resistance welding
(c) arc welding
(d) TIG welding
(e) MIG welding.

50. In arc welding, temperature of the following order may be generated
(a) 1000°C
(b) 1500°C
(c) 5500°C
(d) 8000°C
(e) 10,000°C

51. The phenomeon of weld decay occurs in
(a) cast iron
(b) brass
(c) bronze
(d) stainless steel
(e) carbon steel.

52. Laser welding finds widest application in
(a) heavy industry
(b) structural work
(c) process industry
(d) electronic industry
(e) all of the above.

53. Two sheets of different materials but same thickness can be spot welded by
(a) adjusting the current
(b) adjusting the time duration of current
(c) adjusting the pressure applied
(d) changing the size of one electode
(e) all of the above.

54. Arc blow occurs in
(a) gas welding
(b) gas cutting
(c) arc welding when straight polarity is used
(d) arc welding when reverse polarity is used
(e) welding stainless steel.

55. In thermit welding, the iron oxide aluminium oxide are mixed in the proportion of
(a) 1 : 1
(b) 3 : 1
(c) 1 : 3
(d) mixture is of different oxides
(e) none of the above.

56. Metal deposited on to the workpiece from the electrode
(a) is forced across the arc
(b) falls because of gravity
(c) is attracted towards the workpiece due to the positive polarity of the workpiece
(d) is attracted towards the workpiece due to the negative polarity of the workpiece
(e) is attracted due to electromagnetic action.

57. Which of the following carbon steels is most weldable
(a) 0.15% carbon steel
(b) 0.30% carbon steel
(c) 0.50% carbon steel
(d) 0.75% carbon steel
(e) 1.00% carbon steel.

58. Fluxes are used in welding in order to protect the molten metal and the surfaces to be jointed from

(a) oxidation (b) carburising
(c) dirt (d) distortion and warping
(e) unequal temperature distribution.

59. The carburising flame as compared to oxidising flame is

(a) more luminous (b) less luminous
(c) equally luminous (d) unpredictable
(e) none of the above.

60. The most commonly used flame in gas welding is

(a) neutral (b) oxidising
(c) carburising (d) all of the above
(e) none of the above.

61. The temperature of the inner luminous cone of neutral flame is the order of

(a) 1000°C (b) 2000°C
(c) 2500°C (d) 3500°C
(e) 5900°C

62. Oxygen to acetylene ratio in case of neutral flame is

(a) 0.8 : 1.0 (b) 1 : 1
(c) 1.2 : 1 (d) 2 : 1
(e) none of the above.

63. The temperature produced in oxyhydrogen flame as compared to oxy-acetylene flame is

(a) same (b) more
(c) less (d) unpredictable
(e) none of the above.

64. Filler material in welding should have

(a) same composition as the parent metal to be welded
(b) same melting temperature as the parent metal to be welded
(c) same composition and same melting temperature as the parent metal to be welded
(d) same composition as of electrode
(e) melting temperature much lower than the melting temperature of the parent metal.

65. Low hydrogen electrodes are baked prior to use in order that

(a) proper strength in obtained
(b) welding is free from arc blow
(c) welding is free from moisture pick-up
(d) current required is minimum
(e) electrode does not crumble during use.

66. In MIG welding, the metal is transferred in the form of

(a) a fine spray of metal (b) molten drops
(c) weld pool (d) molecules
(e) very fine metal.

67. In which welding process the parts are heated to welding temperature and then mechanical force is applied at the ends to unite the metals

(a) TIG (b) thermit
(c) atomic hydrogen (d) plasma
(e) forge.

68. In inter gas arc welding following is used for welding magnesium
(a) non-combustible electrode in combination with helium and DC current
(b) combustible electrodes and argon in combination with AC current
(c) straight polarity DC current
(d) carbon dioxide, because of its excellent penetration and high speed
(e) none of the above.

69. Submerged arc welding is
(a) a process which uses a mixture of iron oxide and granular aluminum
(b) accomplished by maintaining a hot molten metal pool between plates
(c) a process in which arc is maintained under a blanket of flux
(d) all of the above
(e) none of the above.

70. In a DC arc welding operation, the voltage arc length characteristic was obtained as $V_{arc} = 25 + 5\,l$ where the arc length l was varied between 5 mm and 7 mm. Here V_{arc} denotes the arc voltage in volts. The arc current was varied from 400 A to 500 A. Assuming linear power source characteristic, the open circuit voltage and the short circuit current for the welding operation are
(a) 45 V, 450 A (b) 75 V, 750 A
(c) 95 V, 950 A (d) 150 V, 1500 A *[GATE-2012]*

ANSWERS

1. (b)	**2.** (a)	**3.** (a)	**4.** (b)	**5.** (a)	**6.** (a)	**7.** (a)
8. (c)	**9.** (a)	**10.** (c)	**11.** (d)	**12.** (b)	**13.** (b)	**14.** (c)
15. (c)	**16.** (b)	**17.** (c)	**18.** (a)	**19.** (c)	**20.** (d)	**21.** (a)
22. ( )	**23.** (e)	**24.** (c)	**25.** (d)	**26.** (b)	**27.** (d)	**28.** (b)
29. ( )	**30.** (e)	**31.** (d)	**32.** (d)	**33.** (c)	**34.** (c)	**35.** (d)
36. (e)	**37.** (a)	**38.** (c)	**39.** (c)	**40.** (e)	**41.** (b)	**42.** (b)
43. (b)	**44.** (c)	**45.** (c)	**46.** (e)	**47.** (d)	**48.** (e)	**49.** (a)
50. (c)	**51.** (d)	**52.** (d)	**53.** (d)	**54.** (c)	**55.** (b)	**56.** (a)
57. (a)	**58.** (a)	**59.** (a)	**60.** (a)	**61.** (d)	**62.** (b)	**63.** (c)
64. (c)	**65.** (c)	**66.** (a)	**67.** (e)	**68.** (a)	**69.** (c)	**70.** ( )

QUESTIONS

1. Define the plasma arc welding process.
2. How many types of plasma arc welding?
3. Differentiate between plasma arc welding (PAW) and Tungsten Inert gas welding (TIG) process.
4. What is Kerf? define the same.
5. What is the keyhole method of welding?
6. What is the difference between plasma gas and shielding gas?
7. Why is the most common use of plasma torch?
8. Advantages of PAW.

9. What is an Arc? How does arc differ from plasma arc?
10. Describe with sketch the principle and working of electron beam welding.
11. Write the advantages and applications of EBW.
12. Explain how electrons penetrate very deep into narrow weld.
13. Explain with the neat sketch working principle of EBW. *[UPTU 2007-08 MFS]*
14. Explain the keyhole technique of welding use in EBW, and what are the limitation of EBW?
15. What are the possible difficulties encountered during EBW and how it can be remove?
16. Why a vacuum chamber is used in electron beam welding?
17. How are electrons accelerated?
18. Why is the depth to width ratio so grate in electron-beam welding?
19. What is kinetic energy? How is it applied?
20. What are the two types of electron beaming? What are their individual advantages?
21. What are the power source requirements for electron beam welding and cutting?
22. What is the laser beam welding (LBW)? Explain the same with neat sketch.
23. Differentiate LBW (Laser beam welding) with EBW (electron beam welding).
24. Explain the principle of laser and write uses/application of LASER. *[UPTU 2007-09]*
25. Describe the advantage of LBW. *[UPTU 2008-09]*
26. Justify why laser beam welding (LBW) is used in electronic industries.
27. Keyhole technique in LBW (Laser beam welding)?
28. Enlist some of the common lasering material used for laser production, what safety precautions need to be taken against laser beam hazards during LBW?
29. Explain what is the effect of vacuum on the weld produced by LBW.
30. How is depth penetration accomplished with the laser?
31. What are the basic components of the welding laser?
32. Describe, with neat sketch, the mechanism of explosive welding and explain the importance of re-entrant jet also brief write about weld-interface and welding.
33. Enlist the application/uses of explosive welding.
34. Enlist the advantage and limitation of explosive welding.
35. Define the different terms used in explosive welding such as buffer plate, flyer plate. Impact velocity, surface jetting etc.
36. Define the ultrasonic welding with the help of a neat sketch.
37. Write the advantages and limitations of ultrasonic welding.
38. Write the application of ultrasonic welding.
39. Explain the various component of USW.
40. What is ultrasonic wave/vibration and how it is generated?
41. Define the HAZ in welding.

42. Define the metallurgy of fusion weld with the help of neat sketch.
43. What is the effect of heat on welding and HAZ?
44. What is a HAZ?
45. Write the short notes of effect of cooling on formation of HAZ.
46. What is weld decay?
47. What is weld decay how it can be minimize?
48. Enlist the point which are responsible for HAZ cracking.
49. Define heat source efficiency.

❑❑❑

Unconventional Machining Process (UCM)

Chapter 11

11.1 INTRODUCTION

Unconventional machining is also an other machining process in which metal removal take place. Unconventional machining process differ with conventional machining process in term of metal removal. In conventional machining process a large quantity metal removed (in the form of chips) in a single stroke, where as in unconventional machining process metal removal rate is very slow as well as material removed in term of fine particles (Powder Form).

In unconventional machining process only small quantity of material can be removed and in conventional machining process high degree of cutting (shear) stress is required for cutting purpose whereas such condition is not applicable in case of unconventional machining process, even if workpiece is very hard then it can not machined by conventional machining process. For cutting hard material unconventional machining process is adopted.

Unconventional machining process is also known as "Non-traditional machining method **(NTMM)**, and the basic working principal of metal removal that no physical contact between the tool and workpiece and no need to be hardened than the job. Unique feature of UCM process that high degree accuracy and high degree of surface finish can be obtained.

11.2 CLASSIFICATIONS OF UCM PROCESS

Unconventional Machining Process (UCM) can be classified on the following basis:

1. On the basis of energy

(i) Chemical
(ii) Electro-chemical
(iii) Electrothermal
(iv) Mechanical

2. On the basis of metal removal technique

(i) Erosion
(ii) Shear
(iii) Vaporization
(iv) Ionic dissolution
(v) Chemical ablation

Classification of Non-conventional Manufacturing processes

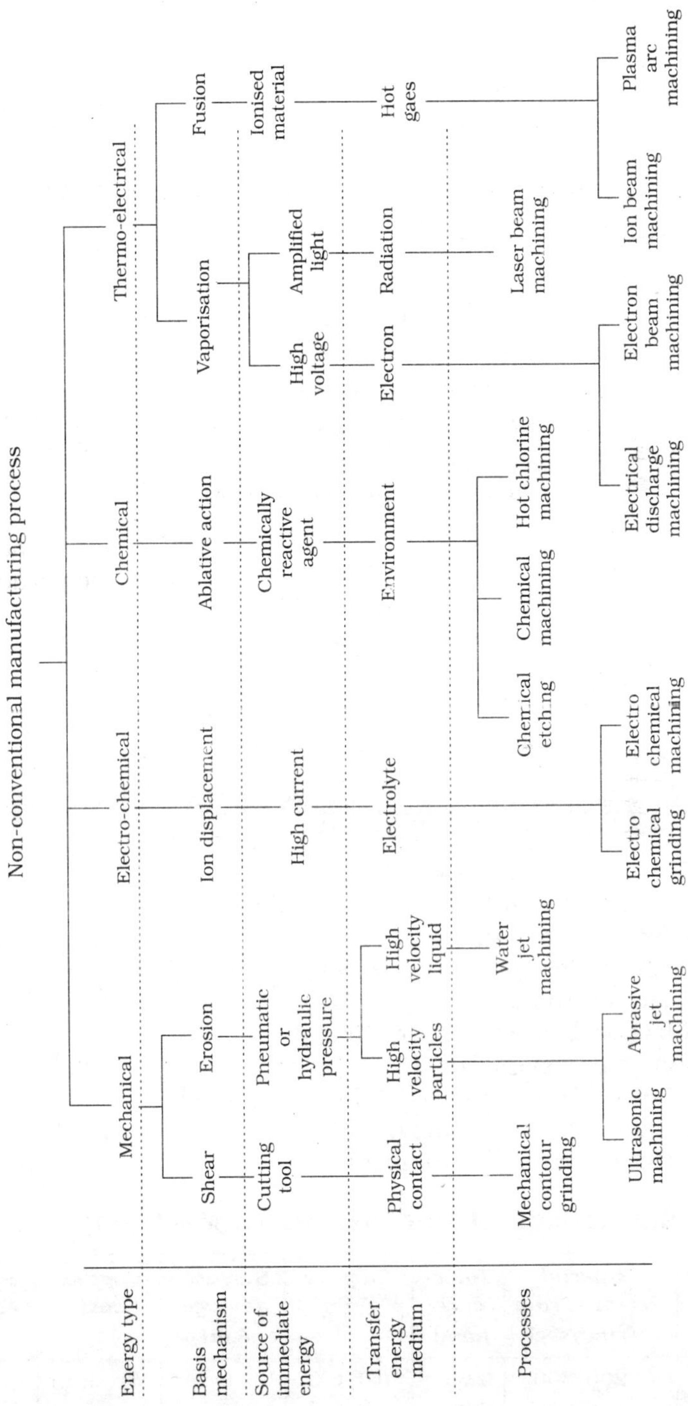

3. On the basis of energy transfer

(i) Physical contact
(ii) Electrons
(iii) High velocity particles
(iv) Electrolyte
(v) Radiation

11.3 NON-CONVENTIONAL MACHINING PROCESS ARE FOLLOWINGS

1. Chemical machining process
   (a) Chemical machining (CHM)
2. Mechanical machine process
   (a) Ultrasonic-machining (USM)
   (b) Abrasive jet machining (AJM)
3. Thermo-electric machining process
   (a) Electrical discharge machining (EDM)
   (b) Plasma arc machining (PAM)
   (c) Ion-beam machining (IBM)
   (d) Laser-beam machining (LBM)
   (e) Electron-beam machining (EBM)
4. Electro-chemical machining process
   (a) Electro-chemical machining (ECM)
   (b) Electro-chemical grinding (ECG)

11.4 COMMON UCM PROCESS

(i) Electro-chemical machining (ECM)
(ii) Electro-discharge machining (EDM)
(iii) Ultrasonic machining (USM)
(iv) Abrasive jet machining (AJM)
(v) Laser beam machining (LBM)
(vi) Electron beam machining (EBM)
(vii) Ion beam machining (IBM)
(viii) Chemical machining (CHM)
(ix) Plasma arc machining (PAM)
(x) Electro-chemical grinding (ECG)

11.4.1 Comparison of Unconventional Machining Process

Process	*Material removal rate (mm^3/sec)*	*Toler-ance* (mm)	*Surface* (μm)	*Surface damage* (μm)	*Capital cost*	*Power (watt) consumption*
ECM	200–300	0.05	0.1–0.3	5	Very high	100000
CHM	15–30	0.05	0.4–0.3	50	Medium	–
EDM	10–20	0.015	0.2–1.3	125	Medium	2700
USM	5–12	0.008	0.2–0.5	30	Low	2400
LBM	1–30	0.03	0.4–1.25	120	Medium	2.0

11.5 ADVANTAGES OF UCM PROCESS

These are several advantages of using UCM over traditional machining process.

(i) Complex shapes which are very difficult to produced by Traditional machining process can be easily produced
(ii) Difficult geometric can be easily produced by UCM process.
(iii) High degree of accuracy and surface finish can be easily achieved by UCM process which is very difficult by conventional machining process.
(iv) High strength alloys can be easily machined by UCM process.
(v) Very hard materials can be easily machined.

11.6 LIMITATIONS OF UCM PROCESS

(i) Process speed is very slow or it is a time consuming process.
(ii) Tool material is not so harder as compared to workpiece.
(iii) Tool forces do not increases as the work material gets harder.
(iv) Set up cost is very high.
(v) Highly skill person is need to perform the operation.
(vi) Material can not be removed in bulk quantity in a single stoke.

11.7 ELECTRO-CHEMICAL MACHINING (ECM) *[UPTU 2011-12]*

11.7.1 Introduction

ECM is very important versatile unconventional machining process in which metal removal take place by passing the current through a electrolyte solution as shown in Fig. 11.1. Basic working principle of electrolysis is based on **Faraday's Law** of electrolyte.

An electrolyte usually **NaCl**, **Sodium nitrate**, **Sodium Chlorate** is passed through a very narrow gap approximate 0.05 to 0.03 mm between the cutting tool (cathode) and the workpiece (anode) as the current (DC) pass through circuit flow of electron start from – ive to + ive side. Tool move towards the workpiece automatically at constant speed by maintaining the constant gap. Tool is the replica of shape that to be produce and metal removal rate basically governed by Faraday's Law of electrolysis, range of power supplies varies from 600 A–40,000 A.

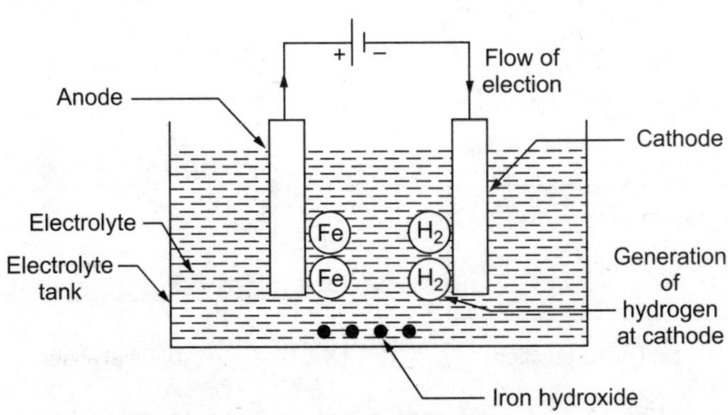

Fig. 11.1 Principle of electrolysis

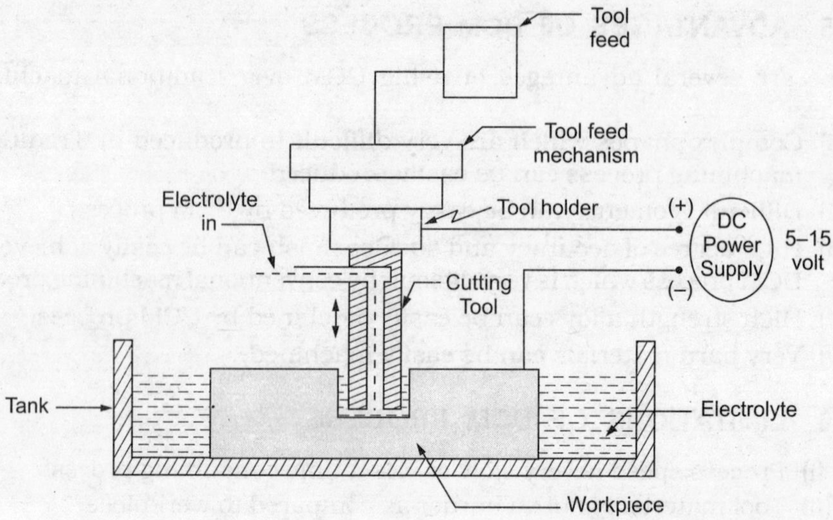

Fig. 11.2 Schetamatic view of electro-chemical maching (ECM) process

A servo motor is used to give the movement of tool. In ECM process proper care must taken as workpiece come in contact with electrolyte may be corresive.

11.7.2 Principle of ECM

It is very oldest method of metal removal and one of the unique feature of **electro chemical machining** is that it is a **stress free** process and it is based on the electrolysis. In this process a DC power source is used, cutting tool work as a cathode while workpiece is used as a anode, as DC power passed through circuit, flow of electron start from cathode to anode in such a way material removal start in the form of particles, reaction take place at anode–

$$Fe \rightarrow Fe^{++} + 2e \quad \text{...(i)}$$

Similarly at cathode

$$H_2O + 2e \rightarrow H_2 + 2OH^- \quad \text{...(ii)}$$

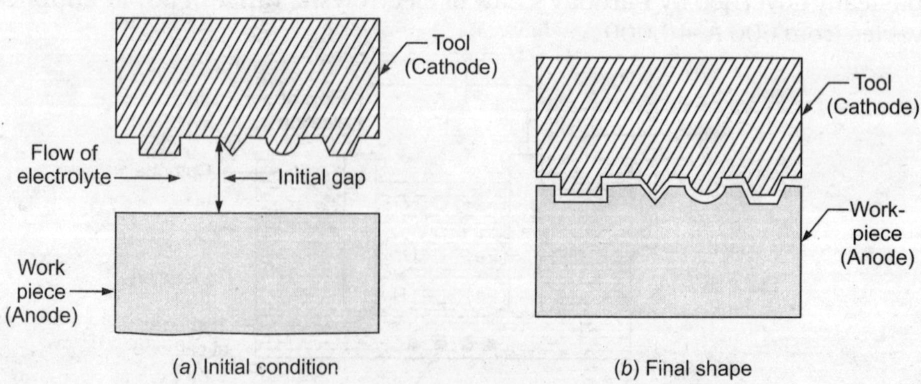

Fig. 11.3 Complimentory shapes generate by ECM

Adding equation (i) and (ii) we get

$$Fe^{++} + 2OH^{-} \rightarrow Fe(OH)_2$$

Net reaction $$Fe + 2H_2O \rightarrow Fe(OH)_2 + H_2$$

There is no physical contact between the workpiece and cutting tool, therefore **no tool wear**, nor plating of the object on the tool so that tool life is very long and one tool produce a large number of components. Voltage and current both are constant in ECM operation.

11.7.3 Characteristic of ECM

Tool material — Copper and brass
Material removal — Electrolysis
Workpiece material — All good conducting metals and their alloys
Process parameter — Current, feed, voltage and electrolyte
Electrolyte pressure — 0.07 to 2.7 MPa
Voltage — 3 to 30 V (DC)
Feed rate — 0.5 to 20.0 mm/min

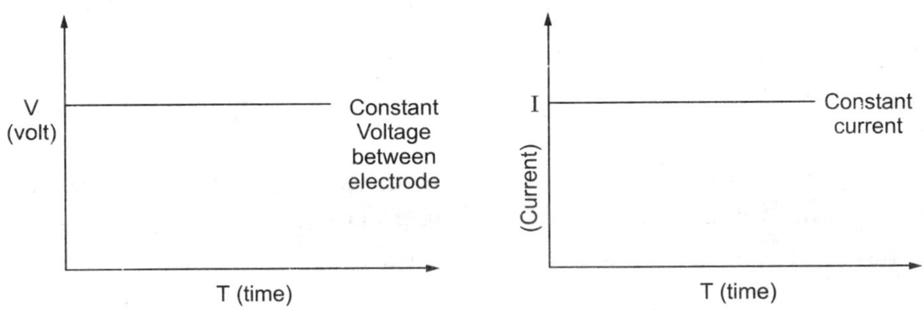

Fig. 11.4 Voltage and current in ECM

11.7.4 Metal Removal Rate in ECM

In electro-chemical machining process metal removal take place according to Faraday's Law of electrolysis which state that chemical change produced during electrolysis is proportional to the current passed and electrochemical equivalence of the anode material. Therefore MRR (material removal rate)

$$MRR = \frac{A \times I}{Z \times F} \text{ kg/sec}$$

(i) $$MRR = \frac{A \times I}{Z \times F \times \rho_a} \text{ mm}^3\text{/sec.}$$

where A = Atomic weight of workpiece material
I = Current passed through system (amperes)
Z = Valency of the work material
F = Faraday's constant
ρ_a = Material (workpiece) density

and Gap resistance is determine by the following method.

(ii)
$$R = \frac{\rho \times h}{A_{gap}}$$

where ρ = Specific resistance of electrolyte

h = Equilibrium gap

A_{gap} = Area of gap

(iii) Tool feed is determine by

$$V_f = \frac{MRR}{A}$$

where A = Area of work surface exposed to electroslyte

MRR = Metal removal rate

11.7.5 Advantages of ECM

(i) No direct contact between the tool and workpiece hence no chance of tool wear.

(ii) During the ECM process no stress generation.

(iii) No cutting forces produce during the process.

(iv) Complex and intricated shapes can be easily machined by this process.

(v) Metal removal rate is quite high in ECM process as compared to other unconventional machining process.

(vi) Dimensional tolerance up to 0.05 mm can be achieved by this process.

(vii) No need of other machining process such as grinding, polishing etc. surface produced by ECM of high degree.

(viii) Frangile parts, which are otherwise not easily machinable can be easily shaped by ECM.

11.7.6 Limitations of ECM Process

(i) Material removal rate is very low as compared to conventional machining process.

(ii) Time consuming process.

(iii) Set up cost is very high.

(iv) Large size part/component can not be machined.

(v) Continuous monitoring is required during process for feed, pressure etc.

(vi) Material which are not conductive can not be machined by this process.

(vii) Some time corrosion may be take place in workpiece or tool.

(viii) Only limited size and shapes component can be machined by this process.

(ix) Designing and fabrication of tool is very difficult.

(x) Very high power consumption is required.

(xi) No sharp edge and corner radius less than 0.2 mm can be produced by this process.

11.7.7 Electrolyte Used in ECM

Various types of electrolytes are used in ECM process and selection of electrolyte basically depends upon the following factors such as:

- type of material being machined,
- feed,
- speed,
- depth of cut etc.

11.7.7.1 A Electrolyte should be Fullfill the following Requirements

(i) It must complete electrical circuit between the workpiece and tool.

(ii) It should cool the workpiece, as during process high heat generation take place.

(iii) It allows desirable machining reaction to take place.

(iv) It should carry away the chip (powder form) from workpiece.

11.7.7.2 Electrolyte should have the following Properties

(i) High specific heat and low viscosity.

(ii) High thermal conductivity.

(iii) Non-toxic and non-corresive.

(iv) Chemical stability.

(v) Inexpensive and easily available.

(vi) Should be resistance to make a passive film on workpiece.

11.7.8 Applications of ECM

ECM process basically used for:

(i) Turning, facing, complex 3–D surfaces

(ii) Grinding

(iii) Trepanning

(iv) Die sinking

(v) Profile of any odd shape contour

(vi) Very narrow slot and holes making

(vii) Multiple hole drilling

(viii) Honing

(ix) Cut off

(x) Jet engine blade cooling holes.

11.7.9 Power Needed for ECM

DC power supply is used for ECM process from 5 to 25 V and current from 500 A to 25000 A is used. To avoid the corrosion from electrolyte, power supply separately used from machine (setup).

Electrolyte	*Materials (alloys)*
HCl or mixture of Brine with H_2SO_4	Nickel based alloys
10% HCl acid + 10% Hydro flouric acid + 10% H_2SO_4	Titanium base alloys
Chloride Solution in H_2O (25% NaCl)	Iron based and its alloys
Sodium Chloride (NaCl)	CO – Cr – W based alloys
Strong alkaline solution	based

Example 11.1

In electro-chemical machining process if current of 1500 A is used, then determine the volume rate of material removal from copper block. If Copper density = 8.96 g/cm^3,
valency = 1 and
atomic weight = 58.93 gm/cm^3 ***[UPTU-2002-03]***

Solution: Given $I = 1500$ A, $F = 96500$

$$Z = 1, \quad \rho_a = 8.96 \text{ gm/cm}^3, \quad A = 58.93$$

$$MRR = \frac{A \times I}{Z \times F \times \rho_a} = \frac{58.93 \times 1500}{96500 \times 1 \times 8.96} \text{ gm/cm}^3$$

$$= \frac{88395}{864640} = \mathbf{0.102 \text{ cm}^3/\text{sec}} \quad \textbf{Ans.}$$

Example 11.2

Estimate the material removal rate (in cc/hr) of an alloy containing –

CO – 18%,
Nickel – 62%, and
Chromium – 20%.

During ECM with current 500 A, density of the alloy is 8.28 gm/cm^3. The following data is available: ***[UPTU-2003-04]***

Metal	*Gram. Atomic weight(A)*	*Valency (Z)*
Cobalt	*58.93*	2
Nickle	*58.71*	2
Chromium	*51.99*	6
		($F = 96{,}500$)

Solution: Given $F = 96{,}500$, $I = 500$ A,

$$\rho_a = 8.28 \text{ gm/cm}^3, \quad t = 1 \text{ hr} = 3600 \text{ sec.}$$

$$MRR = \frac{A \times I}{Z \times F \times \rho_a}$$

$$= \frac{1}{96{,}500} \times \frac{500}{8.28} \times 3600 \left[\frac{58.93}{2} \times \frac{18}{100} + \frac{58.71}{2} \times \frac{62}{100} + \frac{51.99}{6} \times \frac{20}{100}\right]$$

$$= 2.25 \times (5.304 + 18.20 + 1.733)$$

$$= 2.25 \times 25.237 = \mathbf{56.78\ cm^3/hr}$$ **Ans.**

Example 11.3

Determine the material removal rate and electrode feed rate in the electro-chemical machining of an iron surface that is 30 mm × 30 mm in cross-section using castic soda in water as electrolyte, gap is 0.20 mm in between tool and workpiece voltage is 10 V and ρ_a is 3Ω cm.

Solution: Given $A_{gap} = 30 \times 30 = 900 \text{ mm}^2$, $V = 10$ V, $\rho = 3\Omega$ cm

For iron, $Z = 2$, $A = 55.85$, $\rho_a = 7860 \text{ kg/m}^3$

$$R = \frac{\rho \times h}{A_{gap}} = \frac{3 \times 0.2}{900} = 0.00067\ \Omega$$

$$V = I \times R$$

$$I = \frac{V}{R} = \frac{10}{.00067} = 15000\ A$$

$$MRR = \frac{A \times I}{Z \times F \times \rho_a} = \frac{55.85 \times 15{,}000}{2 \times 96{,}500 \times 7860} = 0.000550 \text{ g/sec}$$

$$= \mathbf{550 \times 10^{-6}\ kg/sec.}$$ **Ans.**

11.8 ELECTRICAL DISCHARGE MACHINING (EDM)

Introduction: This process is also known as **spark erosion machining** process as electrical energy is used to remove the material from workpiece by spark discharge without touching. In this process anode and cathode are used and if the both electrode are made by same material then more erosion produce on the positive electrode. For getting the maximum metal removal rate with minimum wear of tool electrode workpiece is used as a positive electrode while tool is used as a negative electrode.

11.8.1 Working Principle of EDM

In electro-discharge machining process material removed by the *spark erosion*. During the operation workpiece and tool both are eroded. Proper gap is maintain between the workpiece and tool. Tool motion is governed by a servo motor. In this process workpiece and tool both are submerged in a dielectric fluid. Workpiece work as a anode while tool work as a cathode. A DC power source is used. Tool is hold in a tool holder and movement of tool is controlled through a reduction gear box driven by a DC shunt motor as shown in Fig. 11.5, which is a reversable so that tool can be with drawn at any time. In electrical pannel there is a provision of one condenser and resistance as supply is passed through the circuit. Condenser charged and voltage will rise to approch the supply voltage. If tool (electrode) touch the workpiece, the condenser is short circuited, causing the supply voltage to appear across the resistance. Once the tool (electrode) comes in accurate position (maintaining gap), sparking generates and condenser rapidly charges and discharged so that a saw tooth wave form is produced across its terminals. Gap should be 0.01 mm to 0.1 mm so that gap voltage is around 70% of the supply voltage for charging the capacitor rank.

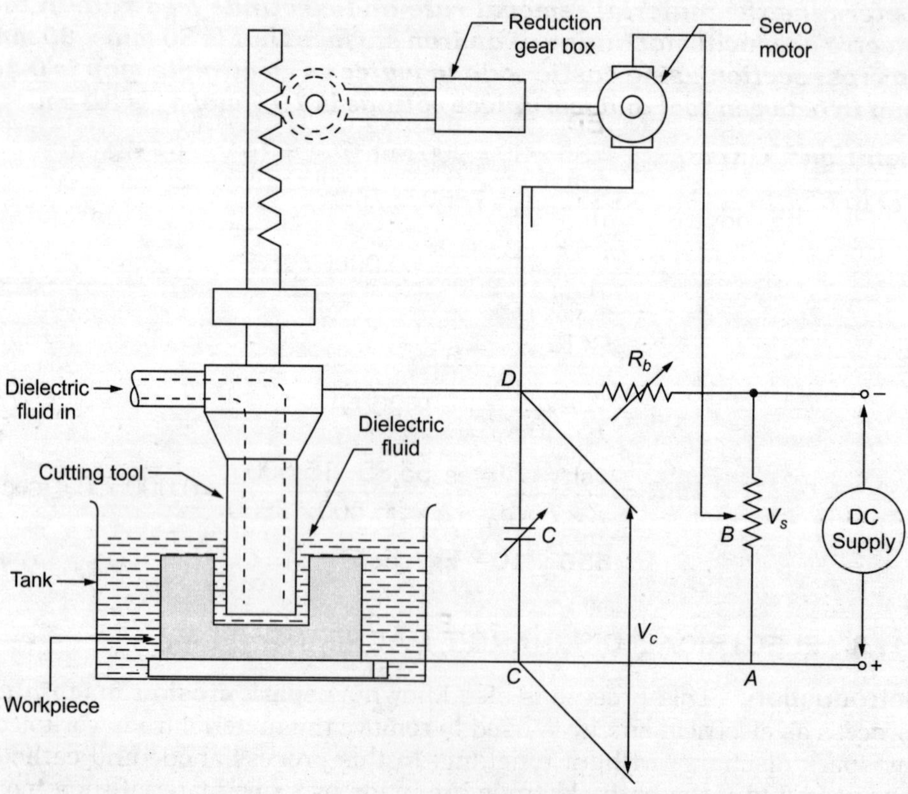

Fig. 11.5 Electrical discharge machining [3]

11.8.1.1 *Feature of EDM*

Tool material — Brass or copper or graphite
Material removal — Spark erosion
Workpiece material — All conducting material and their alloys
Voltage (V) — 40–300
Current (I) — 0.4–400 A

11.8.2 Advantages of EDM

(i) Process is suitable for (extreme) hardened workpiece.
(ii) Machining time is less as compared to conventional machining process.
(iii) Process is very suitable to make a fine (drilled) hole.
(iv) Process is suitable for producing complicate shapes.
(v) All conducting (electrically) material and their alloys can be easily machine.
(vi) No physical contact between the workpiece and tool.
(vii) Process is automated hence no skill operator is needed.
(viii) No cutting force during the process hence high aspect of surface can be machined by EDM.
(ix) As process is not affected by hardness, hence very hard materials can be machined.

11.8.3 Limitations of EDM Process

(i) Process is only suitable for electrical conductive materials.
(ii) More power required as compared to conventional machining process.
(iii) As more energy required for operating the process hence process is more expensive.
(iv) Frequently need to change the electrolyte.
(v) More chance of tool wear.
(vi) Material removal rate is low, hence process is time consuming.
(vii) Corner radius and sharp edges can not be produced by this process.
(viii) In some materials case, more chance of surface cracking.
(ix) Only limited shapes can be produced.

11.8.4 Dielectric Fluid Used in EDM

Dielectric fluid play an important role in EDM process. It is a spark conductor and used to cool the workpiece and tool.

Various types dielectric are used and selection of dielectric fluid basically depends on the various factor such as metal removal rate, types of metal for machining, etc.

A good dielectric fluid should be fullfill the following requirements:

(i) its flash point should be very high so that there is no chance of fire hazard.

(ii) it should have good wetting properties for cooling the workpiece as well as the tool.

(iii) it should have low viscosity.

(iv) it should be capable to remove the swarf (cutting) particles from machining gap.

11.8.5 Condenser Circuit and Metal Removal Rate

11.8.5.1 Distance Capacitance Circuit

This circuit operatate at self-oscillation Fig. 11.6 show a RC circuit. In this process a range of spark from 3000–10000 spark/see produce.

11.8.5.2 Charging

Condenser is charge through a variable distance by passing the current by a DC power source.

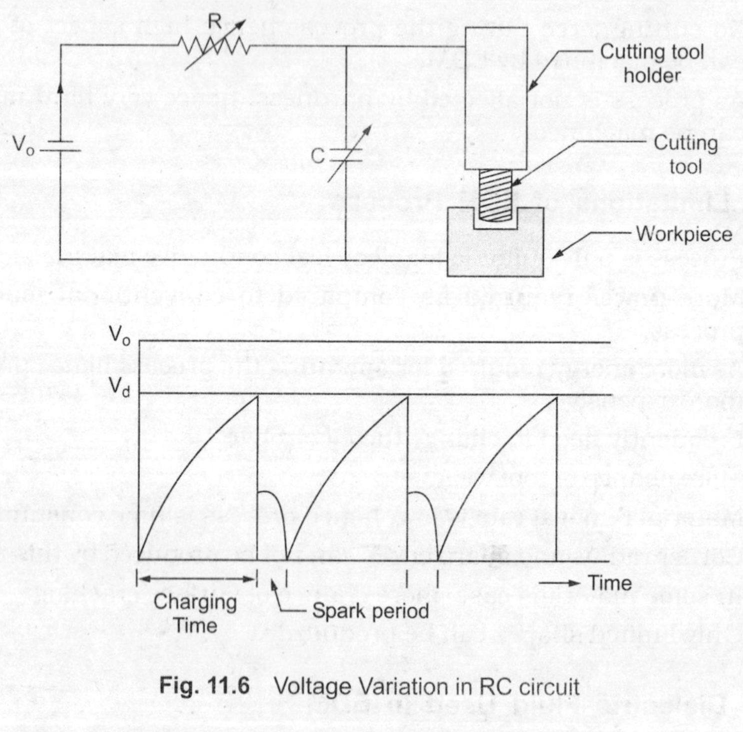

Fig. 11.6 Voltage Variation in RC circuit

11.8.5.3 Metal removal rate

For EDM process material removal rate (MRR) is derectly proportional to the multiplication of frequency of charging and energy delevered per spark. Mathematically it can be expressed as

$$\text{MRR} \propto \frac{1}{2} C V_b^2$$

$$\text{MRR} = \frac{1}{2} K_m f.\ C.\ V_b^2$$

where f = frequency of charging = $\frac{1}{t}$

t = time of condensor charging,

C = capacitance,

R = resistance,

V = charging voltage

V_b = break dwon voltage/discharging voltage

V_s = supply voltage

Since
$$V_t = V_s \left[1 - e^{-\left(\frac{1}{R_c \cdot C}\right)}\right]$$

$$t = R_c \cdot C \log_e \frac{1}{\left(1 - \frac{V_b}{V_s}\right)}$$

Therefore,

$$\text{MRR} = \frac{K_m \cdot V_b^2}{2R_c} \cdot \frac{1}{\log \frac{1}{\left(1 - \frac{V_b}{V_s}\right)}}$$

11.8.6 EDM Wire Cut Machine

Introduction: EDM wine cut machine is shown in the Fig. 11.7. In EDM wire cut machine, a wire act as a electrode and the working principle of wire

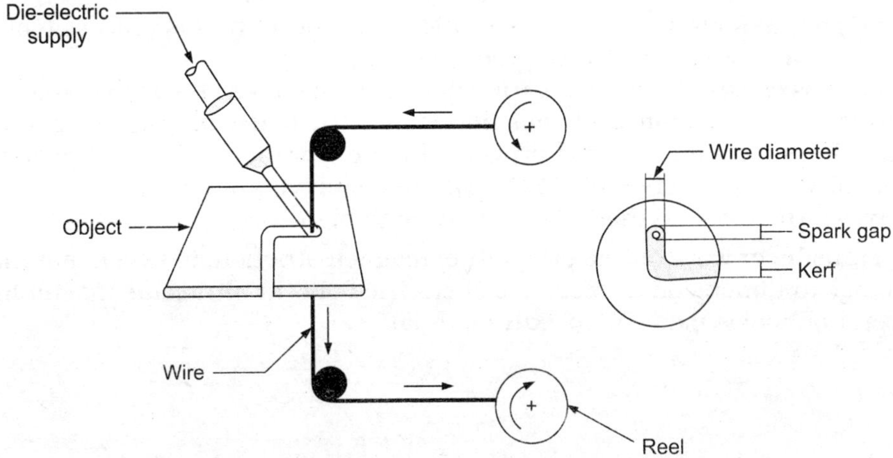

Fig. 11.7 EDM wire cutting

EDM is same as EDM. This wire EDM is suitable to produce a large variety of products such as gear, dies, tool, turbine blades rotors, etc. in wire cut EDM workpiece is clamped on table and table moves in X and Y axis by a NC controlled system, copper or brass wire is continuously moving. During the process jet of dielectric is provided and workpiece is submerged in dielectric. Process consume the time from half an hour to 20 hours.

During the process material cuts due to the vaporization of metal. As there is no involvement of cutting force hence fragile, complex geometrics of parts can be easily machined. Spark is produce one at a time in rapid sucession between electrode and workpiece. Power supplies provide a voltage between wire and workpiece. One of the main limitation of this process that workpiece should be electrically conductive.

11.8.6.1 Advantages of Wire EDM

(i) high degree of surface finish can be achived.

(ii) process is also very suitable for small batch production as programming can be very easily done.

(iii) very less chance of rejection of parts as before performing the operation program can be check.

(iv) Process is very suitable to machine any complicated through hole dies of a electrical conductive material.

11.9 ULTRASONIC MACHINING PROCESS (USM)

11.9.1 Introduction

Ultrasonic machining process is also an unconventional machining process. In this process metal removal take place due to vibratory motion of tool. Ultrasonic sound means a sound which have the hearing frequency above them human earing and the range of ultrasonic sound is usually above then 20,000 Hz.

11.9.2 Working Principle

In ultrasonic machining process a high frequency sound is produced and through that sound high speed vibrations produced. In this process tool is the replica of shape that to be produced.

Process consist of ultrasonic transducer which convert the high frequency sound into vibration and tool holder is attached with the ultrasonic transducer. Motion of tool is provided by the vibration and metal removal take place by the action of abrasive grains (slurry). Abrasive particles feed between the tool and the workpiece through high speed.

Transducer work on the principle of magnetostriction, in which material change in dimension in response of electric field. In ultrasonic machining operation tool is made out of **soft** material.

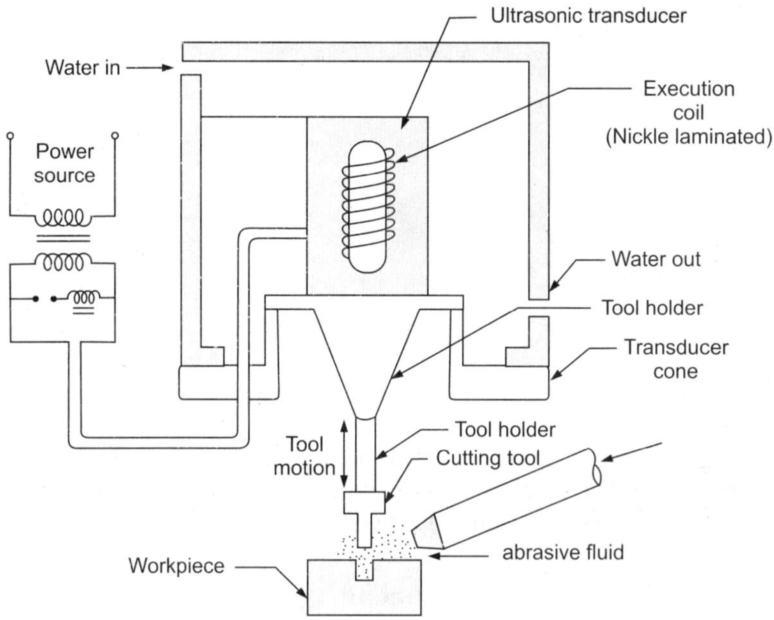

Fig. 11.8 Ultrasonic machining process

11.9.3 Elements of USM

(i) Workpiece material
(ii) Transducer and generator
(iii) Abrasive slurry
(iv) Tool shape and tool material.

(i) Workpiece Material: Process is only suitable for machining of brittle material as material remove in this process by brittle failure. Soft and ductile material can be cut by this process but not economical hence not recommended, even material those may be dissolved in slurry media use not recommended too.

(ii) Transducer: Transducer basically keep in vibrating tool at high frequency which basically consist magnetic coil having the coil wrapped around it.

(iii) Abrasive Slurry: Abrasive slurry play an important role in metal removal, during the process material removed due to the cutting action of abrasive particles. For different material different abrasive particles are used at different speed.

Usually **Boron Carbide** is used abrasive slurry as it has the faster cutting action. Even some cases Al_2O_3, ***silicon carbide***, ***Diamond*** dust are also used.

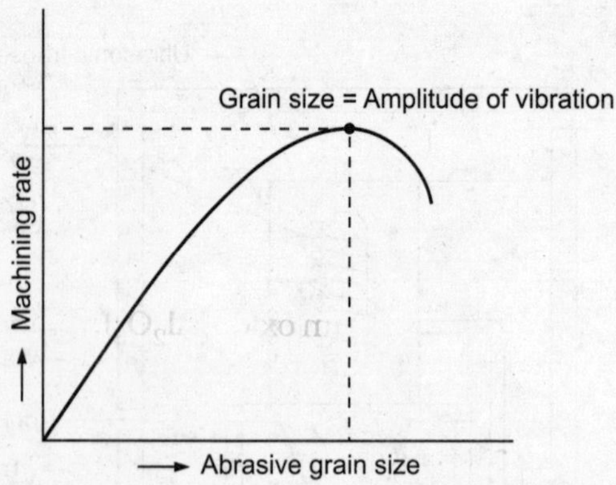

Fig. 11.9 Machining rate as a function of abrasing grain size

Abrasive grain size usually from 200 to 2000. However, abrasive particle size depends upon the surface finish required. Abrasive is suspended in a liquid with 25% to 50% by value of abrasive.

11.9.4 Advantages of USM Process

- Process is suitable for hard and brittle material.
- No noise during the process.
- Machine component/parts produced by this process is free from stress.
- Metal removal cost is low.
- Surface finish of a high degree can be achived by this process.
- Closer dimensional tolerance can be easily achived.
- Process is quite safer, less chance of accident.
- No change in mechanical properties and no change in metallurgy of component/part as no heat generation during the process.
- No thermal stresses..

11.9.5 Limitations of Process

- Process is not suitable for softer material.
- Metal removal rate is low.
- Only limited shapes can be produced.
- Cutting tool is the replica of shapes that to be produce; if product shape change it is mandatory to change the tool.
- Set up cost of system is high.
- Power required to operate the process (system) is high.
- Heavy materials can not be cut in a single stroke.
- Abrasive particles require to change at regular interval.
- One type abrasive can not be use for all types of material.

11.9.6 USM Characteristics:

Tool material	:	Mild steel or brass.
Mechanism of Material removal	:	Fracture of workpiece due to high impact of abrasive particles.
Process element	:	Abrasive particles, ultrasonic transduer feed force, frequency.
Abrasive particle	:	Aluminium oxide (Al_2O_3), diamond dust, silicon carbide, boron carbide.
Gap between tool and workpiece	:	0.2 to 0.5 mm
MRR	:	3 mm^3/sec.
Surface finish	:	0.25 to 0.50 μm R_a
Power supply	:	220 volts, $\simeq$ 12 A

11.9.7 Material Removal Rate

Process can be employed on mostly all materials but process is not recomented for that material having the hardness < 50 HRC. Material removal rate/unit time is inversely proportional to the cutting area of tool provided circulation of the slurry of other operating conditions are remain constant. Material removal rate increases as the abrasive particle size increases. When its size become equal to the amplitude of vibration, then rate drops down.

$$\text{Wear ratio} = \frac{\text{Volume of material removed from workpiece}}{\text{Volume of material croded from cutting tool}}$$

material removal in USM process is a complicated phenomenon, and occuss by fracture and plastic deformation to varying degree.

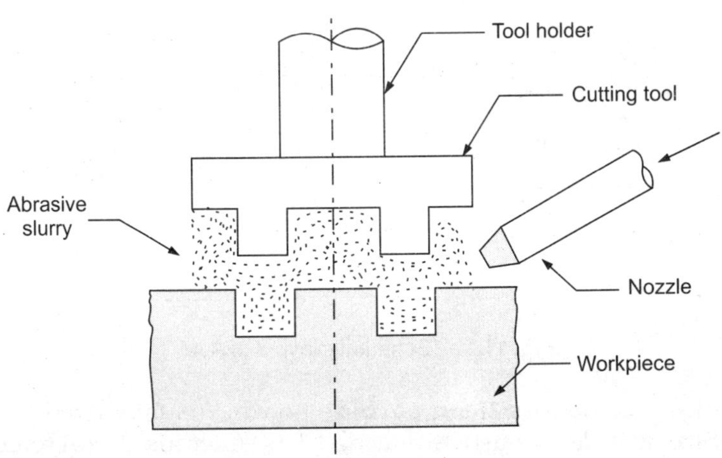

Fig. 11.10 Working principle of metal removal mechanism in USM

11.9.8 Applications

(i) holes and round hole of any shapes which a tool can be made.
(ii) both the conductive and non-conductive material can be machined.
(iii) coining operations for glass, ceramics etc.
(iv) very hard material can be machined.
(v) micro drilling hole of 0.1 mm.
(vi) threading via appropriately rotating and translavting the workpiece.
(vii) jewellary for shaping.
(viii) dentistry.

11.10 ABRASIVE JET MACHINING (AJM)

11.10.1 Introduction

In abrasive jet machining process, abrasive particle mixed with some gas, are fed on the workpiece through a high speed pressure with the help of a nozzle. Figure 11.11 shows a set up for AJM in which abrasive particle mixed with a gas (usually CO_2, N_2 and compressed air) in a mixing chamber in corret ratio. Process is quite similar to sand blasting or shot blasting. A pressure regulator control the flow of abrasive particles.

Abrasive particle impinge on the workpiece at velocities of about 1000–20,000 m/min.

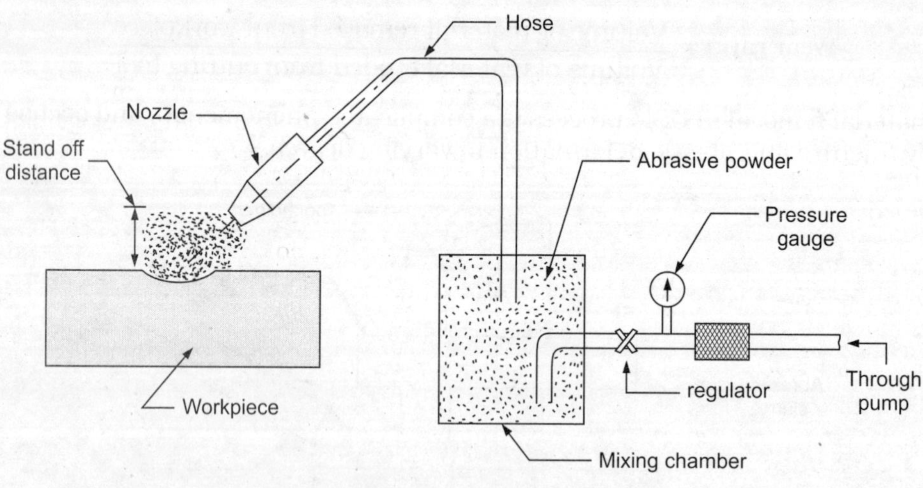

Fig. 11.11 Schematic layout of AJM

Aluminium oxide or ***silicon carbide*** powder usually used for cutting purpose. Size of abrasive particle may be 15–60 microns. Metal removal rate is affected by **size of the particles, distance of nozzle** from workpiece and **speed** of abrasive particles.

11.10.2 Elements of AJM

(i) Type and size of abrasive
(ii) Carrier (medium) gas
(iii) Speed
(iv) Work material
(v) Stand off distance
(vi) Design of nozzle
(vii) Cutted shape

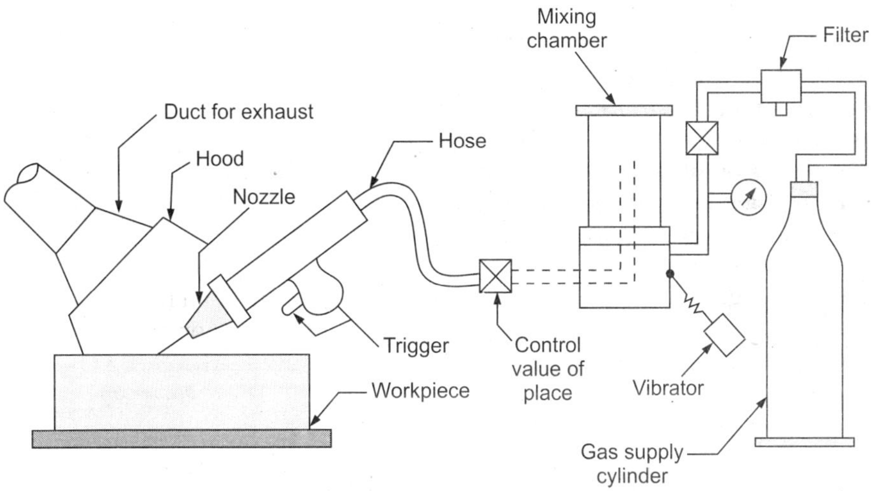

Fig. 11.12 Block diagram of abrasive jet machining [22]

(i) Type and size of abrasive: Carrier gas used in AJM should not be flamable at the time and discharge from nozzle on the workpiece and gas should be such that abrasive powder should be properly mixed and it should be cheap and easily available.

Mostly **carbon dioxide (CO_2), Nitrogen (N_2)** and compressed **air** are used as a carrier gas. On the other hand abrasive particle size depends upon the type of material to be cut, quantity to be cut (range, finish etc.) abrasive having sharp edge and irregular shapes.

(ii) Carrier gas: A carrier gas is used in abrasive jet machining (AJM) process and gas used for this should not flare excessively when discharge from nozzle to the open air. Gas should be cheap, non-toxic and should be easily available.

Commonly Nitrgen (N_2), Carbon dioxide (CO_2), air (O_2), are used as a carrier gas. If air is used as a carrier gas then it should be properly dried for H_2, oil or any other particles. Pressure range should be 2-8 kg/cm^2. High pressure may be result of high nozzle wear while low pressure rate may responsible for low rate of material removal rate (MRR).

(iii) Speed: Due to the kinetic energy of abrasive particles metal removal take place by **erosion.**

In abrasive jet machining the abrasive used for cutting purpose are **Al_2O_3, Silicon carbide,** while dolomide of 200 grit size is used for light cleaning and etching, glass beads of 0.3–0.8 mm is used for light polishing and deburring abrasive particle can be reused as their cutting edges become blunt after cutting.

Abrasive	*Grain size*	*Use*
Al_2O_3	12–25 μ	cutting and grooving
Sodium bicarbonate	28 μ	high finishing at < 50°
Silicon carbon (SiC)	25–40 μ	cutting and grooving
Dolomite	200 mesh	etching and polishing

(iv) Work material: AJM process best suitable for brittle material like ceramics, glass, refrectories etc. M.R.R. depends upon the hardness of workpiece and type of abrasive used during the process.

(v) Stand off distance: ***SOD*** is the distance between the nozzle and the workpiece face (surface). Figure 11.13 Stand off distance effect the MRR as well as the surface finish of part produced by AJM. If the value of stand off distance is very large then surface finish of part is poor. If the gap between the nozzle and workpiece (stand off distance) is low then value of MRR increases. Means on increasing the SOD surface finish is poor. Where as on decreasing the SOD surface finish increases.

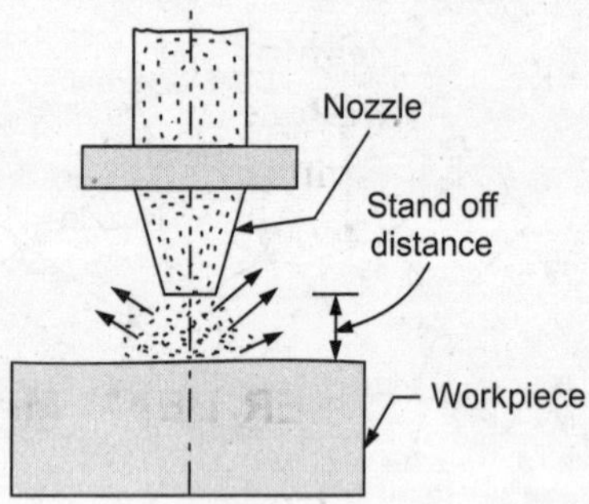

Fig. 11.13

(vi) Design of nozzle: As the abrassive particles flow with carrier gas with in the nozzle then design of nozzle is very important. Its design should be such that it can easily flow the mixture without any wear resistance. commonly **Tungsten carbide** and **sapphire** are used as a nozzle material as both can with stand there properties under high pressure condition.

Average particle size (μm)	*Surface roughness (μm)*
10	0.15–0.2
25	0.4–0.8
50	1.0–2.0

11.10.3 Advantages of AJM

- Process is suitable for brittle and hard material.
- Process is suitable to produce the intricated shape hole.
- Set up cost is not very high.
- Process is suitable to out fragile, heat sensitive parts with out any damage.
- As no physical contact between the nozzle and workpiece hance very little or no heat generation.
- Low power consumption is required.

11.10.4 Limitations of AJM

- Material removal rate is slow as composed to other process hence application of process is limited.
- Machining accuracy is poor.
- Due to high kinetic energy wear rate of nozzle is very high.
- Proper cleaning of workpiece is required before performing the operation.
- Abrasive powder can not be reused.
- Process requires dust collecting system.
- Embedding of abrasive particles in soft workpiece.

11.10.5 Applications of AJM

- To cleaning the surface.
- To produce fines line on the surface of object.
- Removing any previous thermal oxide layer on the surface of object.
- Cutting the various materials like quartz, mica etc.
- For machining semiconductor like germanium etc.

11.11 LASER BEAM MACHINING (LBM) *[UPTU 2011-12]*

11.11.1 Introduction

In LBM workpiece ***melted*** and ***vaporized*** by means of intense, monochromatic beam of light known as laser.

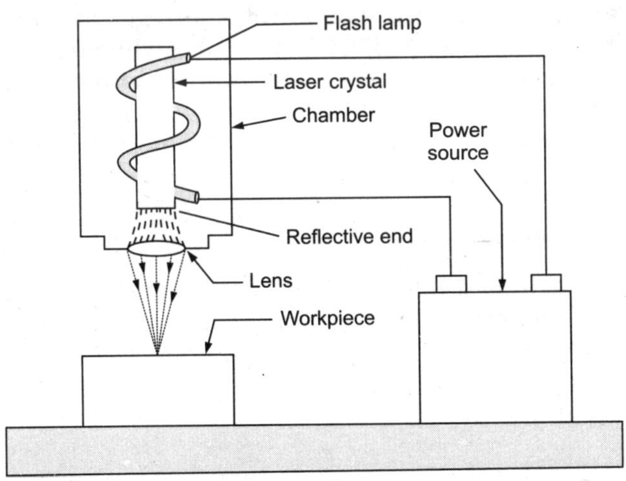

Fig. 11.14 Block diagram of laser beam machining (LBM)

Laser stands: for light Amplification using Stimulated Emission of Radiation.

11.11.2 Working Principle

In LBM the high (sourcecey energy) produce by a crystal is focuse (optical energy) on the workpiece surface. This ray (high) of very high density falls on the surface (very small area) of workpiece and it start to melt and evaporate in a controlled manner.

Laser is a electromagnatic radiation. Process do not require any vacuum and through LBM a large variety of metallic and non-metallic materials can machined by focussing a laser beam on a small portion (spot) 1/100 of a square mm and beam can concentrated in a short flash to a power density of 1,00,000 kW/cm^2 and this energy is sufficient to **melt and vaporized** the workpiece surface (spot).

11.11.2.1 Types of layer

Different type of lasers are used to produce high direction beam of optical & in-fared radiation. However two lasers are most common laser i.e., **solid laser and CO_2 laser.**

11.11.2.1.1 Solid State Laser

In solid lasers commonly **ruby, Aluminium** glass (Nd-glass), and **neodymium-yttrium aluminium gar net (Nd-YAG)** are used. These laser material are fabricated into vodes and their ends are finish for a very high degree optical tolerance through this process energy is injected into the material by intence high flux, which is absorbed by laser material and finally converted into a collimated to laser beam. High intansity light flux is given by xenon tube, and life of laser is appriximate is 10,000–1,00,000 pulse.

11.11.2.1.2 CO_2 Gas Laser

In CO_2 gas laser system usually **Nitrogen (N_2), Carbon dioxide (CO_2) and helium (He)** is used by a glass discharge tube. Nitrogen intermediary eletrical energy and vibrational energy of CO_2 molecules while helium cool the gas mixture. CO_2 laser have several advantage over the former (solid state laser), i.e., high and continuous power output and high concession efficiency. CO_2 laser system is very suitable for cutting organic materials like leather, rubber, paper, plastic etc.

11.11.2.2 Metal Removal Rate

In laser beam machining maximum material removal rate depends upon the thickness of the material and power of laser used. Lower cutting rates give better quality of the cutting surface. Material removal rate is the order of 400 mm^3/hr.
cutting rate is calculated by

$$C = k \times \frac{P}{E \times A \times t_1} \text{(mm/min)}$$

where P = Power of laser (W)

E = Vaporization energy of material (W/mm^2)

k = Constant (material)

A = Area of laser-beam at focal point (mm^2)

t_1 = Thickness of material (mm)

Now energy required to raise a value of material to its vaporization point

$$E = \frac{\rho\, v\, C_p(t_m - t_\infty) + C_v(t_b - t_m) + h_f + h_{fg}}{\eta}$$

where v = volume to be evaporated (m^3)

η = efficiency

ρ = density of material (kg/m^3)

C_p = specific heat at constant pressure

C_v = specific heat at constant volume

t_∞ = ambient temperature

t_b = boiling point

h_f = latent heat of fusion (J/kg)

h_{fg} = latent heat of vaporization (J/kg)

11.11.3 Characteristic of LBM

Material removal take place	: By **metling and vaporization**
tool	: High powered focused laser beam
Medium	: Process **carried** out **Air**
Workpiece material	: Process can be perfrom on **any material**
Process parameter	: (i) Melting temperature of workpiece (ii) Power intensity of laser beam (iii) Diameter of focused laser beam.

11.11.4 Advantages of LBM

(i) Process is suitable for machining of any known material.

(ii) No physical contact between the workpiece and tool.

(iii) As no physical contact between the workpiece and tool hence no large mechanical forces subjected to workpiece.

(iv) Two or more cuts can be performed at a time.

(v) Process can be performed heat shock sensitive material such as ceramic by 50%.

(vi) Process is suitable to cut the complex shapes with high precision.

(vii) Process is very suitable to make the narrow hole with high degree accuracy.

(viii) Process can be performed in any environment such as **air, inert gas,** vacuum and even some cases liquid medium too.

(ix) Full utilization of laser beam.

(x) By this process even two different material can be join.

(xi) High cutting speed in a narrow kerf's also possible.

(xii) Very samll hole can be drilled by laser drilling in a very hard material.

(xiii) Process is suitable for higher productivity with more safety.

(xiv) Heat affected zone is very small.

11.11.5 Limitations of LBM

(i) Skill operator is required to perform the operation.
(ii) Metal removal rate is low.
(iii) Initial cost of equipment is very high.
(iv) Limited application (this workpiece and material removal rate is very low).
(v) Flash lamp life is very short.
(vi) Difficult to control the output energy of laser precisely.
(vii) Thick sheet can not be cut by this process.

11.11.6 Applications of LBM

(i) Microdrilling upto 0.005 mm diameter with high degree accuracy.
(ii) Drilling hole in hard material such as ceramics, tungsten etc.
(iii) LBM process is extensively used in electronic industries.
(iv) For cutting the nano and composite material.
(v) Hole in vanes of jet engine.

11.11.7 Common Application of LASER

Type of laser	*Application*
PCO_2, $CWCO_2$, Nd: YAG, ruby	Metal cutting
PCO_2	Ceramic cutting
PCO_2, Nd:YAG, Nd:glass, ruby	Drilling in metals
Excimer	Drill in plastic
PCO_2, $CWCO_2$, Nd:YAG, Nd:glass, ruby	Metal welding

where P = pulsed, CW = continuous wave

11.12 ELECTRON BEAM MACHINING

11.12.1 Introduction

Electron beam machining (EBM) and Laser Beam Machining (LBM) are thermal processes considering the mechanisms of material removal. However electrical energy is used to generate high energy electrons in case of Electron Beam Machining (EBM) and high-energy coherent photons in case of Laser Beam Machining (LBM). Thus these two processes are often classified as **electro-optical thermal** processes.

Electron beam machining is a latest machining process specially to make a drill of small workpiece. This technology is not used from machining purpose even this technology is also used for welding purpose too. In EBM process high energy electron are concentrated on a small zone of workpeice.

11.12.2 Working Principle

In electron beam machining process a high energy electron beam are concentrated on the surface of the workpiece. Electrons are generated by as source refer as tungsten filament wire.

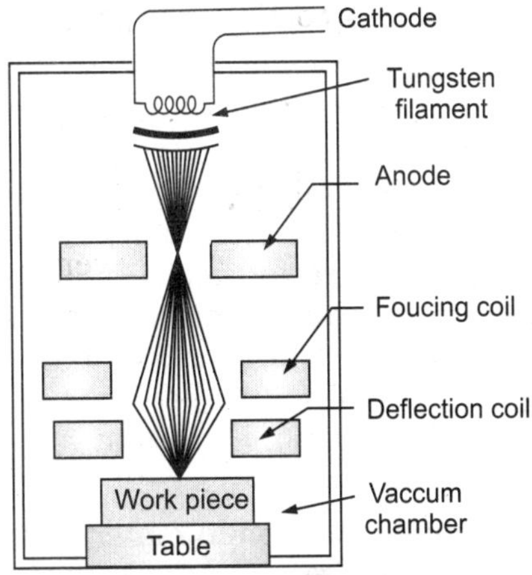

Fig. 11.15 Block diagram of electron beam machining

Vacuum is required for beam machining process and electron are accelerated to the half velocity of light it is very high velocity. As switch is on tungsten filament is heated upto 2500°C in a vacuum (10^{-4} mm of kg), electrons start to generate electrons are passes through anode tube and finally focussed by a magnetic coil. There is a provision of deflection coil for controlling the beam. In electron beam machining process machining of material occurs due to **melting and vaporization** of workpiece.

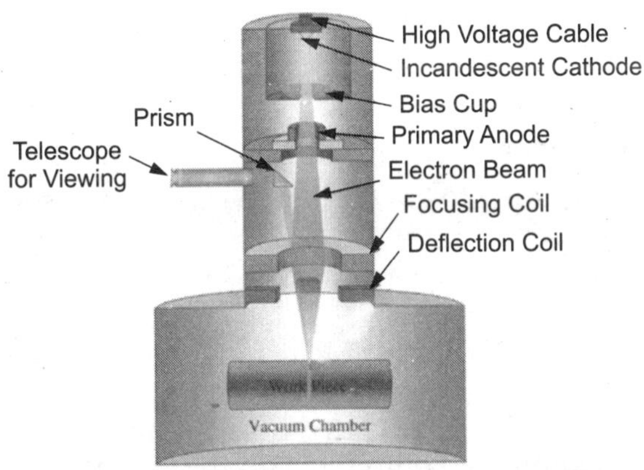

Fig. 11.16 Electron beam machining (EBM)

EBM process is very suitable to produce very fine machining of 0.03 mm out of 0.3 mm sheet and the large of high power current usually 100 – 1000 μA with power output of 100–1150 W.

11.12.3 Characteristic of EBM

Material: Process is suitable Material for all materials
Method of Material Removal: by melting and vaporizing
Medium: vacuum (10^{-5} f Hg) Medium
Consumption of Power: 500 W/mm^3 min.

11.12.4 Electron Beam Machining–Process

Electron beam is generated in an electron beam gun. The construction and working principle of the electron beam gun are complicated. Electron beam gun provides high velocity electrons over a very small spot size. **Electron Beam Machining** is required to be carried out in vacuum. Otherwise the electrons would interact with the air molecules, thus they would loose their energy and cutting ability. Thus the workpiece to be machined is located under the electron beam and is kept under vacuum. The high-energy focused electron beam is made to impinge on the workpiece with a spot size of 10-100 mm. The kinetic energy of the high velocity electrons is converted to heat energy as the electrons strike the work material. Due to high power density instant melting and vaporisation starts and "melt-vaporisation" front gradually progresses.

Finally the molten material, if any at the top of the front, is expelled from the cutting zone by the high vapour pressure at the lower part. Unlike in Electron Beam Welding, the gun in EBM is used in pulsed mode. **Holes** can be **drilled** in **thin** sheets using a single pulse. For thicker plates, multiple pulses would be required. Electron beam can also be manoeuvred using the electromagnetic deflection coils for drilling holes of any shape.

11.12.5 Advantages of EBM

(i) As no direct contact between the workpiece and tool (electron), hence no stress generation.

(ii) Process is suitable to cut any metal/non-metal which exit in vacuum.

(iii) Process is suitable to cut/Drill in a very narrows space.

(iv) Faster process.

(v) Electron beam can travel at two-third speed of light.

(vi) Process not change the properties of workpiece during machining.

11.12.6 Limitations of EBM

(i) Vacuum medium is required to run the operation.

(ii) Initial cost of equipment is very costly.

(iii) As EBM is thermal metal removal process hence more change to distort the part.

(iv) Melting and evaporation of the metal in EBM holes usually make then have slight bell or trumpet shape.

(v) Only upto 4 mm depth material can removed.

(vi) Skill operator is required to perform the operation.

11.12.7 Applications of EBM

(i) To machining of thin materials.

(ii) To engraving the punches

(iii) To drill gas orifice for pressure different device such as used in **aircraft engine**, **nuclear reactors** and many more.

(iv) To hole depth-to-diameter ratio 100 : 1 with multiple pulse

(v) To fabricate wire drawing dies, very small orifice etc.

(vi) To scribe thin films.

(vii) To produce microholes and narrow slats with high precision in any material.

11.12.8 Generation of Contract of Electron Beam

Electrons are produce by a device known as electron gun (Tungsten filament wire). Electron gun consist:

(i) A cathode which is a hot tungsten filament (2500°C) emitting high negative potential electrons.

(ii) An anode at ground potential through which the electron pass.

(iii) A grid cup, negatively based with respect to filament before starting the collide the electron with workpiece a variable strength electromagnetic field (lens) is used to refocused the electron beams. This result in an electron beam using a cross-sectional diameter of 0.01–0.02 mm with a power density of 1–5 billion w/cm^2.

11.12.9 Electron Beam Machining–Equipment

Electron beam gun, which is the heart of any electron beam machining facility. The basic functions of any electron beam gun are to generate free electrons at the cathode, accelerate them to a sufficiently high velocity and to focus them over a small spot size. Further, the beam needs to be manoeuvred if required by the gun.

The cathode is generally made of tungsten or tantalum. Such cathode filaments are heated, often inductively, to a temperature of around 2500°C. Such heating leads to thermo-ionic emission of electrons, which is further enhanced by maintaining very low vacuum within the chamber of the electron beam gun. Moreover, this cathode cartridge is highly negatively biased so that the thermo-ionic electrons are strongly repelled away form the cathode. This cathode is often in the form of a cartridge so that it can be changed very quickly to reduce down time in case of failure. Localized heating by focused electron beam Gradual formation of hole Penetration till the auxiliary support Removal due to high vapour pressure.

Just after the cathode, there is an annular bias grid. A high negative bias is applied to this grid so that the electrons generated by this cathode do not diverge and approach the next element, the annular anode, in the form of a beam. The annular anode now attracts the electron beam and gradually gets accelerated. As they leave the anode section, the electrons may achieve a velocity as high as half the velocity of light.

The nature of biasing just after the cathode controls the flow of electrons and the biased grid is used as a switch to operate the electron beam gun in pulsed mode.

After the anode, the electron beam passes through a series of magnetic lenses and apertures. The magnetic lenses shape the beam and try to reduce

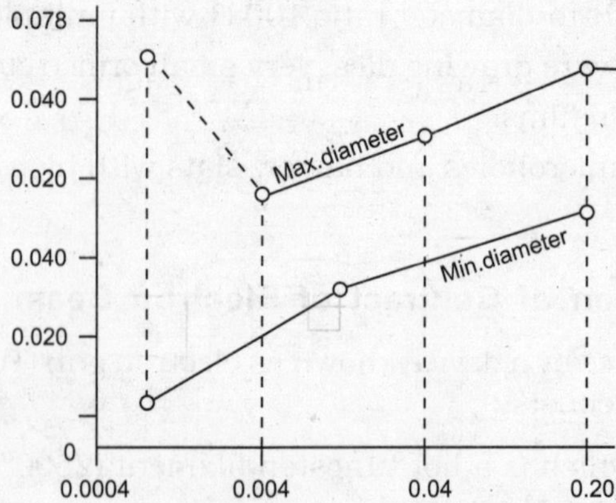

Fig. 11.17 Range of EB performance for steel & Nickel alloys

the divergence. Apertures on the other hand allow only the convergent electrons to pass and capture the divergent low energy electrons from the fringes. This way, the aperture and the magnetic lenses improve the quality of the electron beam.

Then the electron beam passes through the final section of the electromagnetic lens and deflection coil. The electromagnetic lens focuses the lighting system for alignment port for diffusion pump. High voltage supply to cathode. Cathode cartridge bias grid anode vacuum throttle valve magnetic lens aperture electromagnetic coils deflector coils. Port for vacuum gauge telescope for alignment electron beam to a desired spot. The deflection coil can manoeuvre the electron beam, though by small amount, to improve shape of the machined holes.

Generally in between the electron beam gun and the workpiece, which is also under vacuum, there would be a series of slotted rotating discs. Such discs allow the electron beam to pass and machine materials but helpfully prevent metal fumes and vapour generated during machining to reach the gun. Thus it is essential to synchronize the motion of the rotating disc and pulsing of the electron beam gun.

Workpiece is mounted on a CNC table so that holes of any shape can be machined using the CNC control and beam deflection in-built in the gun.

One of the major requirements of EBM operation of electron beam gun is maintenance of desired vacuum. Level of vacuum within the gun is in the order of 10^{-4} to 10^{-6} Torr. {1 Torr = 1mm of Hg} Maintenance of suitable **vacuum** is **essential** so that electrons do not loose their energy and a significant life of the cathode cartridge is obtained. Such vacuum is achieved and maintained using a combination of rotary pump and diffusion pump.

11.13 ION BEAM MACHINING

- An ion beam is a type of particle beam consisting of ions.
- Ion beams have many uses in electronics manufacturing (principally ion implantation) and other industries.

Main components of Ion Beam Machining:

11.13.1 Ion Source

- The ion source has arrays of permanent magnets to produce a multi-cusp magnetic field in regions remote from the plasma grid and the RF antenna.

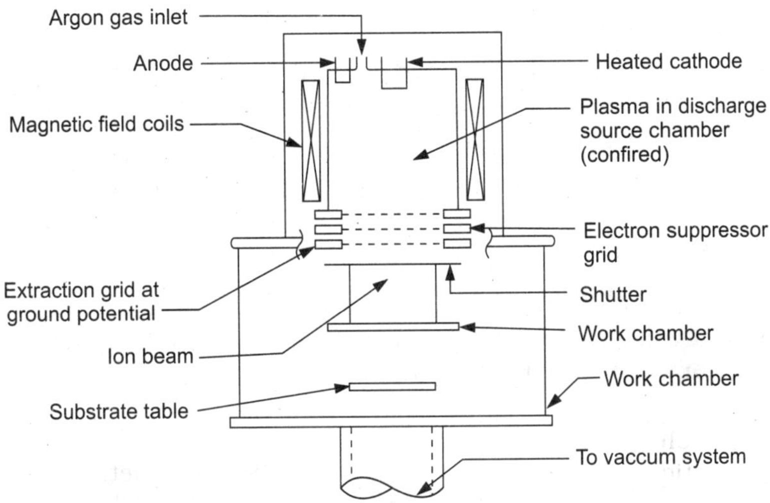

Fig. 11.18 Ion beam machining

- The field confines the plasma by lengthening the path of ionizing electrons and reducing their drift to the walls.

11.13.2 Plasma Source

- Plasma sources generate plasmas.
- Excitation of plasma requires partial ionization of neutral atoms and/or molecules of a medium.
- There are several ways to cause ionization: collisions of energetic particles, strong electric fields acting on bond electrons, or ionizing radiation.
- Following reaction take place.

$$Ar + e^- \rightarrow Ar^+ + 2e^-$$

- Argon ions are thereby produced.
- A magnetic field, obtained from an electromagnetic coil or a permanent magnet, is often applied between the anode and cathode to make the electrons spiral.
- Spiraling increases the path length of the electrons and hence increases ionization.

11.13.3 Extraction Grid

- The ions are removed from the plasma by means of extraction grids.

- The grids are normally made of two or three arrays of perforated sheets of carbon or molybdenum; these materials can withstand erosion by ion bombardment.
- The perforations in each of the sheets are aligned above one another.

11.13.4 Ground Electrodes

- Ground electrodes are used for extracting positively charged ions from the source that combine downstream to form a broad beam.
- Individual electrodes in close proximity to the extraction electrode can be biased to inhibit back streaming of neutralizing electrons close to the source or back to the extraction electrode.

11.13.5 Power Supply

- Output voltages and currents are precisely controlled from the front panel or by remote programming.
- Digital displays on the front panel directly monitor the voltage and current outputs.
- Remote analog signals proportional to each output are provided at the rear panel I/O connector.

11.13.6 Focus of Ion Beam

- Focusing of the Ion Beam is also provided by the construction and shape of the electrodes.
- The suppression electrode produces an inner zero electrostatic field, and an outer electrostatic on a field such that ions entering this outer field are deflected by an amount that is a function of their distance from the edge of the inner field.
- The result is a focused beam having a uniform intensity over a given target area and at a given distance from the lens.

11.13.7 Application of Ion Beam

- Ion beams can be used for sputtering or ion beam etching and for ion beam analysis.
- Ion beam etching, or sputtering, is a technique conceptually similar to sandblasting, but using individual atoms in an ion beam to ablate a target.
- Reactive ion etching is an important extension that uses chemical reactivity to enhance the physical sputtering effect.
- The wafer is then placed in a vacuum chamber, and exposed to the ion beam.
- The impact of the ions erodes the target, abrading away the areas not covered by the photoresist.

- This method is frequently enhanced by bleeding a reactive gas into the vacuum system, which is known as reactive ion etching.
  - Ion-beam Etching
  - Ion-beam sputtering
  - Ion-beam sculpting
  - Basis
  - Broad area ion exposure
  - TEM exposure

Smoothing

- Texturing
- Cleaning
- Shaping, Polishing,

Thinning

- Milling

- Nitrogen implantation has been used to increase wear resistance and give longer life,
  - injection molding screws
  - high speed steel tools
  - a clutch housing tool
  - hip prosthesis
- **Yttrium** gives oxidation and wear resistance.
- **Titanium** and carbon on iron gives lower friction and better wear.
- **Chromium** is used to maintain strength of holes.

11.13.8 Advantage of Ion Beam Machining

- Low temperature processing reduces handling stress problems.
- No dimensional changes.
- Good adhesion of treated surface.
- New alloys possible.
- Can improve corrosion, oxidation, wear, hardness, friction, fatigue.

11.13.9 Limitation of ion beam machining

- Very shallow treatment (< 1 μm).
- High cost.
- Surface can be weakened by radiation effect.

OBJECTIVE TYPE QUESTIONS

1. Metal removal take place in EDM by ***[GATE-2004]***

(a) Erosion of metal
(b) Chemical reaction of metal
(c) Dissolution of metal
(d) Melting and evaporation

2. In which of the following one machining process requires vacuum
(a) Laser beam machining (b) Plasma arc machining
(c) Ultrasonic machining (d) Electron beam machining

3. Match List I (Machining process) with List II (Operating media) and select the correct answer using the code given below the list.

List I	List II
A. Abrasive JAT machining	1. Dielectric
B. Electron beam machining	2. electrolyte
C. Electro-chemical machining	3. abrasive slurry
D. Electro discharge machining	4. Vacuum
	5. Air

(a) A–2, B–5, C–3, D–4 (b) A–4, B–5, C–2, D–1
(c) A–4, B–2, C–3, D–5 (d) A–5, B–4, C–2, D–1

[IES–2004, GATE–2010]

4. Match List I (Machining process) with List II (Application) and select the correct answer using the code given below the list ***[IES–2005]***

List I	List II
A. EDM	1. Hole and cavity in hard and brittle material
B. LBM	2. Micro-drilling and micro-welding of material
C. USM	3. Shaping of hard metals or reshaping of cemented carbide tools
D. ECM	4. Shaping of cemented carbide tool

(a) A–1, B–2, C–3, D–4 (b) B–1, C–3, D–2, A–4
(c) A–2, B–1, C–4, D–3 (d) A–3, C–1, B–2, D–4

5. Match 4 pairs from List I and II. ***[GATE-1998]***

List I	List I
A. ECM	1. Plastic Shear
B. EDM	2. Erosion/Brittle fracture
C. USM	3. Corrosive reaction
D. LBM	4. Melting and Vapourization
	5. Ion displacement
	6. Plastic shear and ion displacement

(a) A–2, B–6, C–3, D–1 (b) A–4, B–1, C–2, D–6
(c) A–2, B–3, C–4, D–6 (d) A–3, B–6, C–2, D–4

6. Match the following components with appropriate machining process. ***[GATE–1992]***

List I	List II
A. Square hole in a high strength alloy	1. Milling
B. Square hole in a ceramic component	2. Drilling
C. Blind hole in a die	3. ECM
D. Turbine blade profile on high strength alloy	4. Jig boring
	5. EDM
	5. USM

(a) A–5, B–6, C–4, D–3
(b) A–2, B–4, C–6, D–1
(c) A–6, B–5, C–3, D–1
(d) A–3, B–6, C–5, D–1

7. Machining process in which the workpiece is immersed in an electrolyte solution is known as
(a) Electro-chemical machining
(b) Laser beam machining
(c) Electro discharge machining
(d) Chemical machining.

8. Tool is made in EDM process
(a) Tungsten carbide
(b) Copper or brass
(c) Stainless steel
(d) Diamond
(e) None of these

9. In electro-discharge machining process, workpiece and Tool (electrode) are submerged in
(a) Chemical solution
(b) Dielectric fluid
(c) Electrolyte solution
(d) None of the above

10. In ECM process the gap between workpiece and tool is usually
(a) 2.0 mm
(b) 0.25 mm
(c) 0.80 mm
(d) 0.12 mm

11. Match the following from List I and List II

List I	List II
A. EDM	1. Electric spark
B. USM	2. Mechanical
C. LBM	3. Radiation
D. ECM	4. Electrical

(a) A–1, B–4, C–2, D–3
(b) A–4, B–1, C–2, D–3
(c) A–1, B–2, C–3, D–4
(d) A–3, B–4, C–3, D–2

12. Erosion of metal in EDM is
(a) Continuous
(b) Proportionate to the number of spark
(c) None of the above
(d) Not known

13. Ultrasonic machining is based on
(a) Uniform machining
(b) Chemical
(c) Vibractory waves of high frequency
(d) None of these

14. In which machining process slurry is used
(a) EDM
(b) LBM
(c) USM
(d) EBM

15. In USM process slurry is fed by
(a) Manual system
(b) High power pump
(c) Any of the above
(d) None of these

16. Match the following non-traditional maching process with the corresponding material removal rate mechanism. *[GATE–2011]*

(Machining Process)	(Mechanism of metal removal)
P. Chemical machining	1. Erosion
Q. Electro chemical machining	2. Corrosive reaction
R. Electro-Discharge machining	3. Ion displacement
S. Ultrasonic machining	4. Fusion and vaporization

(a) P–2, Q–3, R–4, S–1
(b) P–2, Q–4, R–3, S–1
(c) P–3, Q–2, R–4, S–1
(d) P–2, Q–3, R–1, S–4

17. In which of the following machining process ionize gas is used
(a) USM
(b) EBM
(c) PAM
(d) LBM **[NTPC–2011]**

18. In spark erosion machining process, the gap between tool and workpiece is filled with
(a) A liquid of die electric
(b) Photo etchant
(c) Acid solution
(d) Brine solution

19. The machining process in which tool is known as electrode
(a) ECM
(b) USM
(c) EDM
(d) EBM

20. Turbine blades are machined by **[GATE–1997]**
(a) EDM
(b) ECM
(c) USM
(d) AJM

21. The following non-conventional method of machining essential requires electrolyte
(a) IBM
(b) ECM
(c) EDM
(d) LBM

22. Tool in the case of ultrasonic machining is made by
(a) Brass or copper
(b) Diamond
(b) Stainless steel
(d) HSS

23. Ultrasonic machining is not suitable for
(a) Lead
(b) Stainless steel
(c) Brittle material
(d) Plastics

24. As tool and work are not in contact in EDM process hence
(a) no wear of tool
(b) no relative motion occurs between then
(c) no force occurs between tool and work
(d) no power is consumed during metal cutting **[GATE-2003]**

25. In ECM, metal removal is due to
(a) ion displacement
(b) erosion
(c) fusion
(d) corrosion **[GATE-2001]**

26. Selection of electrolyte for ECM is
(a) non-passivating electrote for stock removal and passivating electrolyte for finish control
(b) depends on current density
(c) passivating electrolyte for stock removal and non-passivating electrolyte for finish control
(d) based on tool work electrodes **[GATE-1997]**

27. During ultrasonic machining the metal removal is achived by
(a) high frequency solid waves
(b) high frequency eddy current
(c) rubbing action between tool and workpiece
(d) hammering action of abrasive paticels **[GATE-2005]**

28. Match List-I with List-II and select the correct answer using the code given below *[GATE-2008]*

(Conconventional machining process)	**(Basic Process)**
List-I	**List-II**
A. Electro-polishing	1. Thermal
B. Electro chemical machining	2. Mechanical
C. Abrasive jet machining	3. Electro-chemcial
D. Electro discharge machining	4. Chemical

Code	A	B	C	D
(a)	4	3	2	1
(b)	2	1	4	3
(c)	4	1	2	3
(d)	2	3	4	1

29. Average frequency of spark in EDM is
(a) 1000
(b) 3000
(c) 7000
(d) 10000 spark per second

30. In ECM, the gap between the tool and workpiece is nearly kept as
(a) 0.25 mm
(b) 0.8 mm
(c) 1.0 mm
(d) 1.25 mm

31. Which UCM process uses abrasive slurry
(a) spark erosion machining
(b) electro-discharge machining
(c) ultrasonic machining
(d) electro-chemical machining

ANSWERS

1. (d)	**2.** (d)	**3.** (d)	**4.** (d)	**5.** (d)	**6.** (a)	**7.** (a)
8. (b)	**9.** (b)	**10.** (b)	**11.** (c)	**12.** (b)	**13.** (c)	**14.** (c)
15. (c)	**16.** (a)	**17.** (c)	**18.** (a)	**19.** (c)	**20.** (a)	**21.** (c)
22. (a)	**23.** (c)	**24.** (c)	**25.** (c)	**26.** (d)	**27.** (d)	**28.** (a)
29. (d)	**30.** (c)	**31.** (c)				

QUESTIONS

1. What do you mean by unconventional machining process (UCM)?

2. Classify the unconventional machining process.

3. Compare the different unconventional machining process (UCM)?

4. Write the advantages and limitations of unconventional machining process (UCM)?

5. Explain the mechanism of metal removal in electro-chemical machining (ECM) process. *[UPTU– 2002-03, 2006-07]*

6. Write the advantages and limitations of ECM process.

7. Define the electro-discharge machining (EDM).

8. What is electro-discharge machining (EDM)? Obtain expression of material removal rate in EDM process in term of process parameter.

[UPTU– 2002-03, 2006-07]

9. How are non-conventional machining process differ from conventional machining process? Classify the differ type of non-conventional machining process. ***[UPTU– 2005-06]***
10. What is the principle of EDM? Explain the function of dielectric fluid in EDM. ***[UPTU– 2007-08]***
11. Explain the working principle of EDM with a neat sketch.
12. What are the possible errors cause in EDM?
13. What are the important parameter which control the material removal rate in EDM?
14. Write the advantages, limitations and applications of EDM.
15. What is the principle of working of electro-chemical machining (ECM)? Compare ECM with EDM. What are the advantages, disadvantages and applications of ECM? ***[UPTU– 2010-11]***
16. Explain the EDM process in brief. What are the advantages and disadvantages of electrical discharge machining (EDM) process. When will you go use reverse polarity in EDM process and what are its advantages and disadvantages? Also explain the term wear ratio. ***[UPTU– 2010-11]***
17. What are the characteristics required for a good electrode material in EDM?
18. What is abrasive jet machining (AJM) process? Explain the effects of stand-off distance and abrasive grit size on material removal rate in AJM. ***[UPTU– 2002-03]***
19. Giving a neat sketch explain the working of ultrasonic machining process also discuss its scope of application. **[UPTU– 2002-03]**
20. Giving a neat sketch explain the working of laser beam machining. Also give scope of LBM applications. ***[UPTU– 2002-03, 2006-07]***
21. Briefly explain the working of an USM machine showing important element. ***[UPTU– 2004-05]***
22. What are the important parameters that control the material removal rate in AJM? Briefly explain any two. ***[UPTU– 2004-05]***
23. With the help of a neat sketch, explain the principle and working of abrasive jet machining (AJM)? List the variable and limitations of AJM process. ***[UPTU– 2005-06]***
24. What is Laser? Explain the principle of LBM. ***[UPTU– 2007-08]***
25. With the help of neat sketch, explain the working principle of EBM? ***[UPTU– 2007-08]***
26. What is the working principle of electron beam machining (EBM). What are its advantages, limitations and applications?
27. What type of laser generally used in LBM?
28. Compare the unconventional machining process in term of material removal rate and applications. ***[UPTU– 2008-09]***
29. What are the principle feature of ECM process? What is rate of electrolyte in ECM process? ***[UPTU– 2008-09]***
30. Describe the working method of USM. Give the function of a transduces and tool cone in USM. ***[UPTU– 2008-09]***
31. Compare between plasma arc machining (PAM) and Ion beam machining (IBM). ***[UPTU– 2008-09]***

32. Write a short note on any one of the following giving their uses and limitations. ***[UPTU– 2008-09]***

(a) Dielectric media
(b) Capacitance of the circuit
(c) Resistance of the circuit

33. Explain the role of any two of the following in EDM process.

(a) Dielectric media
(b) Capacitance of the circuit
(c) Resistance of the circuit ***[UPTU– 2003-04]***

34. Explain the application of the following electrode material in EDM copper and graphite. ***[UPTU– 2004-05]***

❑❑❑

Chapter 12

Unconventional Welding Process

12.1 LASER BEAM WELDING

Laser beam welding is also another type radiant energy welding process in which coalescence is generated by the heat obtained from application of a concentrated coherent light beam falling on the surface that to be joint. Laser beam welding like electron beam welding can weld any material. Heat affected zone is very narrow zone due to very high energy intensity and high speed of LBW can also be achieved.

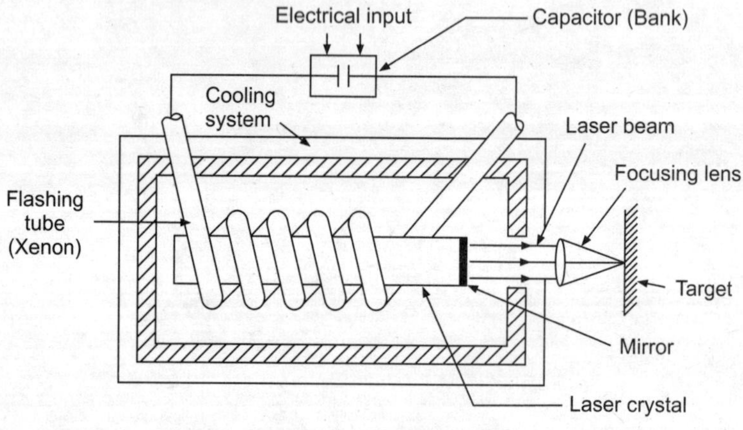

Fig. 12.1 Laser beam welding

Visible light consist a number of radiations of different wavelength. Wave travel in random direction and are not in phase. Radiation must be converted to the same wavelength and brought into phase using laser.

LASER is the an abbreviation of **Light Amplification By Stimulated Emission of Radiation.**

Three type of lasers are available, i.e.,

- **solid laser**,
- **gas laser**, and
- **semiconductor laser**.

But solid state laser ND : YAG **(yttrium aluminium gas net)** crystal absorbs the energy and convert it into a single wavelength beam of infrared light with a diameter of few mm. Output of YAG laser is limited. Current required to operate the system is 400 W average power which gives weld penetration about 1 mm for deeper penetration, we require high level output, i.e., 15 K watt. In commercial high power laser used for welding YAG crystal is replaced by a tube filled with a mixture of CO_2, N_2 and He gases. A high voltage developed across electrode placed inside the tube make the gase fluorescence.

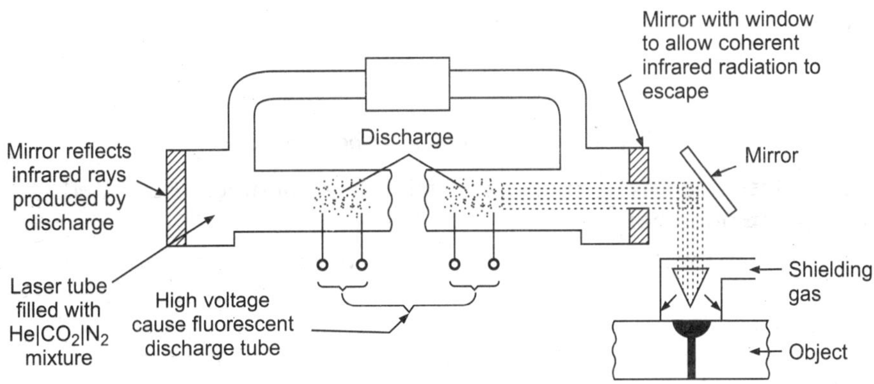

Fig. 12.2 CO_2 Laser's outline diagram

12.1.1 Working Principle

The focal spot (thousands of an mm in diameter) is targeted on the weld joint surface or by focal length selection above or below. At the surface the enormous concentration of light energy is converted to thermal energy object. Surface melting occurs and progresses through the weld joint by thermal conductance. For welding, beam energy is maintained below the vapourization temperature of the weld joint material. For hole drilling or cutting vapourization is required. As weld joint penetration is dependent on conducted heat. The thickness of materials to be welded is generally less than 0.080 inches if the ideal metallurgical and physical characteristics of laser welding are to be realised.

Benefits are **narrow welds**, **no distortion**, **minimal heat affected zones** and **excellent metallurgical quality**.

As with electron beam the intense, concentrated energy produces melting and coalescence before a substantial heat affected zone can develop. As the welds are narrow and are low volume there is a minimal reservoir of heat for conductance into the adjacent area. When materials to be welded are thick and particularly if they have high thermal conductance (such as Al) this important metallurgical advantage of minimal heat affected zone can be destructively affected. It is claimed that since the source of energy is light of a specific wavelength contaminants in the weld pool or on the facing surfaces of the joint may be preferentially vapourized by their particular light absorbing characteristics resulting in a kind of weld purification. The

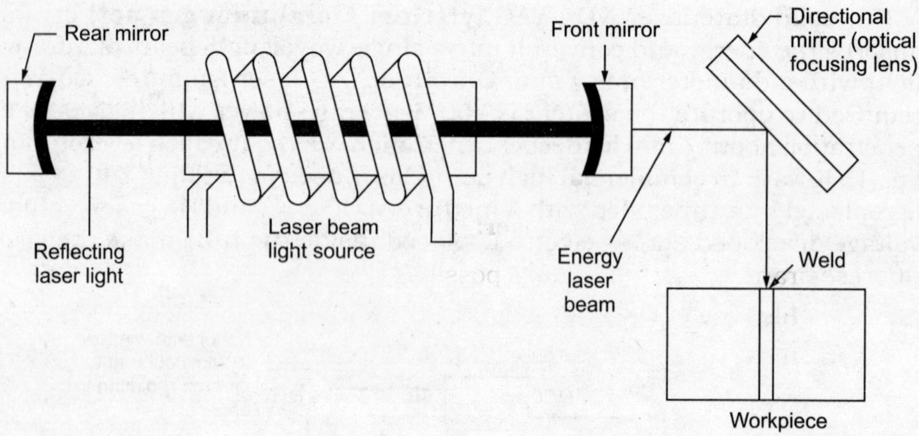

Fig. 12.3 Setup of a laser beam welding

excellent fatigure strength of laser welds is sometimes attributed to this purifying phenomenon.

Energy distribution across the beam is produced by the design of the resonant cavity, including mirror curvatures or shape and their relative arrangement. This combination results in photon oscillation within the cavity producing specific output beam energy distributions or patterns. These patterns are labelled *transverse energy modes* (TEM) and have specific identifying numbers.

It is important to note before proceeding that the function of all lasers whether they be gas (**carbon dioxide**, **helium**, **neon**, etc.) or other lasing sources is based on the principle of the excitation of atoms by means of intense light, electricity, electron beam, chemicals, etc., and the spontaneous and stimulated emission of photons. Depending on the lasing source, output frequencies differ widely and are capable of a great number of applications. These range form welding to critical surgery, resistance trimming, communication, etc.

12.1.2 Types of Laser

Basically two primary types of lasers used in welding and cutting:

- Gas lasers (CO_2 type)
- Solid state lasers (Nd: YAG type)

1. **Carbon Dioxide Laser:** This type of the laser consist a high purity carbon dioxide with helium gas while nitrogen is used as a lasing medium. Important characteristics of CO_2 laser are:

   (a) Infrared (10.6 μm).

   (b) Transmission of beam only by missor (not by fibre optic).

   (c) Cutting laser range 0.5–10 kW.

   (d) High cutting speed by CO_2 laser.

   (e) Reflective material are not easy to cutting welding, i.e., aluminium copper.

(f) Non-metallic material can be easily cut by CO_2 laser.

2. **Nd: YAG Laser:** Nd: YAG (yttrium aluminium garnet) laser basically consist transmitting mirror, power supply source, resonator, reflecting. It is a **solid** type laser and among the solid laser it is very famous. Nd: YAG laser is used in industries as its good thermal characteristics.

Characteristics of Nd: YAG laser:

(a) Transmission of beam is possible by optical fibre.

(b) This type of laser is available upto 2 kW (energy).

(c) This type of laser is not suitable for cutting of non-metallic material

(d) Excellent performance for recision work but cutting speed is not so good as CO_2 laser.

(e) Wavelength of laser is satisfactory absorb by metallic material.

12.1.3 Advantages of Laser Beam Welding

(i) Welding operation can be performed inside plastic housing/case or transparent glass.

(ii) A large variety of material can be joint by this process.

(iii) No provision of high electric current and its effect as no electrons are used.

(iv) Low total thermal input.

(v) Can be transmitted over long distance with a minimum loss of power.

(vi) Complicated shape's object can be welded easily by this process.

(vii) No vapourization of metal.

(viii) Beam can be focused with greater accuracy an object as it is light.

(ix) Heat affected zone are very narrow.

(x) With the help of this process heat treatment material can easily welded without affecting their heat treated condition.

12.1.4 Disadvantages

(i) Rapid cooling rate may cause cracking in certain metals.

(ii) High initial cost investment.

(iii) Optical surface can damaged.

(iv) Maintenance cost is very high.

(v) Trained and expert operator is required to performing welding.

12.1.5 Applications

(i) Cutting and welding of material is possible.

(ii) Small electronic components and integrated circuitry in electronic industries.

(iii) Welding of *nuclear plant* components.

(iv) Transmission component for a car.
(v) For welding of high melting point metals.
(vi) A large dissimilars such as Cu, Ni, Al, SS, Tn, Zr etc. can easily weld by this process.

12.2 ULTRASONIC WELDING

Ultrasonic welding is also another type of radiant welding process (Radiation process) as it is a radiation welding process therefore it is consider as a advance welding technique through ultrasonic welding many metals could be combined that could not be weld by any other joining process. Ultrasonic effect upon the molten metal generates a smaller grain size, giving the metal more strength.

In ultrasonic welding as shown in Fig. 12.4 as metallic tip vibrating with ultrasonic frequency (capacity of vibration above the human hearing capacity is known as ultrasonic) is produce to weld a thin plate to a thicker one supported on an anvil range generally used of frequency from 20,000 Hz to 60,000 Hz. Ultrasonic welding combines pressure and high-frequency vibration motion to form a solid-state bond.

Ultrasonic welding is capable to join such combination as:

(i) Aluminium to steel
(ii) Aluminium to tungsten
(iii) Aluminium to molybdenum, and
(iv) Nickel to brass.

With the help of ultrasonic welding different melting temperature material can be strongly joint.

Ultrasonic welding equipment consist of two main part, i.e., a **power source** and a **ultrasonic transducer**. Power source convert 50 Hz main supply to a high frequency power and that is transform of a transducer to magnetic flux and lastly kinetic motion which is amplified by a velocity transformer. An

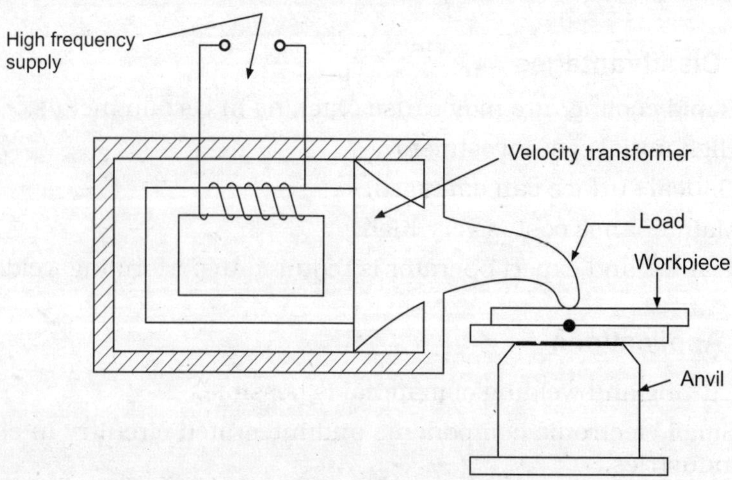

Fig. 12.4 Typical setup for ultrasonic welding arrangement

ultrasonic weld is competed in about as much time as it tares to strike an arc in arc welding.

Ultrasonic transducer are basically two types:

(i) Piezoelectric crystal
(ii) Magnetostrictive transducer.

Piezoelectric crystal vibrates at a natural resonant frequency as a current flows through them and the magnetostrictive transducer is composed of laminations of nickel, or a nickel alloys which is surrounded by a coil of wire quartz is a typical piezoelectric crystal. Crystal may also be **ammonium dihgrosphaten**, **Rochelle salt**, **barium titanate ceramic**. Generally there is a coupler or horn to funnel the vibration to a tip which is placed directly over the weldment. Pressure is also applied in the ultrasonic welding.

12.2.1 Welding Equipments

Ultrasonic welding consist the following equipment:

(i) A high frequency converter
(ii) A transduces-coupling system
(iii) An anvil
(iv) A force application device
(v) Timer
(vi) Control system for electrical, electronic and hydraulic/premeiotic system.

(i) Frequency Converter: A frequency converter is used for producing high frequency electrical power at the design frequency of the welding system or process. USW equipment may be driven by electron tube frequency converter, motor-alternator or silicon-controlled rectifier converter.

(ii) Transducer Coupling System: It consist a transducer that convert electrical power into elastic vibratory power and a coupling system include the sonotrode **tip** which conduct vibratory power to weld zone.

(iii) Anvil: It is used for supporting to object and it provide the necessary reaction to clamping force. An anvil must not permit compliance of the workpiece with the applied vibration to avoidant energy losses anvil assembly should be bolted from welder frame.

(iv) Force Application Device: It applies static load normal to the plane of weld surfaces. It is

(a) spring-actuated for very small machine
(b) prismatically–actuated in medium size machine
(c) Hydraulically actuated with large unit

(v) Timer: Timer basically used to control the weld interval in spot type, ring and line welding or a translating and rotating mechanism for continues seam welding.

12.2.2 Working Principle of Ultrasonic Welding

Working Principle: Ultrasonic welding is a simplest type welding process. Workpiece should be welded free from any **oxide flim, coating, moisture** and this is possible only due to the vibration generate during the welding process. Workpiece that to be welded should be properly clamp between the welding tip and an anvil in the process. Anvil is used just to support the workpiece both tip and anvil are faced with high speed steel.

In ultrasonic welding a frequency converter is used which convert 50 cycle line power into a very high power (**electrical**). Ultrasonic welding system consist a transducer which convert electrical power into ultrasonic vibration energy and this vibration energy transfer into joint through welding tip which is attached with transducer tip oscillates in the plane of the joint interface frequency and the pressure exerted on the material depends on the type and size of the rail and type of welding machine.

During in ultrasonic welding due to vibration, vibration combined with static clamping force, induce dynamic shear stresses in the workpiece then plastic deformation take place at the joint interface any coating (oxide) or any other oxide coating are shattered and dispersed so that intimate contact and bonding of the workpiece surface take place.

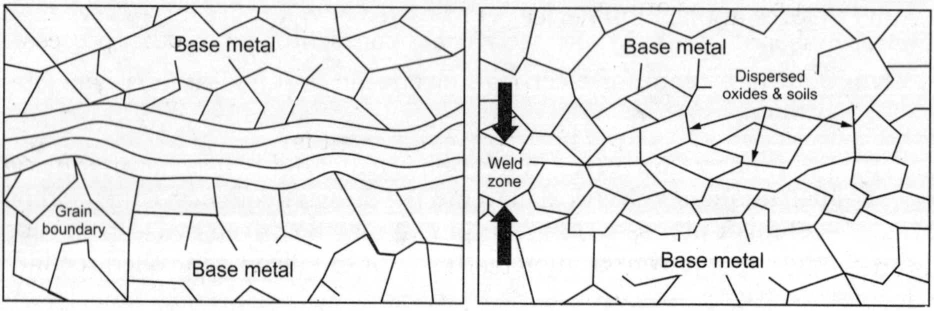

Fig. 12.5 Workpiece before and after ultrasonic welding.

12.2.3 Materials Suitable for Ultrasonic Welding (Application)

The materials which are most suitable for ultrasonic welding are **non-ferrous** metals and their alloys of any combination. Therefore, applications that involved bounding materials such as **copper**, **aluminium** and **brass** are very common. Materials such as **lead**, **zinc** and **tin** cannot be welded.

12.2.4 Applications of Ultrasonic Welding

Ultrasonic welding is used in every type of industries. It is used for fabrication of small **moter armature**, **aluminium** and **gold**, **lead wire** connections to transistors of diodes, helicopter access doors, dissimilar metal joints in solar collectors. However the unique application of the process is, welding of containers of explosive such as nitroglycerine, pyrotechnics (fire works) and reactive chemical of aluminium to stainless steel.

12.2.5 Advantages of Ultrasonic Welding

(i) Low temperature is involved in ultrasonic welding.
(ii) As low temperature encountered in ultrasonic welding metallurgy of parent metal does not completely change.
(iii) Stronger joint is produced in ultrasonic welding.
(iv) Welding aluminium to glass is possible (welding of glass is not possible by other manner).
(v) No need for slag removal as it removes automatically due to vibration.
(vi) It is a faster welding process and can achieve proper penetration.
(vii) Equipment is similar reliable and only moderate skill operator is required to operate the machine.
(viii) Thick and thin section can be joined together.
(ix) Minimum surface deformation results.
(x) Dissimilar metal having different melting point can be joint easily.

12.2.6 Limitations of Ultrasonic Welding

(i) Main disadvantage of ultrasonic welding that it is restricted to aluminium.
(ii) This process is only applicable only up to 3.2 mm thick.
(iii) Ultrasonic welding is not economically competitive when other process can be used to do the same job.
(iv) Material being welded tend to weld with the tip and anvil.
(v) Due to fatigue loading life of the equipment is short.

12.3 ELECTRON BEAM WELDING *[UPPSC 2010]*

It is one of the latest type of welding technique in which welding operation is performed under the vacuum chamber. An electron beam welding machine has three basic components.

(i) Argon which produces a controlled beam of electron.
(ii) A vacuum chamber with associated pumping equipment.
(iii) A unit which either travels the beam along the joint line or moves the work under the gun.

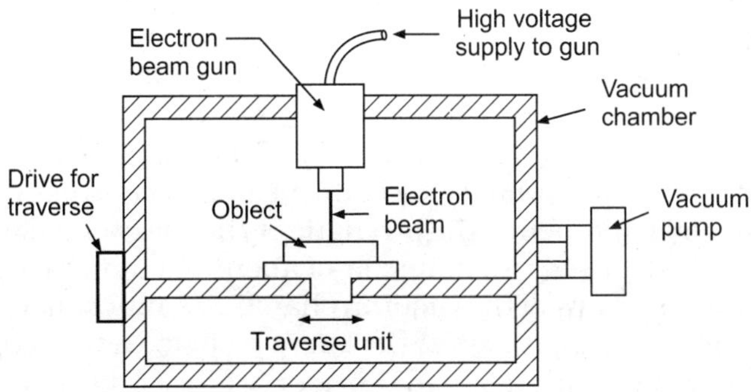

Fig. 10.6 (a) Basic components of an electron beam welding machine

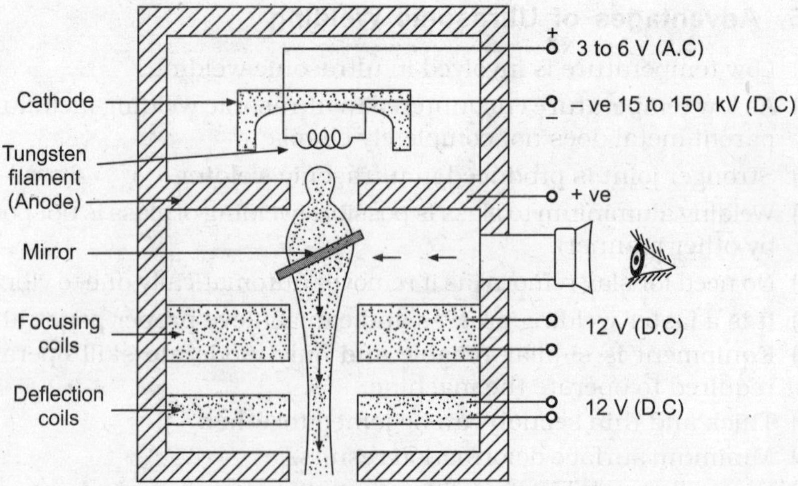

Fig. 12.6 (b) Main components of EBW gun used for welding

12.3.1 Definition

Electron beam welding is define as a fusion welding process in which coalescence is produced by the heat obtained from a concentrated beam composed primarily of very high velocity electron. In EBW kinetic energy of beam due to high velocity convert into thermal energy these by causing the workpiece metal to melt and fuse.

12.3.2 Electron Gun

Different types of gun have been used in commercial equipment for electron-beam welding but in the main they, are based on the working principle of cathode ray tubes and thermionic valves found in television set. Electron generated by the heating of tungsten filament wire which is mounted in a cup shaped electrode. Which help to control the flow of electron anode consist a hole in center and available in the form of disc. Two range of voltage are used. In low voltage system, the difference between the anode and cathode usually lies from 15 to 30 kV while a range from 70 to 150 kV is common for high voltage machine. Several advantages can be achieved by both system. But the resultants welds are essentially similar as the voltage pass through the gun, electrons move rapidly from cathode to anode. Some electron collected there and some pass through the hole in the centre of the anode and continue to travel until they meet a solid object when they give up their kinetic energy in the form of heat. gun contains magnetic coils which deflect the electron towards the central longitudinal axis of the beam. By controlling the current flowing in the coils and hence the strength of the magnetic field, the beam can in this way be focused to a diameter of about 0.5 mm. Viewing windens can be fitted into the walls of the vacuum chamber. A much more satisfactory method is to sight along the axis of the beam by using a telescopic eyepiece and angled mirrors.

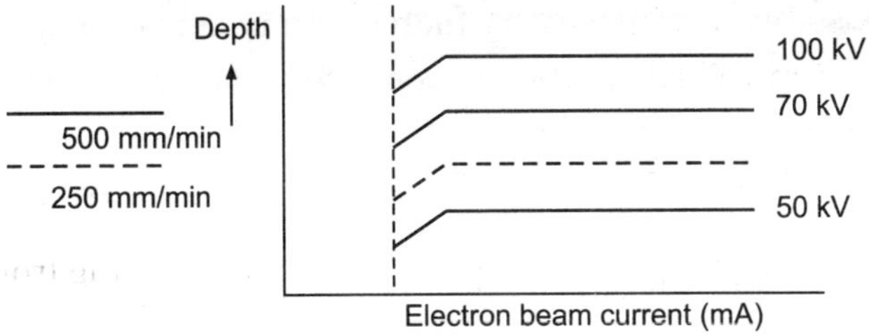

Fig. 12.7

12.3.3 Vacuum Chamber

It is another important component of EBW, in this electrons are projected through air they losses their energy in collision with O_2 and N_2 atoms. Therefore it is essential operate in a vacuum at a pressure of 10^{-4} to 10^{-2} torr (0.013 to 1.3 N/m^2) if beam travels at any distance. Vacuum chamber must be so large to accommodate the work and gun. It must also be designed such that to protect personnel against radiation hazards. When electron beams collides with the surface of object some part of energy released is converted to X-rays. There should not be allowed telescope, as they can harm operator in vicinity and they are contained by applying a lead lining to the chamber.

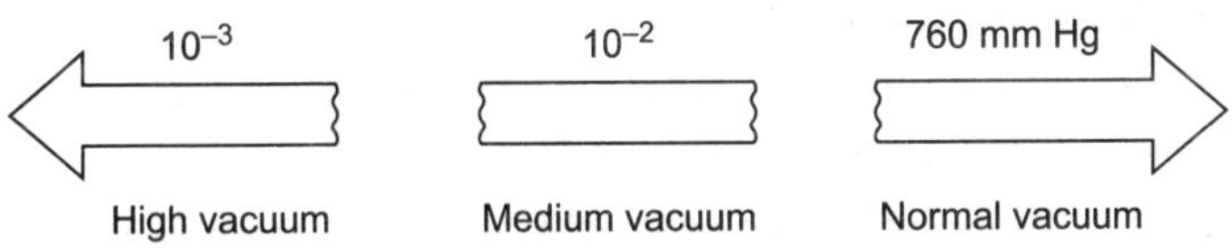

Fig. 12.8

12.3.4 Procedure

Heat source in the EBW for melting joint, is focused beam of high velocity electrons. Electron beam upon impinging the object release heat by converting their kinetic energy. These electrons are accelerated towards, anode as large potential difference that exists between them. Higher the potential difference higher would be acceleration. Current levels are low ranging between 50 mA to 1000 mA depending on the accelerating voltage, the electron would travel at the speed of 5000 to 200000 km/sec. Depth of penetration of weld metal depends on their electron speed. Which in turns depends upon accelerating voltage. When high velocity electron beam strike on the workpiece all kinetic energy is converted into heat energy as these electron penetrate the metal. Material is melted and forming a key hole around the beam. This when solidified, forms the joint.

12.3.5 Advantages of EBW

(i) High quality (excellent) weld can be produced at higher speed.

(ii) Heat affected zone (HAZ) and fusion zone are very narrow.

(iii) A very minor distortion in weld object as the energy input is in a narrow concentrated beam.

(iv) It is possible to achieve very high depth penetration in a single pass.

(v) Precise control of operation is possible and machine can be transport one place to another places for performing welding process.

(vi) In some case of weld plate thickness no edge preparation is required.

(vii) No generation of flux, slag on weldment after welding therefore no need of cleaning object after welding.

12.3.6 Limitations of EBW

(i) All types of metal cannot be welded.

(ii) Cost of equipment is very high.

(iii) Time and equipment needed to generate vacuum for welding of every new object.

(iv) A skill person is required to perform the welding operation.

(v) X-ray generate during process which is harmful for operator skin and eyes.

(vi) Very large and complex shapes cannot be welded by this process.

12.3.7 Applications of EBW

(i) For welding of refractory and reactive metals.

(ii) Low distortion components can be welded which are used in *automobile aerospace shattle, aeroplanes*, etc.

(iii) *Gear cluster* for transmission system and driven and shaft.

(iv) Intricate valve arrangements made from corrosion-resistant alloys and pressure capsules.

12.4 PLASMA ARC WELDING

Plasma arc welding is also another type of arc welding process. In plasma arc welding, plasma is the temporary state of gas. Gas get ionized after passing the electric current through it and gas becomes a conductor of electricity. In

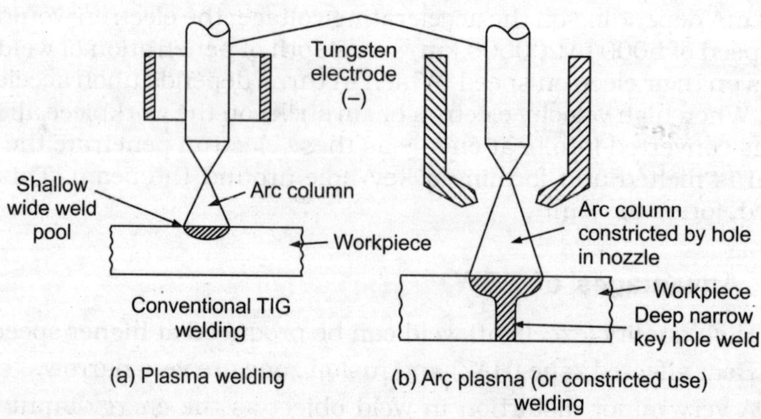

Fig. 12.9 Plasma arc welding

ionized state gas atoms break into electrons (–) and ions (+) energy of the plasma jet and temperature dependent upon the electrical power employed to create arc plasma.

Plasma arc welding is a constricted arc process where arc is constricted with the help of a water cooled small diameter nozzle which squeezes the arc, increases pressure temperature and heat.

Plasma arc welding process are basically two types:

(i) Non-transferred arc
(ii) Transferred arc.

12.4.1 Non-transferred Arc

Arc is formed between the electrode (–) and the water cooled constricting nozzle (+) arc plasma comes out of the nozzle as a flame arc is independent on workpiece and workpieces is not a part of circuit. High density metal coating can be produced by this process. A non-transferred arc is initiated by using a high frequency unit in the circuit.

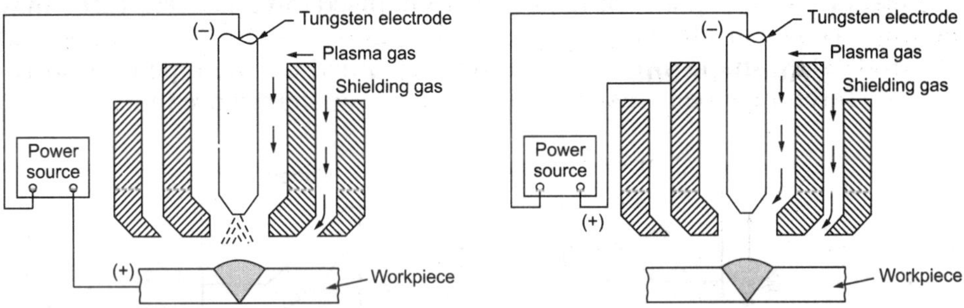

(i) Transferred arc (ii) Non-transferred arc

Fig. 12.10 Arc in plasma welding

2. Transferred Arc. In this process arc is generated between the electrode (–) and the workpiece (+) arc is transferred from electrode to workpiece. Transferred arc can be used for welding at high arc travel speed for arc initiation a current limiting resistor is put in the circuit which permit a flow of about 50 ampere between the nozzle and electrode and a pilot arc is establish between the electrode and the nozzle.

12.4.2 Uses of Equipment

(i) Power source
(ii) High frequency generator and current limiting resistor
(iii) Plasma torch
(iv) Shielding gases
(v) Current and gas decay controller.

1. Power Source. A DC power source (generator/rectifier) having dropping characteristic and open circuit voltage of 70 volt used for plasma arc welding rectifier preferred over DC generator.

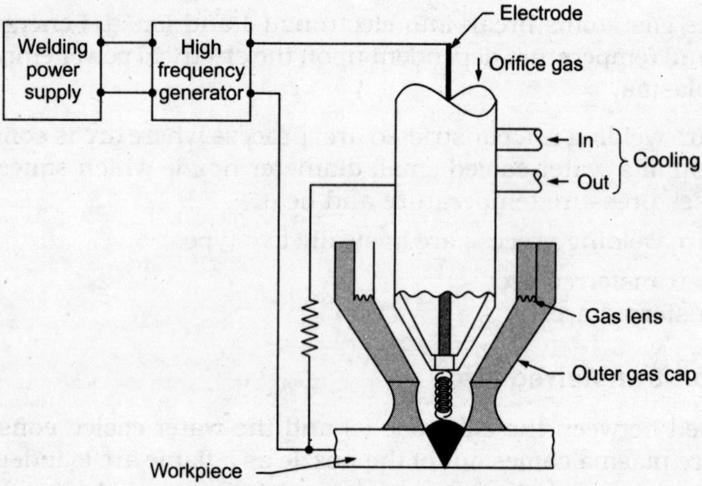

Fig. 12.11 Plasma arc welding

2. Plasma Torch. A hand operated or mechanised torch is used. Torch may be water cooled or air cooled. Water cooled increases the life of nozzle and electrode size, shape and type of torch depended on the type of weld required. Torch may be transferred arc or non-transferred arc type.

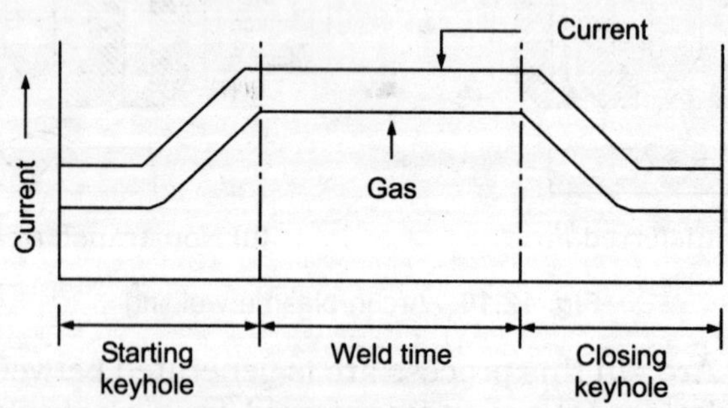

Fig. 12.12

3. Shielding Gas. Two gas or mixture of gases are used orifice gas at lower pressure and flow rates forms the arc plasma. Argon, He, Ar + H_2 and Ar + He mixture is used as inert gas or gas mixture. Argon is widely used. Helium is preferred as a broad heat input pattern and flatter cover pass is desired, for cutting purpose a mixture of argon and hydrogen (10 – 25%) or that of N_2 and H_2 is used.

12.4.3 Process

Process of work material cleaning and filler metal addition same as Tungsten Inert Gas welding (TIG). Temperature experienced in a tungsten arc used for TIG welding are of the order 11000°C. If the arc is surrounded by a water-cooled copper tube, it is constrained and temperature in arc column rest to about 2000°C.

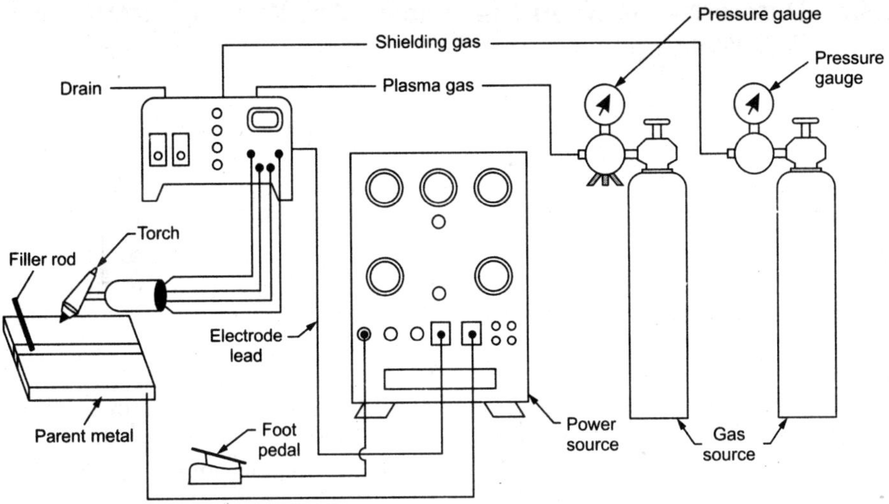

Fig. 12.13 Setup for plasma arc welding (PAW)

12.4.4 Advantages of Plasma Arc Welding

(i) It permits faster metal deposition rate and high arc travel speed as compared to TIG.
(ii) Uniform penetration is possible.
(iii) Stability of arc.
(iv) Excellent weld quality.
(v) Plasma arc welding can produce radiographic quality weld at high speed.
(vi) It can weld steel pieces up to about half inch (½″) thick, square butt joint.

12.4.5 Limitations of PAW

(i) Special protection is required as generation of infrared and ultraviolet radiations produce.
(ii) Consumption of inert gas is very high.
(iii) It can easily weld only upto 25 mm thickness.
(iv) More chance of electrical hazards are associated with this process.
(v) Welder need car plugs because of unpleasant, disturbing and damaging noise.

12.4.6 Applications

(i) Welding of a titanium upto 8 mm thick.
(ii) Nickel (Ni) and high Ni alloys.
(iii) For melting, high melting point metals.
(iv) Carbon steel, stainless steel, Ni, Cu, brass, monal, Inconal, Al, Titanium, etc.
(v) Welding of rocket (case) motors.

12.4.7 Difference between the Plasma Arc Welding (PAW) and TIG Welding

[UPTU 2006-07]

Plasma arc welding	*TIG welding*
1. Two gases (inert) are uses one for plasma gas and other for shielding purpose.	**1.** Only one gas used, which forms plasma as well as shields the arc and molten weld pool.
2. It uses a constricted arc.	**2.** It uses a non-constricted arc.
3. Electrodes remains with in nozzle therefore chance of tungsten inclusion and electrode contamination is zero (nil).	**3.** Chance of tungsten inclusion and electrode contamination if carelessly welded.
4. Very high temperature of the order of about 11000°C can be achieved.	**4.** Temperature of the order about 4000°C can be achieved.
5. Deeper penetration be achieved.	**5.** Penetration obtain by TIG is not so deeper.
6. Input heat distribution in the job are improved.	**6.** Input heat distribution is not so improved in TIG as PAW.
7. No more filler metal required.	**7.** More filler metal required.
8. Faster metal deposition rate can be achieved.	**8.** Metal deposition rate is not so faster.
9. Inert gas consumption is very high.	**9.** Inert gas consumption is very low.
10. Welding equipment cost is very high.	**10.** Welding equipment cost is not so high.
11. Cutting of hard and brittle material is possible.	**11.** Cutting of hard and brittle material is not possible.

12.5 EXPLOSIVE WELDING

Explosive welding is also another type of solid state welding. In explosive welding, welding can be done by making one part strike over the other part at a very high but subsonic velocity. This can be done by the using of explosive usually of the **ammonium nitrate** base. Process is completed with in the microseconds.

Explosive welding produce strong metallurgical bonds between metals combination which cannot be achieved by any other joining method. Explosive welding offer an excellent alternative for joining dissimilar metals and alloys with varying physical and metallurgical properties.

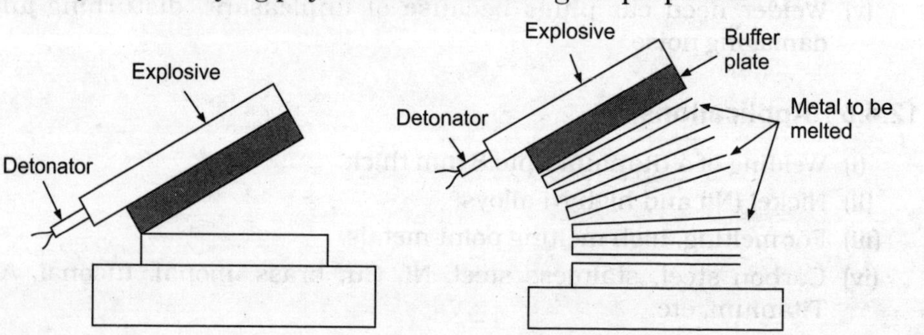

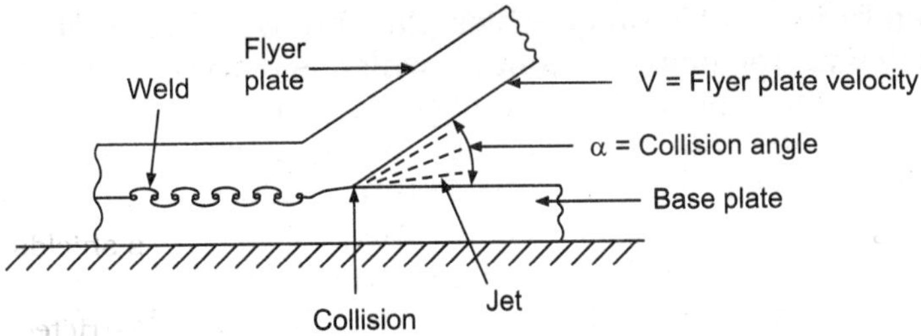

Fig. 12.14 Basic feature of a explosive point welding

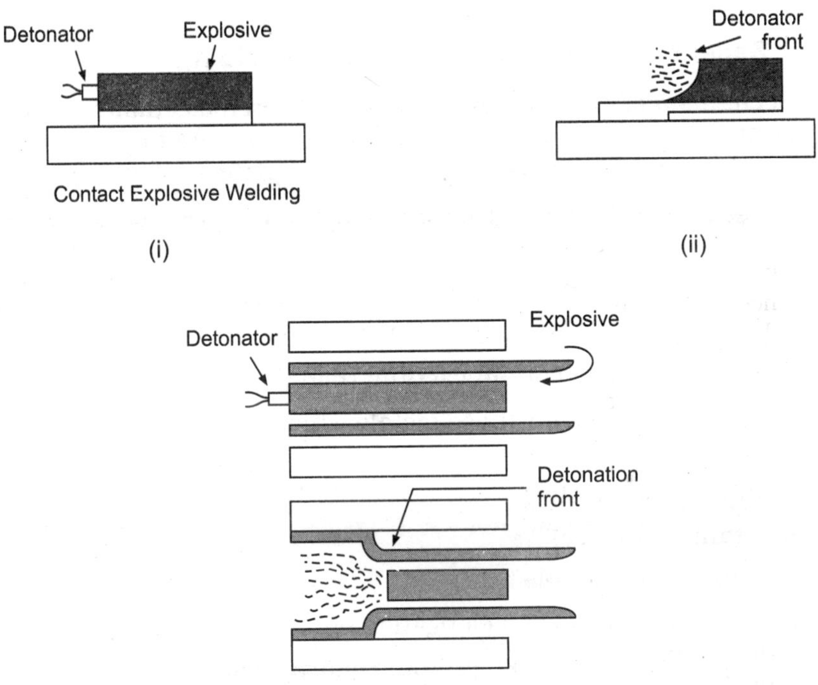

Fig. 12.15 Explosive welding application

12.5.1 Development of Explosive Welding

Previously explosive welding was consider as a form of cold pressure welding but it is now not so although cold pressure welding are used as early as 3000 B.C. for welding sponge iron. A major limitation of pressure welding is that a severe deformation is needed to obtain a bond. Welding by directly applying pressure demands that material surfaces deformed in order to expose clean, oxide free surface in explosive welding. Plastic deformation is restricted to the interface or the area between the two faces of metal.

12.5.2 Procedure

Flyer plate to be joint with the parent plate. Buffer is above the flyer plate and made by rubber, cardboard for protection the top surface of flyer plate from make from detonation of explosive charge. Plate (flyer) should be an inclination from 1 to 10° range. Thick plate know as flyer plate. Now charge is exploded with the help of detonator which is placed at the one end of the flyer plate, as the explosive ignited flyer plate moves toward the target with the velocity of 150 to 500 m/sec. Pressure is also produced between the flyer plate and the parent plate of order 700 to 7,000 N/m^2. Under such high pressure and high velocity metal flows ahead of the joining front acting like a fluid jet in such a way interlocking bond is formed. Interlocking bond is very important aspect of an explosive welding. By the interlocking the strength of joint formed.

Explosive welding needs a large area and it can be performed outdoor.

12.5.3 Salient Feature About Explosive Welding

Basic mechanism of explosive welding is based on molecular bonding as a result of high velocity impact. High velocity is generated by detonation of explosive. For a typical explosive welding of two plates.

- Process can be performed at room temperature in air, water or vacuum too
- Surface contaminants tend to be blown off the surface
- Impact pressure are millions of PSI
- If process is carried out in vacuum then no sound and blast
- Process is suitable for those metals which are prone to brittle joints such as:
  (i) aluminium to steel
  (ii) titanium to steel
- Process is not suitable for:
  (i) brittle metals having elongation 6%
  (ii) impact value 10 ft. Lb (V-notch charpy test)
- If two materials comes closer contact with each other then they can be bond at the level of molecule
- It does not happened normally as surface contamination
- Normal welding overcomes this problem by melting materials so that they mix in liquid phases.
- Important characteristics are:
  (i) critical velocity, and
  (ii) critical angle, collision angle
- When parameters match up, the surfaces form a liquid jet starting at the point they impact, and is directed away from the welded seam.
- The plates have an airgap between, and plastic, liquid or granular explosives are placed on the plate. The backer plate is supported by an anvil. (for light backer sand can be used where as for stronger bucker concrete/steel can be used).
- Cladding plate is supported by the tack weld supports at the edge.

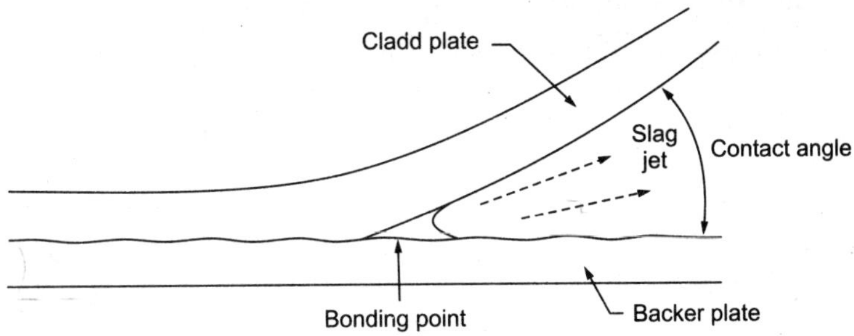

Fig. 12.16

- Small gap is required between the plate and buffer (rubber and plexiglass) for the case of a high velocity explosive. Angled interface is used only for the case of high velocity explosive.
- The detonation velocity should not greater than 120% of the sonic velocity in the metal.
- **Different explosive forms are such as:**
  1. cast shapes
  2. plastic flexible sheet
  3. power/granular
  4. pressed shapes
  5. cord
- As there is a max. velocity in welding therefore due to thermal effect joint become weaker.
- To efficiently use explosives the plate separation is 1/2 to 1 times the cladding plate thickness.
- Explosive with high velocity (4500–7500 m/sec) are:
  (*a*) TNT
  (*b*) PETN
  (*c*) Composition B
  (*d*) RDX
  (*e*) Primacord
  (*f*) Composition C_4
  (*g*) Detasheet
- Range of medium velocity explosive (1500–4500 m/sec) are:
  (*i*) Nitroguonidine
  (*ii*) Dynamites
  (*iii*) Diluted PETN
  (*iv*) Ammonium nitrate
  (*v*) Ammonium perchlorate
  (*vi*) Amatol
- **Types of there bond:**
  1. *Straight, direct metal-to-metal:* Best type of bonding but difficult to obtain when collision velocity less than critical velocity.
  2. *Wavy:* The interface is strong and the interface has waves.
  3. Straight (continues layer).

12.5.4 Weld Interface and Weld Parameter

A buffer may be required between the explosive and the prime component to protect the surface of component. The manner in which the explosive is detonated is extremely important explosive must provide uniform detonation so that collision velocity will be uniform from the start to the finish of the weld. A flat interface is formed when the collision velocity is below some critical value for the particular combination of metal. Increasing the stand off distance increases the angle of approach between the flyer plate and the target plate. This result in increased size of the wave for successful explosive welding the angle of impact is 5 to 25°.

12.5.5 Terms Used in Explosive Welding

1. **Buffer Plate:** A buffer plate is a rubber spacer which is in between the flyer plate and the explosive. It acts as a damping device.
2. **Flyer Plate:** Plate which is to be welded with the target plate is known as flyer plate. There is a buffer above the flyer plate which may be rubber, card based or any similar material to pivoted the top surface from damage from detonation of explosive charge.
3. **Stand-off Distance:** It is the distance which is maintained between the flyer plate's lower edge and the upper edge of target plate and it is from 1/4 to 1/2 of the thickness of the flyer plate.
4. **Surface Jetting:** When the flyer strike with the target plate then a local deformation and fusion takes place which result in joining of the two plates. This phenomenon is called *surface jetting*.
5. **Critical Velocity:** Velocity of the flyer plate which is enough to course the surface jetting is called the *critical velocity*.
6. **Impact Velocity:** Velocity by which the flyer plate strikes the target plate is called impact velocity. Impact velocity depends upon the ratio of the weight of the explosive to that of the weight of flyer plate and also on the contact angle, for every material there is a minimum velocity below which welding does not take place for example, copper can not be welded below 120 m/sec velocity and aluminium at velocity less than 225 m/sec.

12.5.6 Advantages of Explosive Welding

(a) Process is very simple.
(b) Very large surface can be welded (bonded) by this process very easily.
(c) Heat treated metal can be join by this process without affecting their properties.
(d) A large no. of different thickness plate can be explosively *clad* together.
(e) No chance of heat affected zone (HAZ).
(f) Good explosive bonds typically have a strength equal to greater than that of the weaker of the two metal joins.
(g) Better mechanical properties of joint can be achieve by this process.

12.5.7 Limitations

(a) Noise and vibration are generated during the welding process when explosive ignited.

(b) Storage of explosive is also very big problem because explosive can ignited by chance.

(c) This explosive process (*cladding*) is not suitable for brittle substance. There should be some ductility in material.

(d) If thickness of material pass above the 60 mm then large quantity of explosive is required for welding which is not easy process.

(e) Berllium, glass, ceremics, tungsten, boron cannot be explosive welding.

12.5.8 Applications: Main Applications

(a) (i) Cladding of metal

(ii) Welding

(iii) Joining

(b) A large numbers of dissimilar metal can join/cladding successfully with the explosive welding that is:

(i) Tungsten to steel,

(ii) aluminium to steel,

(iii) aluminium to stainless Steel,

(iv) aluminium to nickel and its alloys,

(v) titanium alloys to Cr-Ni alloys,

(vi) copper to stainless steel, stainless steel to nickel, etc.

(c) Cladding process in-storage tank (fuel), electrical industries, ship-building industries, aerospace industries, etc.

(d) Pressure vessels and heat exchange tube is the example of explosive cladding.

(e) Tube and pipe upto 2.0 meter length can be *clad* by this process.

(f) Metal wire, composite material can be fabricated by explosive.

(g) Explosive *cladding* is finding use in the die-casting industries for nozzle, die-cast biscuits and other components.

Experiments

Experiment 1

OBJECT

To make a threaded bolt of given dimension on lathe machine.

TOOL'S USED

Lathe machine, Threading tool, Vernier caliper, Steel rule, and Knurling tool.

RAW MATERIAL USED

$\Phi \times 32$ mm × 105 mm long M.S. Rod.

THEORY

Threading mechanism: The main mechanism of cutting threads on lathe machine is to establish a specific relationship between the longitudinal movement of the tool parallel to the axis of rotation of the job. This will determine the pitch or lead of the thread. This is achieved with the help of lead screw and a split nut. The two halves of the split nut are fastened to the carriage. When the nut is closed on the lead screw, it acts as complete nut and the carriage starts moving as the lead screw rotates. The lead screw is geared to the spindle and the proper speed ratio between the two is set by mean of a gear change box. Therefore as the lead screw rotates, the carriage will move a predetermined distance (depending upon the pitch or lead of the thread) per revolution of the job. It is require that the split nut must be engaged a exact predetermined time, for taking successive cuts, so that the tool enters the helical groove of the cut previously produced otherwise the tool may remove some of the desired thread. This is achieved with the help of a thread dial, which is mounted on the carriage and is driven by the lead screw through a worm gear. The face of the thread dial is graduated in to an even number of full and half divisions, whenever the lead screw rotates and the split nut is not engaged, the thread dial rotates. The split nut must be engaged when a particular line on the dial face coincides with the zero line for cutting even number of threads, the split nut should be engaged when any line on the dial coincides with zero line and for cutting odd number threads when any numbered line coincide with zero line.

Knurling: It is the operation of plastically displacing metal in to a particular pattern for the purpose of creating a hand grip or roughened surface on a workpiece.

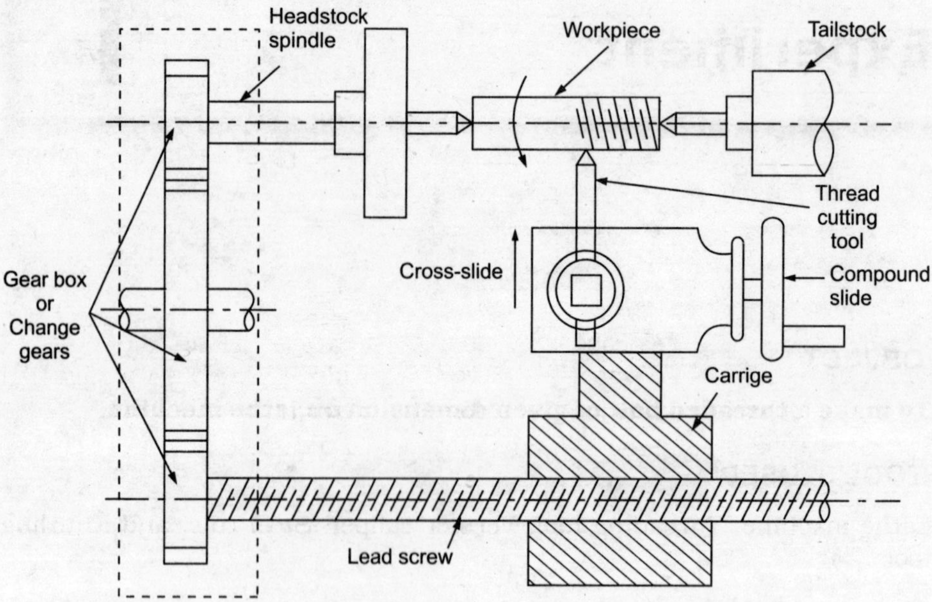

Fig. 1.1 Lathe machine set-up for thread cutting operation

PROCEDURE

1. Hold the job on lathe chuck and hold the tool on tool post.
2. Face the job and turn by 1 mm depth of cut.
3. Knurled the job in the length 45 mm with knurling tool.
4. Start cutting thread; the tool is fed inward until it first scratches the surface of the job. The graduated dial on the cross slide is noted or set to zero. The split nut is then engaged and the tool moves over the desire job length (55 mm). At the end of the tool travel it is quickly withdrawn by means of cross slide. The split nut is disengaged and the carriage is returned to the starting position for the next cut. These successive cuts are continued until the thread reaches its desired depth. The depth of first cut is usually 0.25 to 0.40 mm. This is gradually decreased for the successive cuts until for the final finishing cut; it is usually 0.027 to 0.075 mm.

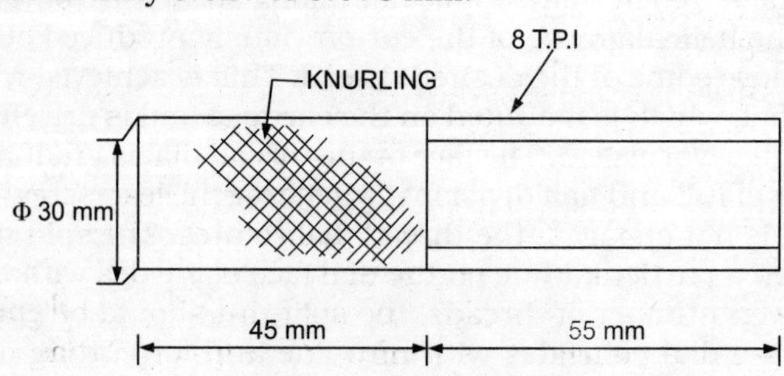

Fig. 1.2 Knurling operation

PRECAUTIONS

1. Job and tool should be hold rigidly.
2. Coolant should be use in turning and threading.
3. Change gear should be proper according to given pitch or lead.
4. Apron should be use.
5. Do not check the dimension while job is revolving.
6. Chuck key should be immediately removed from the chuck after tightening in work piece.
7. Wear goggles at time of machining.

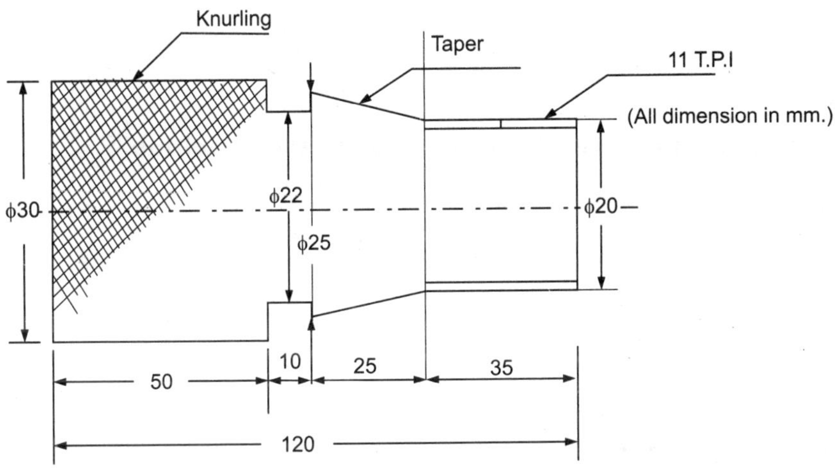

Fig. 1.3 Knurling, grooving, Taper turning and thread cutting operations.

Experiment 2

OBJECT

To make a block of given dimensions on shaper machine.

TOOL'S USED

Shaper machine, tool, Vernier caliper, Steel rule.

RAW MATERIAL USED

50 × 50 × 50 M.S. Block.

THEORY

Quick return mechanism or ram drive: The mechanical ram drive is a slotted arm quick return motion mechanism. The slotted arm is pivoted at its lower end to the column and its upper end is connected by a link to an adjustable ram block, which is secured to the ram. Power from an electric motor is transmitted to the bull gear pinion. A pin fits freely in to a hole in bull gear block, which can slide in the slot of the rocker arm. As the bull gear rotates it carries the pin in a circular path with the help of crank. This makes the rocker arm to swing back and forth about its stationary pivot. Thus the rotary motion of the bull gear is converted into the reciprocating motion of the ram with the cutting tool. The length of the ram stroke depends upon the radius of the crank pin on the bull gear, which can be adjusted. The larger the crank radius, the longer the ram stroke will be and vice-versa. It's clear that the return stroke is quick so as to minimize the return ideal time. The bull gear is massive in size so that it also performs the function of the flywheel. This makes the reciprocating movement of the ram smooth and without any jerks.

Feed mechanism: We know that the table feed is intermittent and is accomplished on the return (non-cutting) stroke when the tool has cleared the workpiece. The cross feed is given to the table with the help of a cross feed screw. The spring loded reversible pawl gets its motion from the slotted disk through a connecting rod and pawl arm or rocker arm. The slotted disk is driven from the bull gear through a spur gear drive and rotates at the same speed as the bull gear. Because the pawl is straight on one side and slanted on the other, it moves the ratchet wheel in on direction only. A spring pushes against the pawl to keep it contact with the ratchet wheel. As the slotted disk rotates the connecting rod oscillate the pawl about the cross feed screw, so that it moves the ratchet wheel by engage with it and hence rotates the cross feed screw which moves the table. The straight side of the pawl pulls of pushes the gear for positive movement, while the slanted side merely slides over the teeth preparatory to taking another bite. A knob at the top of the pawl enables the operator to rotate it 180° to reveres the direction of feed or 90° to stop it altogether. Adjusting the eccentricity or offset of the driving pin in the slotted disk controls the rate of feed.

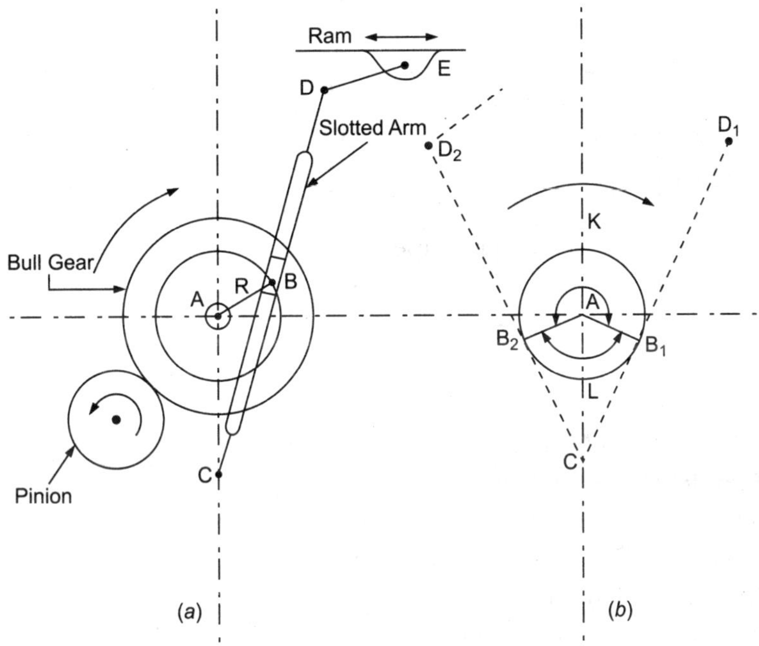

Fig. 2.1 Mechanical ram drive (Shaper)

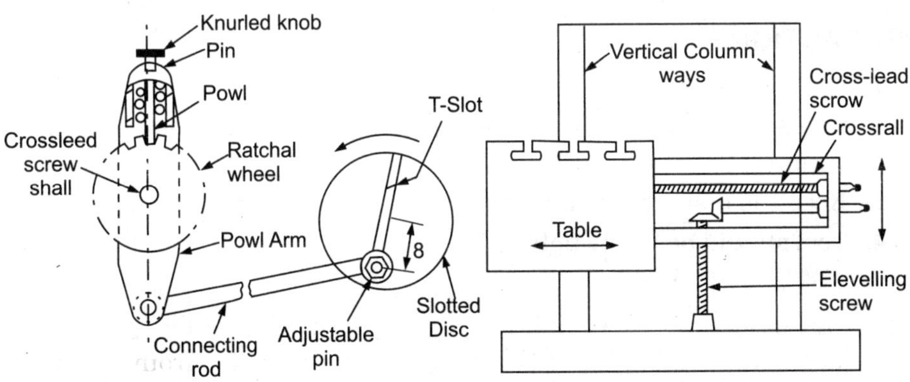

Fig. 2.2 Feed mechanism

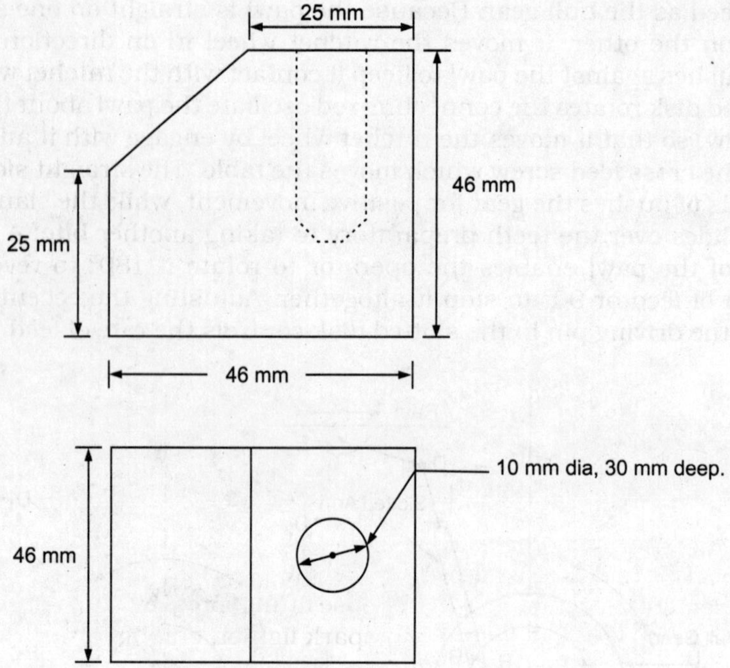

Fig. 2.3 Block made by Shaper

OPERATIONS

1. Hold the job in vice.
2. Hold the tool in tool post.
3. Set the stroke length and depths of cut, and start the machine.
4. Make all the sides according to drawing (46 × 46 × 46).
5. Check all the dimension of the job as per drawing.
6. Make the taper on the job.
7. Hold the job on the drill machine and make hole of 10 mm dia and 30 mm deep.

PRECAUTIONS

1. Workpiece should be firmly hold in vice.
2. Workpiece should not be check for dimensions while machine is running.
3. Keep the distance from machine.
4. Tool should be firmly tight.

Experiment 3

OBJECT

To make a corner joint by oxy- acetylene welding.

TOOL USED

O_2 cylinder, C_2H_2 cylinder, hose and hose fitting, pressure regulator, welding torch, filler rod, goggles, gloves and spark lighter, chipping hammer.

MATERIAL USED

70 × 70 × 6 mm. two M.S. plates.

THEORY

Oxy-Acetylene welding is a process where in coalescence is produced by heating with a gas flame or flames without the application of pressure and with the use of filler metal. The welding heat is obtained from burning a mixture of Oxygen and Acetylene gas. The gases are mixed in the proper proportion with in a welding torch, which provides control of the welding flame. The mixture is burnt at the blowpipe nozzle or tip. Acetylene (C_2H_2)

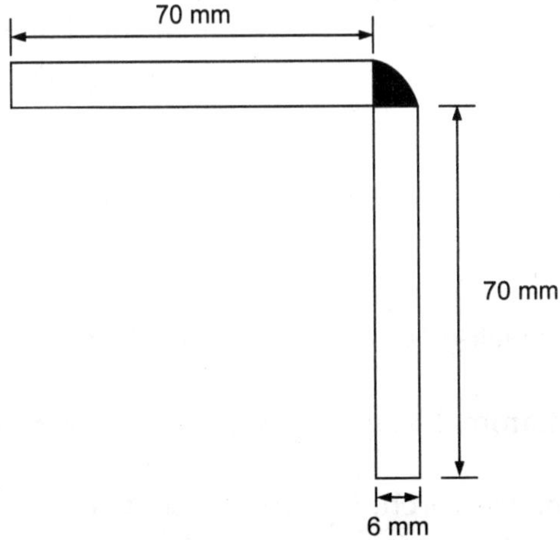

Fig. 3.1 Corner joint

produces higher temperature, (in the range of 3200°C). It contains more available carbon and releases heat when its components (C + H) dissociate to combine with O_2 and burn.

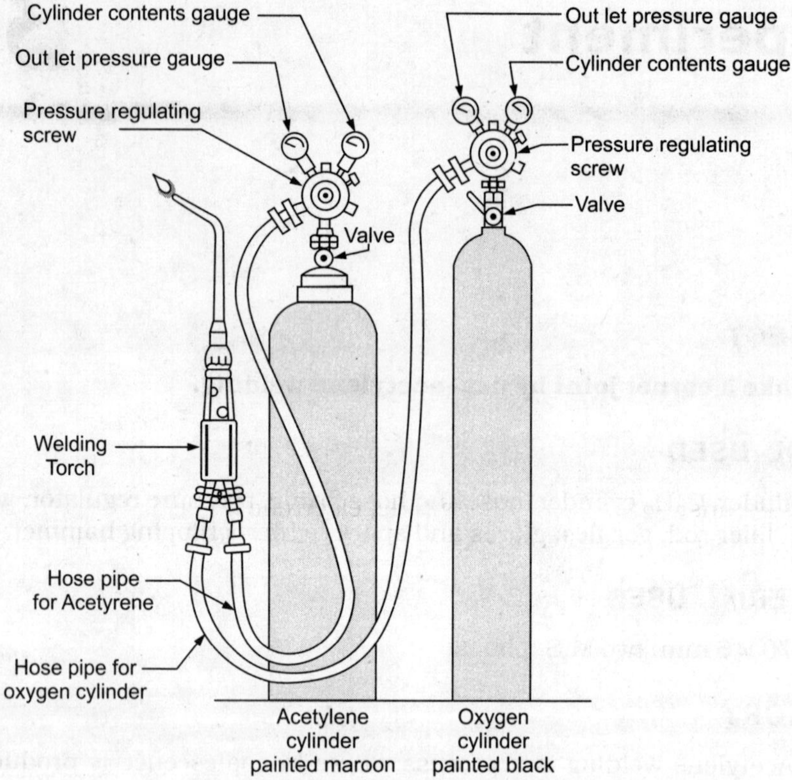

Fig. 3.2 High pressure gas welding equipments

OPERATIONS

1. First fix the job in bench vice and fileing to remove the irregular surface.
2. Now for lighting the torch, O_2 cylinder is opened before the C_2H_2 cylinder. After igniting the torch adjust the flame of the torch.
3. After adjusting the torch flame, put the workpiece on the welding bench and starts the welding process.
4. After welding on both sides remove the extra filler material from joints and clean it by sand paper or emery paper.

PRECAUTIONS

1. Use a proper welder's eye shield with dark glass when looking into flame.
2. Remove all flammable material from near the cylinders or welding area.
3. When lighting the torch, O_2 cylinder is opened first.
4. Loose clothes should always be avoided.

Experiment 4

OBJECT

To make a T-joint by electric arc welding.

TOOL USED

Welding transformer, welding electrode, electrode holder, chipping hammer, file, goggles, gloves,etc.

MATERIAL USED

70 × 70 × 6 mm. two M.S. plate.

THEORY

Electric arc welding is a welding process where in coalescence is produced by heating with an arc or arcs, with or without the application of pressure and with or without the filler materials.

In EAW the arc is struck between an electrode and the work. To strike an arc, the electrode is brought in contact with the work at the point where the welding is to be started, after connecting the work to the welding circuit. After a light contact, the electrode is immediately with drawn to a distance of from 2 to 4 mm from the work. Only a comparatively low potential difference is required between the electrode and the work to strike an arc.

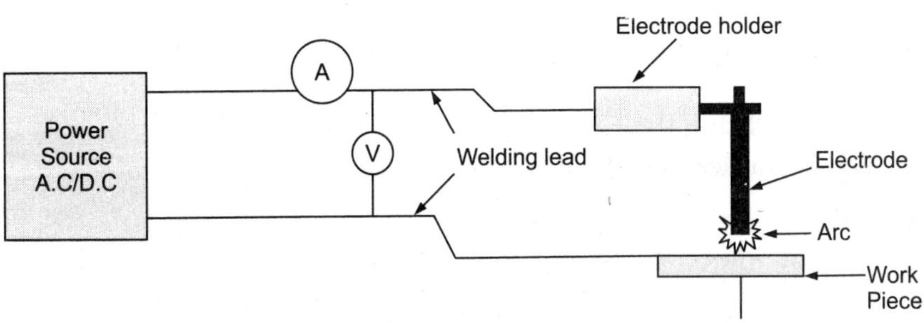

Fig. 4.1 Electric arc welding set up

From 40 to 45 V is usually sufficient for D.C. and from 50 to 60 V for A.C. The voltage falls after the arc is established. A stable arc can be maintained between a metal electrode and the work metal with a voltage of 15 to 30 V is needed to strike an arc between non-consumable electrode and the work. The temperature of electric arc depends upon the type of electrodes between which it is struck. It is about 3200°C on the cathode and about 3900°C on the anode for carbon electrodes and 2400°C and 2600°C respectively for metal electrodes. The temperature may reach 5000°C to 5500°C in the center of arc.

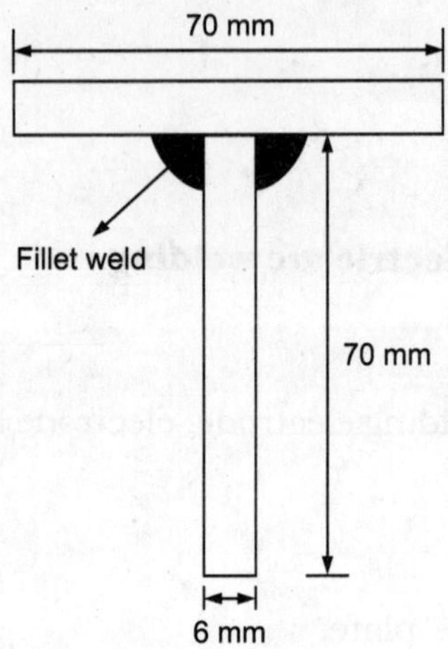

Fig. 4.2 T-joint

OPERATIONS

1. First fix the job in bench vice and fileing to remove the irregular surface.
2. Put the workpiece on the welding bench in desire position.
3. Starts the welding process.
4. After welding on both sides remove the extra filler material from joints and clean it by sand paper or emery paper.

PRECAUTIONS

1. Use a proper welder's eye shield with dark glass when look in to arc.
2. Remove all flammable material from near the cylinders or welding area.
3. When lighting the torch, O_2 cylinder is opened first.
4. Loose clothes should always be avoided.

Experiment 5

OBJECT

To make a job (Lap joint) by seam welding.

TOOLS USED

Seam welding machine, tong, gloves.

RAW MATERIAL USED

Two piece of sheet. (100 × 100 × 1) mm.

THEORY

Seam welding is a one of the type of resistance welding. Seam welding is also known as continuous spot welding and it is differ with spot welding in manner that a leak proof joint is produced in seam welding. In seam welding coalescence is produced by the heat obtained from resistance of the work to electric current in a circuit of which the work is a part, and by the application of pressure and without the use of filler metal. The amount of heat generated at the contracting area of the elements to be welded, is determined from Joule's law,

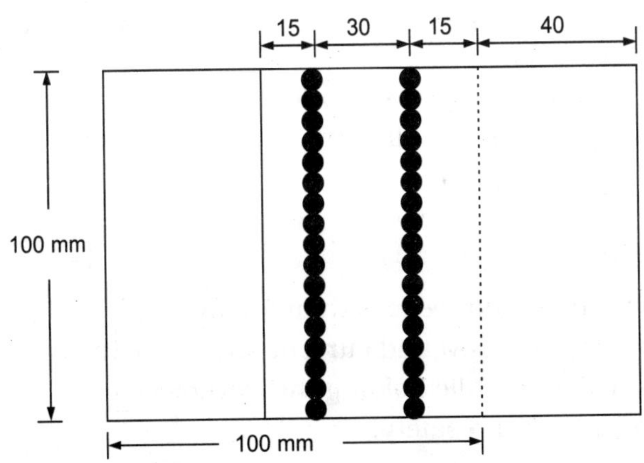

Fig. 5.1 Seam welding

$$Q = I^2Rt, \text{ joules}$$

I = Current in ampere.

R = Resistance of circuit at the contacting area of the elements in ohm.

t = Time during which the current flows, in seconds.

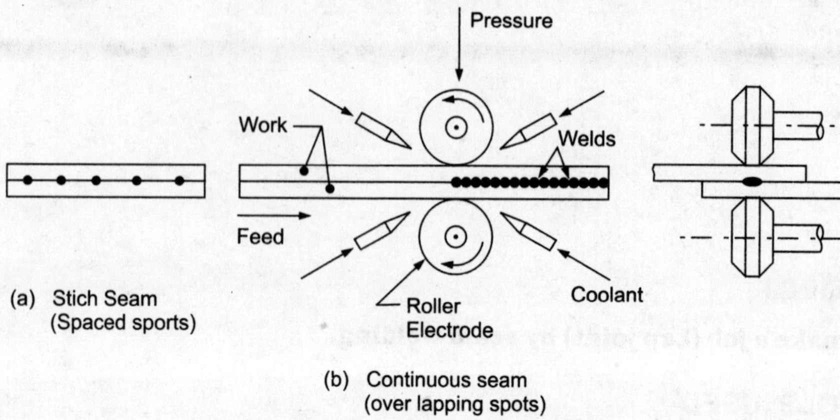

Fig. 5.2 Resistance seam welding

WORKING PROCEDURE

Steps of operation

1. First fix the workpieces in bench vice.
2. Remove the surface in regularties from the workpieces and makes dimensionally as per given drawing.
3. Make the line for seam welding.
4. Adjust the voltage-by-voltage regular according the thickness of metal sheet or workpieces.
5. Start the machine and compressors,
6. Put the workpieces as per marking between the roller disc electrode.
7. Apply the pressure, by-pressure regulator and apply it by foot.
8. Pass the workpiece through the rollers by the speed range 0.5 to 3.5 m/min.
9. Apply the force as high as 5 to 6 kN.
10. After completed the welding release the pressure.
11. Remove the workpiece.

PRECAUTIONS

1. Pressure should not be less than 50 kN.
2. Voltage should be low and current should be high.
3. Always use tong while holding the workpieces.
4. Use gloves for hand safety.

Experiment 6

OBJECT

To make a job (lap joint) by spot welding.

TOOL

Spot welding machine, tong, gloves.

MATERIAL USED

Two pieces 100 × 100 × 1 mm sheet.

THEORY

Spot welding is a one of the type of resistance welding. It is a welding processes where in coalescence is produced by the heat obtained from resistance of the work to electric current in a circuit of which the work is a part, and by the application of pressure and without the use of filler metal. The amount of heat generated at the contracting area of the elements to be welded, is determined.

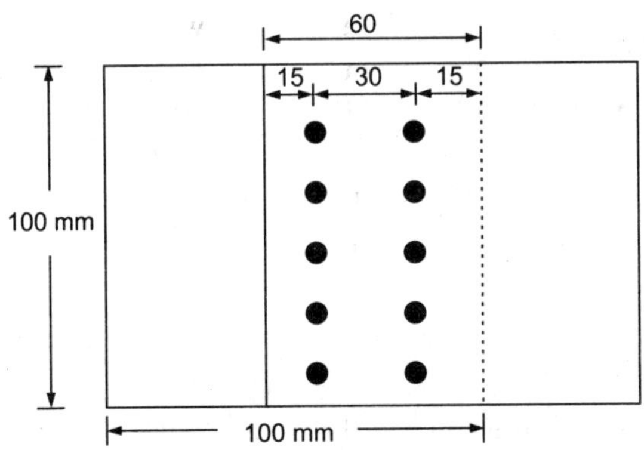

Fig. 6.1 Spot welding

From Joule's law,

$$Q = I^2Rt, \text{ joules}$$

I = Current in ampere.

R = Resistance of circuit at the contacting area of the elements in ohm.

t = Time during which the current flows, in seconds.

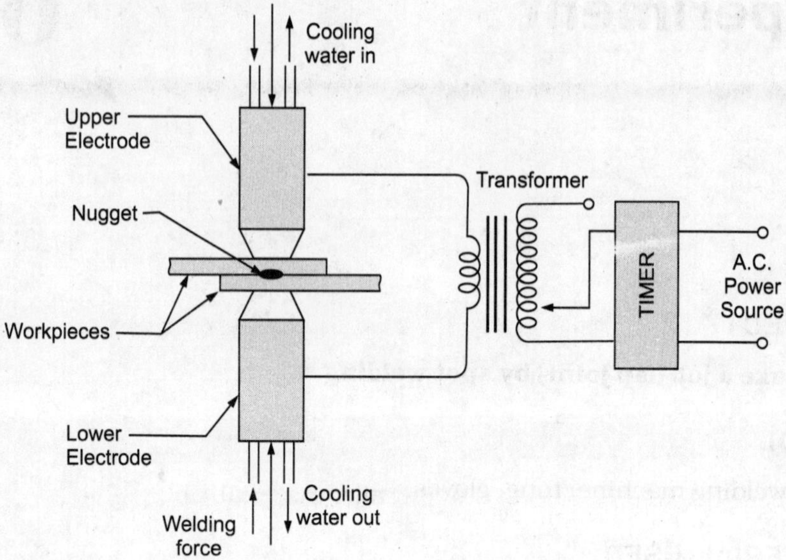

Fig. 6.2 Resistance spot welding

OPERATIONS

1. Position the workpiece and squeeze between the electrodes.
2. Apply a low voltage current to the electrodes. (0.5 to 10 V).
3. Hold until the proper temperature is attained, (upto 400°C).
4. Release current, continue pressure, (upto 2 kN).
5. Releases pressure and remove work.

PRECAUTIONS

1. Pressure should not be less than 50 kN.
2. Voltage should be low and current should be high.
3. Always use tong while holding the workpieces.
4. Use gloves for hand safety.

Experiment 7

OBJECT

To make a butt joint by TIG welding.

TOOLS USE

TIG welding machine, file, try square, filler material, chipping hammer, face shield.

MATERIAL USE

70 × 70 × 40 mm. M.S. plate — two pieces.

THEORY

In TIG welding coalescence is produce by heating with an electric arc between a suitable electrode and the work. Shielding is obtained from an inert gas such as ***helium***, and ***argon***. Pressure is not use and filler material may or may not be use. In this welding non-consumable metal electrode's used. Its melting point is 3400°C. The arc temperature may be in the range of 10,000°K.

The welding head {gun or torch} consists of a light weight handle with provision for holding a stationary ***tungsten electrode***. In the welding head the shielding gas flows by or along the electrode through a nozzle at the end of the gun in to the arc region. This is done to keep the electrode cooler and permits higher currents to be used. Most guns are also water-cooled. In the

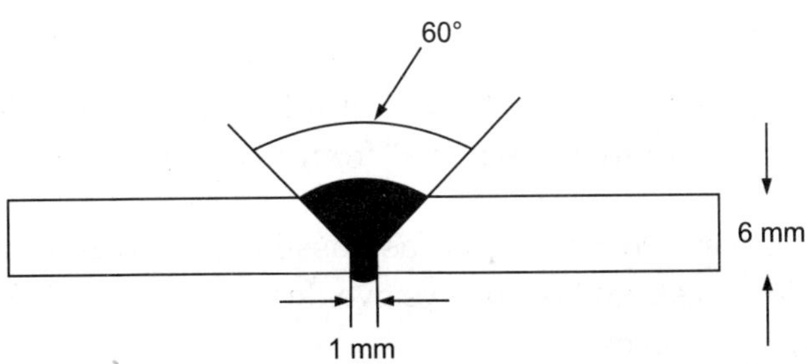

Fig. 7.1 Welding bead by TIG machine

tungsten electrode, 2% thorium and zirconium are added to improve electron emission, arc stability, arc striking and current carrying capacity, and to increase the melting point of tungsten. Electrode diameters of 0.8 mm to 5 mm are commonly used with welding speed ranging from about 8 mm/s to over 50 mm/s.

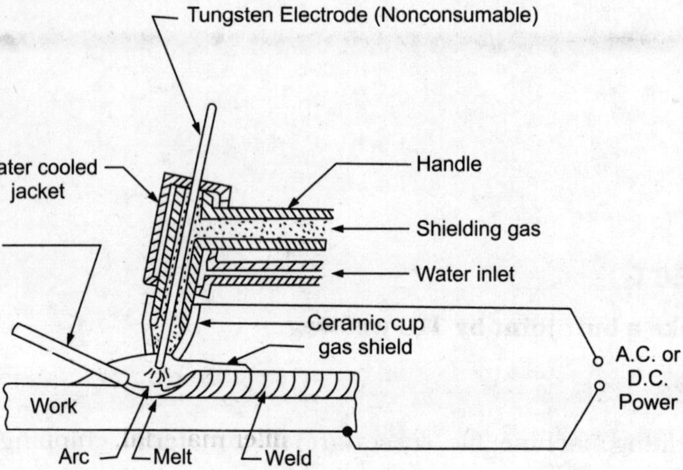

Fig. 7.2 TIG welding machine setup

WORKING PROCEDURE

1. First fix the workpiece in bench vice and prepare the edges for welding as per thickness of workpiece. Such as *V*-shape *U*-shape, single–*V* straightness of edges, etc.
2. Start the machine and open the pressure regulator Valve of argon gas cylinder as per given destruction.
3. Adjust the current and voltage for given workpiece as per thickness of workpiece.
4. Put the workpiece on the welding table in desired position.
5. Start the welding process after creating the arc between electrode and workpiece.
6. After welding on both side remove the slag from the joints.
7. Clean the job by emery paper.

PRECAUTIONS

1. Use a proper welder's eye shield with dark glass when look in to arc.
2. Remove all flammable material from near the cylinders or welding area.
3. Keep your skin covered as far as possible from the arc and gases.
4. Loose clothes should be always avoided.
5. Wear leather gloves.

Experiment 8

OBJECT

To make a lap joint by MIG welding.

TOOL'S USE

MIG welding machine, file, try square, filler material, chipping hammer, face shield.

MATERIAL USE

70 × 70 × 40 mm. M.S. plate – two pieces.

THEORY

In MIG welding coalescence is produce by heating with an electric arc between a suitable electrode and the work. Shielding is obtained from an inert gas such as ***carbon dioxide***, ***helium***, and ***argon***. Pressure is not use and filler material used. In this welding consumable metal electrode's used. Electrode is in the form of wire and fed into the arc with the help of variable speed motor at a preset rate that is determined by the arc voltage. The wire feed increase in the arc voltage caused by an increase in arc length. This process has greater efficiency then TIG and is more rapid.

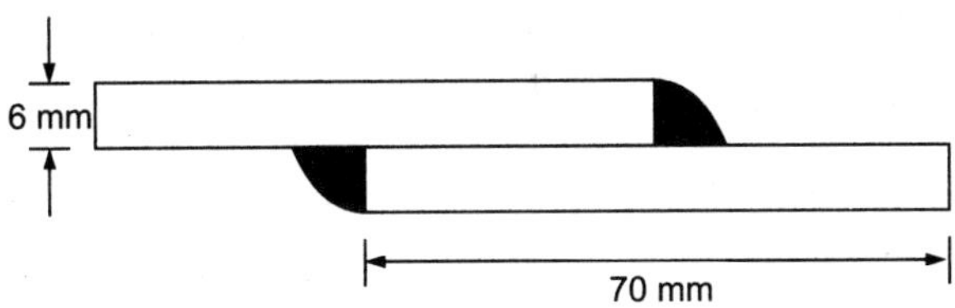

Fig. 8.1 Lap (double) joint

OPREATIONS

1. First fix the job in bench vice and fileing to remove the irregular surface.
2. File both edges of metal pieces through 70° as shown in figure or as per given in Drawing.

3. Put the workpiece on the welding bench in desire position.
4. Starts the welding process (open the pressure regulator valve of carbon dioxide gas cylinder as per instruction).
5. Adjust the current and voltage,
6. Start the welding process by feeding the electrode in the from of wire from electrode holder.
7. After welding on both sides remove the extra filler material from joints and clean it by sand paper or emery paper.
8. Clean the job by emery paper.

PRECAUTIONS

1. Use a proper welder's' eye shield with dark glass when look in to arc.
2. Remove all flammable material from near the cylinders or welding area.
3. Keep your skin covered as far as possible from the arc and gases.
4. Loose clothes should always be avoided.
5. Wear leather gloves.

Experiment 9

OBJECT

To make a job of given dimension on capstan lathe machine.

TOOL'S USED

Capstan lathe machine, Threading tool, Vernier caliper, Steel rule and Knurling tool.

MATERIAL USED

A Extruded bar of Mild Steel.

THEORY

Capstan lathe (ram type): The ram-type turret lathe shown in Fig. 9.2 has the turret mounted on a slide or ram which moves back and forth on a saddle clamped to the lathe bed. Turrets of this type are usually six sided and index on the return stroke. The ram is advanced or retracted by a capstan type hand wheel. Counter-clockwise rotation of the hand wheel advances the ram; clockwise rotation retracts the ram. Adjustable stops are provided to preset the forward stopping position of each turret index. The turret is bored out to accept commercial turret tooling on the lathe by putting an adjustable boring bar in the spindle of the lathe.

Knurling Process: It is the operation of plastically displacing metal in to a particular pattern for the purpose of creating a hand grip or roughened surface on a workpiece.

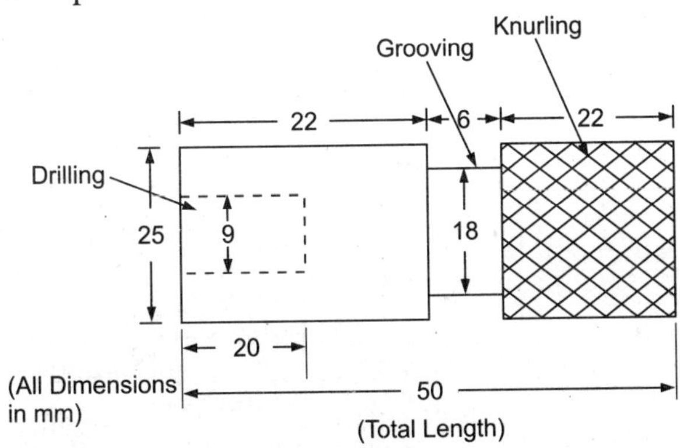

Fig. 9.1 Specimen produced by capstan lathe

PROCEDURE

1. Hold the job on lathe chuck and hold the tool on tool post.
2. Face the job and turn by 1.0 mm depth of cut.
3. Make a groove of 10 mm width and 5 mm depth on given location of the job as in the fig.
4. Knurled the job in the length 20 mm with knurling tool.
5. Make a ϕ 12 mm hole on the one face of the job with the help of drill on turret.

PRECAUTIONS

1. Job and tool should be hold rigidly.
2. Coolant should be use in turning, knurling and drilling also.
3. Apron should be used.
4. Do not check the dimensions while job is revolving.
5. Wear goggles at the time of machining.

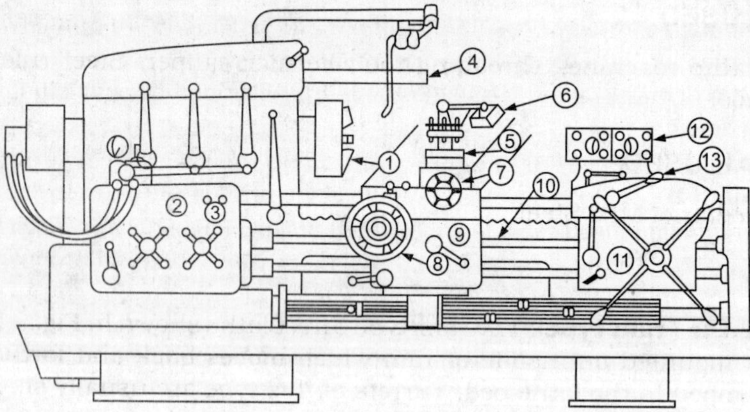

Fig. 9.2 *Turret lathe: 1. Chuck 2. Gear box 3. Electrical push button starter 4. Overarm support 5. Front tool post 6. Rear tool post 7. Hand wheel for cross-slide 8. Hand wheel for longitudinal feed 9. Carriage 10. Feed bar 11. Saddle 12. Turret head 13. Lever for locking the slide*

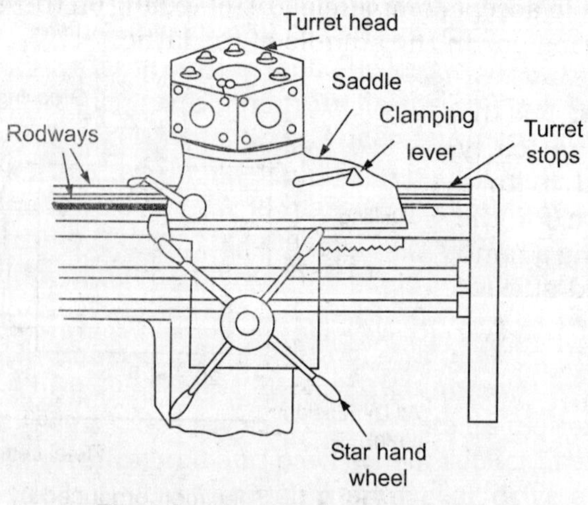

Fig. 9.3 Mounting the turret head

Experiment 10

OBJECT

To calculate the chip thickness Ratio (*r*) of a given mild-steel specimen.

MATERIAL USED

Mild-steel Specimen.

MACHINE AND TOOL'S USED

Center lathe m/c, Vernier calipers, Cutting tool.

THEORY

Fig. 10.1 shows the cutting process.

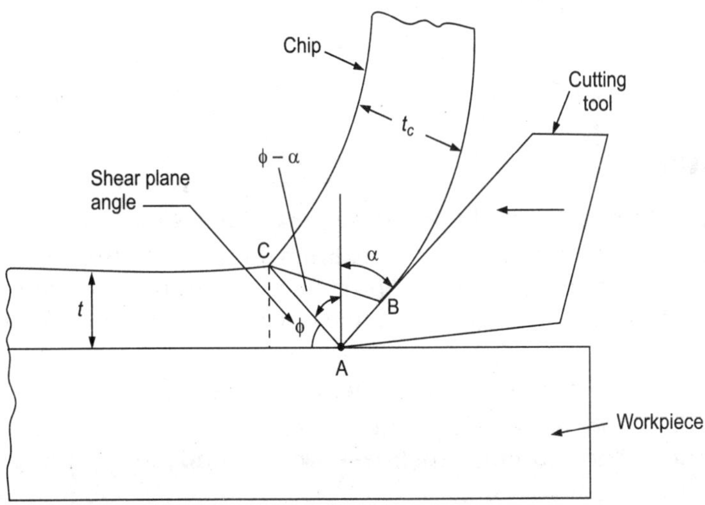

Fig. 10.1 Shear plane and Shear plane angle

$$t = AB \sin \phi$$

$$t_c = AB \cos (\phi - \alpha)$$

where t = depth of cut

t_c = chip thickness

the chip thickness ratio r, which is also termed as cutting ratio, would be

$$r = \frac{t}{t_c} = \frac{\sin\phi}{\cos(\phi - \alpha)} = \frac{1}{\cot\phi \cos\alpha + \sin\alpha}$$

$$\cos\phi \cos\alpha = \frac{1 - r\sin\alpha}{r}$$

$$\tan\phi = \frac{r\cos\alpha}{1 - r\sin\alpha}$$

Experimentally the chip thickness ratio r could be determined by measuring the average thickness of the chips produced under given conditions of feed and speed. From this it is possible to evaluate the shear angle using the above equation. However, direct measurement of chip thickness is difficult, because of the roughness on the outside of the chip. For this purpose an indirect measurement is followed wherein the length of a chip l_c.

Equivalent to known length of an uncut chip is measured. Then, considering the fact that the depth is the same, the average chip thickness t_c would be given as

$$t_c = \frac{t_l}{l_c}$$

where l = length of the uncut chip to obtain an exact size of uncut chip length l, we may introduce a small saw cut parallel to the axis on the workpiece so that the uncut chip size is

$$l = \pi D$$

where, D = diameter of workpiece.

PROCEDURE

1. Clamp the specimen of given dimension on lathe machine.
2. Check the diameter (D) of specimen before machining, start the machine and give a desired depth of cut to the specimen and take the chips. Stop the machine and measure the machined diameter of specimen (d).
3. Check the thickness of chip (t_c) with the help of vernier.
4. Calculate the uncut chip thickness, with the help of uncut chip thickness and cut chip thickness we can find easily chip thickness ratio (r).

OBSERVATION

S. No.	Diameter of the specimen (D) in mm.	Machined dia. of specimen (d) in mm.	Uncut chip thickness (t) =(D-d)/2 in mm.	Cut chip thickness (t_c) In mm.	Chip thickness ratio (r)= $\frac{t}{t_c}$
1.					
2.					
3.					
4.					
5.					

PRECAUTIONS

1. Job and tool should be hold rigidly.
2. Coolant should be use in turning.
3. Apron should be used.
4. Do not check the dimensions while job is revolving.
5. Used goggles at the time of machining.
6. Check the zero error in vernier caliper before measuring the dimensions.
7. Check dimensions properly and carefully.

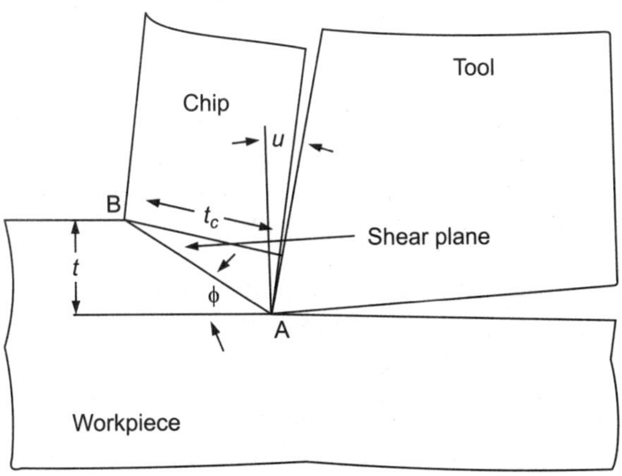

Fig. 10.2 Orthogonal cutting with in shear plane

Experiment 11

OBJECT

To prepare a blank and punch of given size on an aluminium sheet.

MATERIAL USED

Hand press, aluminium sheet, etc.

THEORY

Blanking: This is the operation carried out on processes and consist of cutting the outside contour of a stamping. Production of sheet metal blanks of flat shapes requires a single action press equipped with tools comprising a punch. A corresponding die, stripper to keep the sheet and for spacing successive cuts. The operation of cutting inside the contour i.e., holes and slots is called piercing.

Fine Blanking: This process is a press working technique. It utilizes special press and precession tools and dies to produce parts in final assembly shapes. Parts produced have clearly sheared surface over the entire material thicknes. In conventional methods one-third thickness is sheared and balance is fractured which has to be shaved, milled etc. The dimensional accuracy obtained is also high.

Punching: This is also press operation and consist of cutting holes of various shapes in sheet. This operation is generally continuous and at a time two or more holes may be punched.

Trimming: It consists of removal of cutting away of excess material left around the parting lines in the previous operation.

Notching: This is the operation of removing small amount of metal from the edges of workpieces by the notching action usually obtained with the help of nibbler.

Lancing: It is a special form of piercing operation in which the entire contour is not cut, the blanked material remains attached with the sheet.

OPERATIONS

1. At first required size of the aluminium sheet is cut.
2. It is clamped to the fixture or die securely.
3. After that punch is moved down by turning wheel to complete the operation.

PRECAUTIONS

1. Always keep a safe distance from press.
2. To secure tightly the aluminium sheet with the fixture.
3. Not to touch any part of the device except the wheel used for moving up or down the punch.

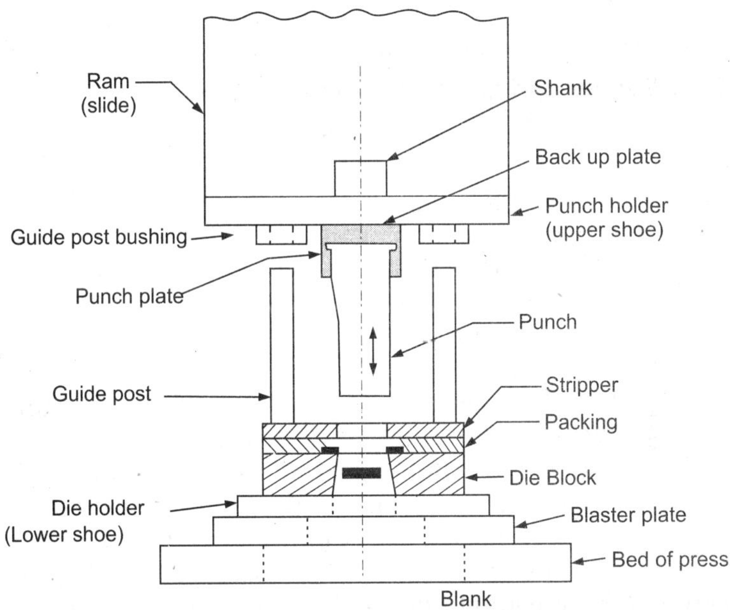

Fig. 11.1 A simple cutting die

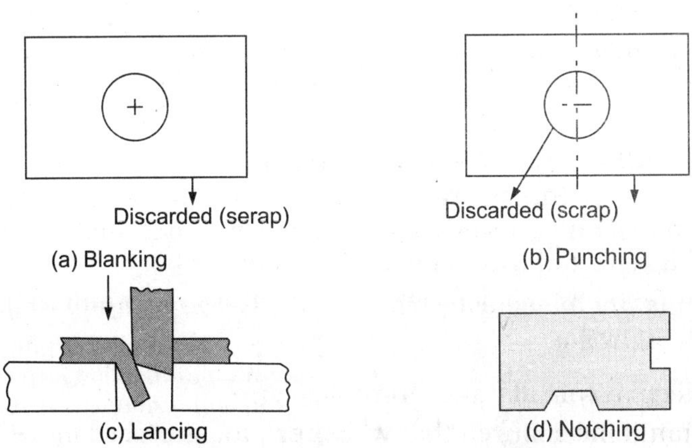

Fig. 11.2 Sheet metal operations

Experiment 12

OBJECT

To make a 90° band of given specimen of tube.

THEORY

Bending is the cold working process, involving plastic deformation in which the total surface area remain constant. Bending of pipe, tube, rolled or extruded shape involves practices and theories of simple supported beam, cantilever and become many of the problems of forming and deep drawing. In bending flow of metal occurs in plastic range and there is a permanent change in shape thinning of outer wall of a bent tube or other hollow section and corresponding thickening of the inner wall of the bend are noticeable effect of plastic flow.

Same common type of bending operations are:

Stretch Bending: The process consists of gripping a sheet of material or a rolled or extruded section at each end in suitable jaws and starching it over a die made to the required contour until complete forming has been achieved by extending and thinning the metal over a profile.

Roll Bending: This operation is carried out at a series of roll stations and a flat sheet of strip.

It transformed into desired section progressively in stages. The metal gauges or section thickness remains constant and only bending operation takes place in each stage.

Draw Bending: The operation is used for bending hollow section. The machine used for this operation has rotating die to which the piece to be bent is clamped and is drawn against a roller. A mandrel is used to eliminate the following effect of tubular section.

PRECAUTIONS

1. To keep a safe distance from pipe vice.
2. It should be ensured that **whether** pipe is placed in between support and roller.
3. Pipe should be filled with sand properly.

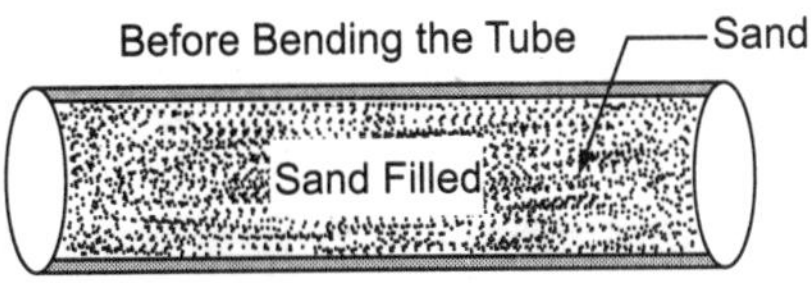

Fig. 12.1 (a) Tube before bending

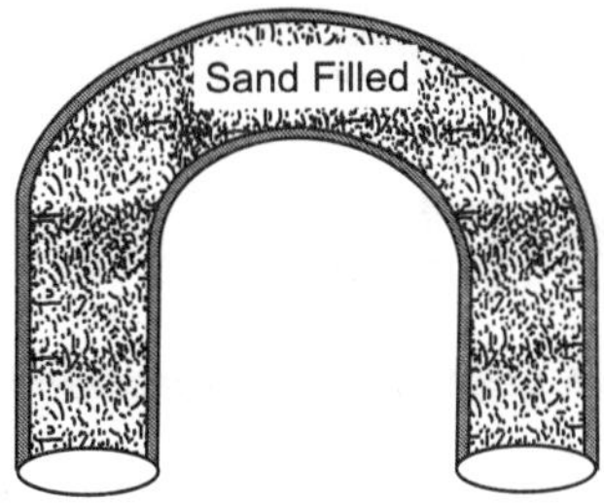

Fig. 12.1 (b) Tube after bending

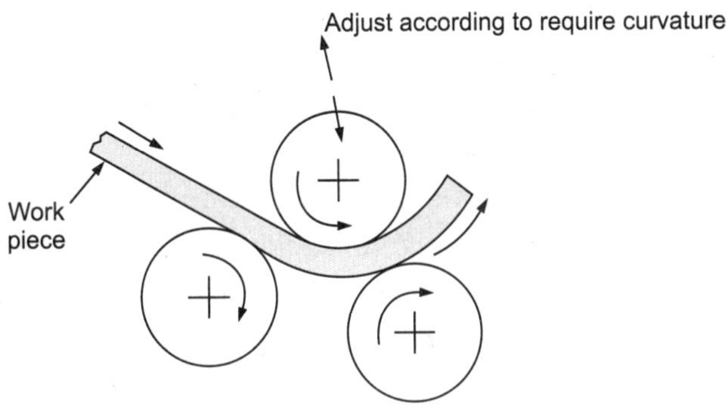

Fig. 12.2 Roll bending

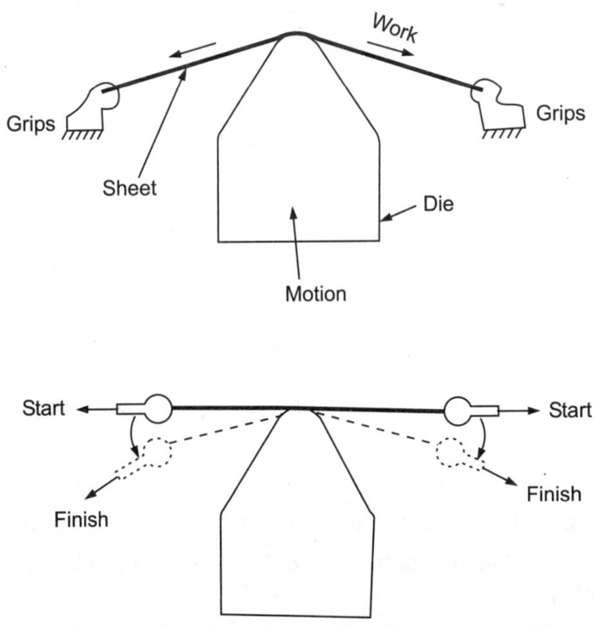

Fig. 12.3 Stretch bending

Bibliography

[1] B. Avitzur, *Metal Forming: Process and Analysis*, McGraw Hill Book Co.

[2] Bhattacharya, A., "Metal Cutting, Theory and Practice, Central Book Publishers, Calcutta (1884).

[3] Boathroyd, G, "*Fundamental of Metal Machining* and *Machine Tools*", Scripta Book Co., Washington 1975.

[4] Davies. A.C., "*Welding*" Cambridge University Press, UK.

[5] Ghosh and Malik, "*Manufacturing Science*", East-west Press, New Delhi.

[6] Gourd, L.M., "*Principle of Welding Technology*", Viva Books, New Delhi.

[7] HMT, "*Production Technology*", Tata McGraw Hill, New Delhi, 1982.

[8] Hould Coroft, P.T., "*Welding Process Technology*", Cambridge University Press, London, 1977.

[9] Juneja. B.L., Sekhon. G.S., "*Fundamental of Metal Cutting* and *Machine Tools*", New Age International Publishers, New Delhi.

[10] Kalpakjlan, Scrope and Schmid, Steven, R, "*Manufacturing Engineering* and *Technology*," Pearson Education (Singapore), Indian branch, Patparganj, New Delhi.

[11] Kaushish, J.P., "*Manufacturing Process*", Printice-Hall of India (P) Ltd, New Delhi.

[12] Khanna, O.P., "*A Text book of Welding Technology*", Dhanpat Rai & Sons, New Delhi.

[13] Khanna, O.P., "*A Textbook of Production Technology*", Dhanpat Rai & Sons, New Delhi.

[14] M.C. Shaw., "*Metal Cutting Principles*" 3rd edition 1954, M.I.T. Cambridge Massachusetts, USA.

[15] Martellotti, m" *A Treatise on Milling and Milling Machine.*" The cincinnati milling machine Co. USA, 1952.

[16] P.H. Black., "*Theory of Metal Cutting*", McGraw Hill Book Co. U.S.A.

[17] Raghuvansi. B.S., "*Manufacturing Process*" Dhanpat Rai & Co. New Delhi.

[18] Rajput, R.K., "*Manufacturing Technology*" – Laxmi Publications (P) Ltd, New Delhi.

[19] Rao, P.N., "*Manufacturing Technology*", – Volume-II, Tata McGraw Hills, New Delhi, 2009

[20] Rao. P.N., "*Manufacturing Technology*," Volume-I, Tata McGraw Hill Publishing company, New Delhi, 2009.

[21] Rizvi, S.A. and Ali, Wajahat, "*Advanced Welding Technology*", Kataria & Sons, New Delhi, 2010.

[22] Rusion off, S.E, "*Manufacturing Process,*" The Times of India press Bombay 1965.

[23] Amstead, B.H, Begeman, M. "*Manufacturing Process*". John Wiley and Sons, USA, 1994.

[24] Sen G.C, & Bhattacharyya A, "*Principle of Machine Tools*", New Central Books Agency, Calcutta.

[25] Sharma. P.C., "*A Textbook of Production Engineering*"., S.Chand & Company, New Delhi.

[26] Singh Balbir and Singh C.K., "*A Textbook of Production Engineering*", Standard Publishers and Distributors, New Delhi.

[27] W.A J. Chapman., "*Work Shop Technology*", Part III, Edward Arnold (Publishers) Ltd.

[28] W.G. Patton, "*Modern Manufacturing Process & Engineering*", Prentice Hall, New India, 1970.

[29] "*A Treatise on Milling and Milling Machines*" The Cincinnati milling machine Co. U.S.A 1951.

[30] Rizvi, S.A. "*Mathematical Modeling of Surface Roughness on Turning of EN8 on CNC machine*". M. Tech dissertation, 2005.

PREVIOUS YEAR EXAMINATION PAPER

B. TECH

(SEM. V) ODD SEMESTER THEORY
2013-2014

MANUFACTURING SCIENCE-II

Time : 3 Hours *Total Marks : 100*

Note: Attempt **all** questions. All questions carry equal marks.

1. Attempt any **four** parts of the following: **(4 × 5 = 20)**

(a) What are the various types of chips? Under what conditions is each formed ?

(b) Explain the Merchant's force circle diagram and derive the shear angle relationship $2\phi + \beta - \alpha = \pi/2$, where ϕ is the shear angle, β is the friction angle and α is the rake angle. State the assumptions of such a diagram.

(c) What are cutting fluids? Discuss some of the cutting fluids used during machining.

(d) When the rake angle is zero during orthogonal cutting, show that

$$\frac{\tau_s}{U_c} = \frac{(1 - \mu r)r}{1 + r^2}$$

where symbols have their usual meanings.

(c) Draw the tool geometry of a single point cutting tool and show the different angles.

2. Attempt any **two** parts of the following: **(2 × 10 = 20)**

(a) What is defference between Capstan and Turret lathe? What are the standard and special tools used on these machines? How are these tools different to engine lathe tools?

(b) Discuss the various criteria used for optimizing the cutting conditions. A cylindrical bar is to be turned. The maximum allowable feed is 0.2 mm/revolutions and at this feed rate Taylor's tool life equation for a tool-work combination is found to be $VT^{0.25} = 75$. The labor cost involved in each regrinding of the tool is ₹ 5.00. On the average, it takes about 2 minutes to change the tool. Find the cutting speed that will lead to maximum production rate. Derive the formula used.

(c) What are the main differences between a shaper and a planer? Discuss the different drive mechanisms used in shaper with the help of suitable diagram.

3. Attempt any **two** parts of the following: **(2 × 10 = 20)**

(a) What are the various factors to be considered in the selection of a grinding wheel? Discuss each in detail.

(b) (i) Why surface finish is important for many application? Illustrate your answer.

(ii) Explain the Lapping process. State its uses, limitations and advantage.

(c) Explain the factors which affect the surface finish in plain milling operation.

A steel block of 20 mm width is being milled using a straight slab milling cutter with 20 teeth, 50 mm diameter, and 10° radial rake. The feed velocity of the table is 15 mm/min and the cutter rotates at 60 rpm. If a depth of cut of 1 mm is used, what will be the power consumption? Assume shear-strength (k) of steel is 250 N/mm^2 and tool-chip interface friction $\mu = 0.3$.

4. Attempt any **two** parts of the following: **(2 × 10 = 20)**

(a) Explain with neat sketch the working and applications of the following:

(i) Coated and uncoated electrode welding.

(ii) Atomic hydrogen welding.

(b) (i) Describe the submerged are welding process with the help of a suitable diagram. What are the advantages and applications of this process?

(ii) Compare electro-slag welding process with that of submerged arc welding from standpoint of heat liberated, joint preparation and welding position.

(c) Describe the oxy-acetylene welding equipments. Draw the different types of flames used in gas welding. How would you identify these flames?

5. Attempt any **two** parts of the following: **(2 × 10 = 20)**

(a) What are the main parameters to be considered while selecting a particular unconventional machining process any why?

(b) (i) How metal removal in EDM is achieved? Discuss any one spark generator used in EDM.

(ii) Draw a neat sketch of the electrode feeding mechanism used in EDM.

(c) (i) What are the applications of ultrasonic machining? Why can very hard material be better cut by the ultrasonic process than soft ones?

(ii) Compare ECM with EDM. Why isn't ECM as widely used as EDM?

Index

I

A

Abrasive grain 202
- grade/hardness 204
- manufacturing 203
- properties 202
- structure 204
- types 202, 203

Abrasive jet machining 372
- advantages of 374
- applications of 375
- elements of 373
- limitations of 375

Acetylene 254
Acetylene generator 254
Alloying elements 39
Arc welding process 262
- advantage of 268
- application of 269
- equipment required 266
- limitation of 269
- types of 262

Arc welding
- history of 245

ASA system (American system) 11
Atomic hydrogen welding 290
- advantages of 291
- applications 292
- disadvantages of 292
- equipments 291
- procedure 291

B

Brazing 308
- advantages of 310
- applications of 311
- different type of 310, 312
- disadvantages of 311
- joints 309
  - different type of 310
- metals that can be joint by 310
- operation 309

C

Cemented carbides steel 37
Ceramics-alumina 38
Chip(s) 13
- formation mechanism 12
- types of 14
- velocity 23

Chromium 40
Chuck adapter 145
Cobalt 39
Collets 145
Combined drills and countersinks 187
- classifications 188
- nomenclature of 188
- process parameters 189

Cubic boron nitride 39
Cutting fluid(s) 25
- composition of a 27
- effects in machining 30
- functions of a 26, 27
- maintenance and disposal 31
- requirements of a good 27
- selection criteria 30

Cutting tool 4
- materials 31
- multi point 10
- single point 5

D

Diamond 38
Distortion 329, 336
- longitudinal shrinkage 331
- main causes for 329
- transverse shrinkage 331
- types of 329

Drill machine 176
- main parts of a 176
- types of 177

E

Electrical discharge machining 363
- advantages of 365
- condenser circuit and metal removal rate 366
- dielectric fluid used in 365

feature of 365
limitations of 365
wire cut machine 367
working principle of 364
Electro-chemical machining 357
advantages of 360
applications of 361
characteristic of 359
electrolyte used in 361
limitations of 360
metal removal rate in 359
power needed for 362
principle of 358
Electrogas welding 297
principle of operation 298
Electron beam machining 378
advantages of 380
applications of 380
characteristic of 379
equipment 381
generation of contract of electron beam 381
limitations of 380
process 380
working principle 379
Electron beam welding 399
advantages of 401
applications of 402
electron gun 400
limitations of 402
procedure 401
vacuum chamber 401
Electroslag welding 295
advantages 297
equipments used in 297
limitations 297
working principle 295
Explosive welding 406
advantages of 410
applications 411
development of 407
limitations 410
procedure 407
salient feature 408
terms used in 410
weld interface and weld parameter 410

F

Fits 228
between mating parts 230
classification of 230
Flank 13
Flank wear 42
Flash butt welding 287
advantages of 288
limitations 288
metal welded by 288
some applications 289
working principle 287
Flux cored arc welding 302
advantages and applications 304
disadvantages 304
equipments 302
principle of operation 302
weldable metals 304
Flux cored arc welding 302
disadvantages 304
principle of operation 302
weldable metals 304
welding equipment 302
Fly cutting 163
Friction welding 299
advantages of 301
limitations of 301
materials that can be welded 301
parameters 301
welding parameter 301
Fusion weld
metallurgy of 319

G

Gas welding 251
acetylene generator 254
advantages of 260
applications of 260
arrangement of oxy-acetylene 256
backfire/popping 255
flashback 256
limitations of 260
operating of cutting torch 261
oxy-acetylene welding 251
oxy-acetylene/gas-cutting 261
oxygen 254
safety/precautions of acetylene cylinder 255
technique 258
welding torch 257
Granding wheels 195
care and safety 209
components 198
selection of a 207
shapcs 209
types of 197
cylinder or wheel ring 197
diamond 198
dish cup 198
saucer 198
straight 197
straight cup 197
tapered 197
Grinding 195

mechanics of 203
operations 209
process 195
Grinding machine 210
types of 210
centerless 213
surface 210
Grinding wheel wear
mechanism of 206

H

Heat affected zone (HAZ) 320
Hole system 231
Honing 220
advantages of 222
limitations of 222
machine 221
process 220

I

Indexing fixture 147
Inert gas 275
Ion beam machining 382
advantage of 385
application of 384
extraction grid 383
focus of 384
focus of ion beam 384
ground electrodes 384
ion source 383
limitation of 385
plasma source 383
power supply 384

L

Lapping 223
advantages of 225
applications of 225
limitations of 225
method of 224
speed 223
Laser beam machining 375
advantages of 377
applications of 378
characteristic of 377
common application of laser 378
limitations of 378
working principle 375
Laser beam welding 392
advantages of 395
applications 395
disadvantages 395
types of laser 394
working principle 393
Lathe machine(s) 73
backgear 85
common parts of 80
counter shaft 88
head stock 83
leadscrew 86
operations 91
boring 96
drilling 96
facing 91
knurling 97
reaming 96
thread cutting 97
turning 91
selection of 73
special purpose 79
specification of a 79
tailstock 87
types of 75
automatic 78
bench 77
capstan 99
centre 76
engine 76
speed 77
tool room 75
turret 101
working principle of a 73
work holding devices 89
Lee and shaffer's theory 24

M

Machinability 42
factors affecting the 43
Machining economics 44
Manganese 40
Metal cutting 3
process 17
Metal inert gas welding 269
advantage of 272
application 273
equipment required for 270
feature of 273
limitation of 273
metal transfer in 273
principle of operation 270
Metal inert gas welding 269
advantages of 272
applications 273
equipments required for 270
features of 273
metal transfer in 273
power source 271
principle of operation 270
Milling 130
angular 158
climb 170

conventional 170
face 160
feeds for 155
form 162
gang 162
helical 169
key way 164
operation 171
plain 157
sawing and parting 168
straddle 159
T-slot 167
Milling cutters 134
classification of 136
angle 139
end 138
helical 136
metal slitting saw 137
side 137
T-slot 138
woodruff key slot 139
nomenclature 134
selection of 140
size of 141
types of teeth 135
Milling machine 130
classification of 130
mounting and indexing work 140
tools for 134
types of 131
knee-type 131
ram-type 133
swivel cutter head ram-type 133
universal horizontal 132
universal ram-type 133

N

Non-conventional machining process 356

O

Oblique cutting 17
Orthogonal cutting 17
Earnest Merchant's model 20

P

Percussion welding 286
advantages of 286
applications of 287
disadvantages of 287
procedure for 286
Phosphorus 40
Planer 124
classification of 125
construction 126
Plasma arc welding 402
advantages of 405
applications 405
limitations of 405
non-transferred arc 403
process 404
uses of equipment 403
Polishing 225
Projection welding 284
advantages of 285
limitations of 285

Q

Quick-change tooling 146

R

Rake face 13
Resistance welding 277
advantages of 279
applications of 280
limitations of 279
types of 280
variables of 278
Restance built welding 289
Right-hand welding technique 259

S

Seam welding 282
advantages of 284
applications 284
equipment 283
limitations 284
principle of operation 282
types of 283
Shaper machine 112
additional components 117
drive mechanisms 117
crank and slotted link 117
Whitworth quick return 119
principal parts of a 112
shapes produced by 120
Shear angle 19
Shear zone 12
Shielding gas 272
Silicon 40
Slotter machine 120
cutting tools 123
drive mechanism of a 122
main parts of a 121
specification of a 122
types of 122
work holding devices 123
Soldering 305
advantages of 307
alloys 307
applications of 308
joint 306

metals joined by 305
method of 307
steps involved in soldering process 307
types of solders 306
Solid state welding 299
Spindle adapters 145
Spot welding 280
advantages of 280
applications of 281
types of 281
Stainless steel weld
weld decay in steel HAZ of 339
Submerged arc welding 292
advantages of 294
applications 295
aspects of 293
equipments 294
limitations of 294
working principle 293
Surface roughness 237
Surface texture 237

T

Tool life 43
Tool wear 40
abrasion 41
adhesion 41
chemical reactions 41
diffusion 41
mechanism of 41
plastic deformation 42
types of 41
Tugsten inert gas welding 274
advantage of 276
applications of 276
equipments used in 275
principle of operation 274
bead 276
Tungsten 40
Twist drills 181
classifications 181
nomenclature of 183

U

Ultrasonic machining process 368
advantages of 370
applications 372
elements of 369
limitations of 370
material removal rate 371
characteristics 371
working principle 368
Ultrasonic welding 396
advantages of 399
applications of 398
equipments 397
limitations of 399
materials suitable for 398
working principle of 398
Unconventional machining process 354
advantages of 357
classifications of 354
limitations of 357
Unconventional welding process 392

V

Vanadium 39
Vises 147

W

Weld metal zone 322
Weldability 341
factors that affect the 341
testing 341
Welding 245
advantages of 249
application of 250
limitations of 249
Welding defect 331
cold cracking 335
cracks 333
creater 336
distortion 336
excessive penetration bead 332
general sources of 332
hot cracking 334
inclusions 338
incomplete fusion and penetration 337
lack of penetration 332
poor weld bead appearance 339
porosity 336
slag inclusion 337
spatters 339
undercut 333
Welding joints 342
Welding stresses 324
contraction of solid metal 327
longitudinal shrinkage 327
residual stresses 325
solidification shrinkage 327
transverse shrinkage 328
Welding stresses and distortion 324
Work-hardened metals 340

□□□

Win Prizes

ATTENTION READERS

Win Prizes

We request for your frank assessment regarding various aspects of the book as given below :

Manufacturing Science-II

S.A. Rizvi • Dr. S.P. Tewari • Wajahat Ali

Please Fill up the Sheet in Neat and Clean Writing

(i) What topics of your syllabus, important for your examination point of view, are not covered in this book ?

....

....

....

....

(ii) What are the chapters wherein the treatment of the subject matter is not systematic or organised or updated ?

....

....

....

....

....

(iii) Have you come across misprints/mistakes/factual inaccuracies in the book? Please specify the chapters and the page numbers.

....

....

....

....

....

(iv) What other books on the same subject have you found/heard better than the present book? Please specify in terms of price and quality

....

....

....

....

....

(v) Further suggestions and comments for the improvement of this book.

....

....

....

....

....

(PS: If need be please attach a separate sheet to write your views)

OTHER DETAILS

(i) Who recommended you this book ? (Please tick (✓) in the box near the option relevant to you.

☐ Teacher ☐ Friends ☐ Bookseller

....

....

....

....

(ii) Name of the recommending teacher, his designation and address :

....

....

....

....

(iii) Name and address of the bookseller you purchased the book from :

....

....

....

....

(iv) Name and address of your institution (Please mention the University or Board as the case may be)

....

....

....

....

(v) Your name and complete postal address :

....

....

....

....

(vi) Course you are enrolled in :

....

....

....

....

The best assessment will be awarded each month: The award will be in the form of our publications as decided by the Editorial Board.

Please mail the filled up coupon at your earliest to :

S.K. Kataria & Sons®

4885/109, Prakash Mahal, Dr. Subhash Bhargav Lane,
Opposite Delhi Medical Association, Daryaganj, New Delhi-110002
Phone: +91-11-23243489, +91-11-23269324; Telefax: +91-11-23243489
e-mail: katariabooks@yahoo.com; katariabook@gmail.com
Website: www.skkatariaandsons.com